Introduction to Nonlinear Circuits and Networks

Bharathwaj Muthuswamy • Santo Banerjee

Introduction to Nonlinear Circuits and Networks

 Springer

Bharathwaj Muthuswamy
Department of Physics
QuEST Lab
Stevens Institute of Technology
Hoboken, NJ, USA

Santo Banerjee
Institute for Mathematical Research
University Putra Malaysia
Serdang, Malaysia

ISBN 978-3-030-09798-1   ISBN 978-3-319-67325-7   (eBook)
https://doi.org/10.1007/978-3-319-67325-7

This Springer imprint is published by the registered company Springer Nature Switzerland AG
The registered company address is: Gewerbestrasse 11, 6330 Cham, Switzerland

Dedicated to Leon O. Chua, Robert M. Fano, Lan J. Chu, and Richard B. Adler

Preface

The purpose of this course-based text-book is to emphasize introductory concepts from classic ideas in nonlinear circuit theory:

1. The book is completely self-contained and does not assume any prior knowledge of circuit theory. It is simply assumed that the reader has taken a first-year undergraduate (elementary) course in differential and integral calculus, along with elementary physics courses in classical mechanics and electrodynamics (with an exposure to matrix algebra). Hence, this book should be accessible to any motivated individual who has taken the abovementioned courses.
2. The book also covers topics that are not typically found in standard circuit textbooks, such as:

   - **Memristors**. The justification is that although (as the reader will learn from this book) memristors are not used in linear circuit theory, a memristor is the fourth fundamental circuit element. Hence, it is only logical that any text on circuit theory discuss memristors. Thus, ideally our book would have been titled *Introduction to Circuits and Networks*. But then a casual reader might mistake this book for emphasizing only linear circuit theory. However, as this book will show, nonlinear circuit theory (memristor included) is accessible to anyone with the correct background who is interested in fundamental circuit theoretic concepts.
   - **Nonlinear chaotic circuits**. We believe that chaotic circuits elegantly integrate a variety of concepts from circuit theory and hence form a natural repository of "projects" for the reader to understand all the concepts discussed in this book.
   - **Nonlinear operational amplifier circuits, for example, Schmitt triggers**. As will be clear from this book, accurate analysis of Schmitt triggers will help dispel common misconcepts about the nature of hysteresis in electronic circuits and will also help the reader understand the deep concept of modeling.
3. Each chapter has illustrative examples, exercises, and a lab component. We will also have (maximum 20 min) conceptual videos for each chapter and end-of-chapter exercises (lab included) online. The purpose of these supplementary

videos is to (*a*) highlight major concepts in each chapter and (*b*) provide solutions (hints for open-ended problems) to end-of-chapter exercises. This extensive use of electronic aids is the "twenty-first century" approach to nonlinear circuits. In the process, we hope to "transform" fundamental ideas in nonlinear circuits from the classic works of Leon O. Chua and others. Nevertheless, the goal of this book's electronic aids is to supplement, not replace, rigor.

Over the course of a teaching career spanning 10 years at the University of California (UC) Berkeley, Dr. Muthuswamy has coordinated with Dr. Leon O. Chua and others to reintroduce nonlinear circuit theory at an elementary level. Much of the material in this book is thus derived from Dr. Chua's EE100 (Electronic Techniques for Engineering) lecture materials. This course was offered by the Electrical Engineering and Computer Sciences department at UC Berkeley for non-electrical engineering majors. Thus, the approach taken by EE100 (and this book) is a top-down view of circuit analysis where we discuss **general principles and emphasize device modeling**. Therefore, the material in this book can be adopted for an introductory course in circuit theory.

At the University of California, Berkeley, we were able to cover most (excluding memristors) of the material in this book in one semester. The material on chaotic circuits was used as a source of projects. For schools that are based on the shorter quarter system (10 weeks of instruction), we would suggest splitting the material in this book into two courses. The first course could cover Chaps. 1 and 2 (network elements). The second course would cover Chaps. 3–5, where Chaps. 3 and 4 discuss techniques of network analysis followed by Chap. 5 as a source of course projects. Another option would be to cover resistive networks in the first course and dynamic networks in the second course. Specifically:

1. First course—resistive networks: only excluding material on dynamic elements in Chaps. 1 and 2 (Sects. 1.9.3–1.9.5, 2.2.3, 2.2.4) and covering all of Chap. 3.
2. Second course—dynamic networks: cover dynamic elements in Chaps. 1 and 2 that were not covered in the first course, followed by Chaps. 4 and 5.

Hence, the way we have organized the chapters is based on the fact that, in circuit theory, the laws of elements are distinct from the laws of networks.

Our goal in writing this book is simple: a student who thoroughly understands the concepts in this book will be well prepared for any follow-up course in circuit theory. For readers who are further interested in advanced concepts, we are planning to write a follow-up volume, *Advanced Nonlinear Circuits and Networks*. A reader who thoroughly understands the material in both volumes will maximize knowledge gained from **any** follow-up electrical engineering course, since our books on circuit theory emphasize both the underlying mathematics and physical experiments.

Hoboken, NJ, USA   Bharathwaj Muthuswamy
Serdang, Malaysia   Santo Banerjee

Acknowledgements

There are a plethora of folks whom we have to thank. From Dr. Muthuswamy's perspective, first and foremost, he would like to thank his MS and PhD advisor Dr. Leon O. Chua for all his support and guidance. Particularly for this book, Dr. Chua guided us in organizing the content. His advice to extract simple yet novel concepts on nonlinear circuit theory from his classic works, *Introduction to Nonlinear Circuit Theory* and *Linear and Nonlinear Circuits*, paved the way for this book. Dr. Muthuswamy was also deeply influenced by conversations with Dr. H. Gopalakrishna Gadiyar and Dr. Ganesan from the Vellore Institute of Technology in the summer of 2014. Their suggestion was to "capture the nonlinear circuits knowledge" of Dr. Leon O. Chua. Thus, capturing that knowledge is the primary purpose of this book.

Ferenc Kovac, Carl Chun, and Dr. Pravin Varaiya from Berkeley have been both professional and personal mentors throughout the years. From Dr. Muthuswamy's academic career in Wisconsin, interactions with Dr. Jovan Jevtic and Dr. Gerald Thomas have been motivational and inspiring. Dr. Sunil T. Mathew from the University of Oklahoma has provided invaluable advice regarding the role of biological memristors. Dr. Muthuswamy would also like to thank his former colleagues at Tarana Wireless for their support when writing a draft version of this book for approval. The College of New Jersey (Dr. Muthuswamy's employer while writing the majority of the book) also deserves special thanks.

We would also like to thank the anonymous reviewers for helping us reformulate the content of this book, so as to make it sustainable. Without Springer's support throughout the writing process, this book would not have been possible. Dan Funke from the College of New Jersey also read through the manuscript and provided valuable feedback.

Dr. Muthuswamy would also like to thank his family for their moral support throughout the process: daughters Shambavi and Thejasvi, father M.G. Muthuswamy, mother Chandra Muthuswamy, brother Karthikeyan Muthuswamy, and sister-in-law Mamta. Last but not the least, Dr. Muthuswamy would like to thank his spiritual advisor, Rajan Kurunthappan and family, for their continued unconditional support over the last two decades.

Mathematical Notation

The mathematical notation used in this book is standard [15]; nevertheless, this section clarifies the notation used throughout the book.

1. Lowercase letters from the Latin alphabet $(a-z)$ are used to represent variables, with italic script for scalars and bold invariably reserved for vectors. The letter t is of course always reserved for time. n is usually reserved for the dimension of the state. j is used for $\sqrt{-1}$, in accordance with the usual electrical engineering convention. Mathematical constants such as π, e, and h (Planck's constant) have their usual meaning. Other constant scalars are usually drawn from lowercase Greek alphabet. SI units are used.
2. Independent variable in functions and differential equations is time (unless otherwise stated) because physical processes change with time.
3. Differentiation is expressed as follows. Time derivatives use Leibniz's ($\frac{dy}{dx}$, for example) or Newton's notation: one, two, or three dots over a variable correspond to the number of derivatives and a parenthetical superscripted numeral for higher derivatives. Leibniz's notation is used explicitly for non-time derivatives. ∂ is the usual symbol for indicating partial derivatives.
4. Real-valued functions, whether scalar- or vector-valued, are usually taken (as conventionally) from lowercase Latin letters f through h, r, and s, along with x through z.
5. Vector-valued functions and vector fields are boldfaced as well, the difference between the two being indicated by the argument font, hence $\mathbf{f}(x)$ and $\mathbf{f}(\mathbf{x})$, respectively.
6. Constant matrices and vectors are represented with capital and lowercase letters, respectively, from the beginning of the Latin alphabet. Vectors are again bolded.
7. In the context of linear time-invariant systems, the usual conventions are respected: $\mathbf{A}$ is the state matrix; $\mathbf{B}(\mathbf{b})$ is the input matrix (vector).
8. Subscripts denote elements of a matrix or vector: $\mathbf{d}_i$ is the ith column of $\mathbf{D}$; x_j is the jth element of $\mathbf{x}$. Plain numerical superscripts on the other hand may indicate exponentiation, a recursive operation, or simply a numbering

depending on the context. A superscripted T indicates matrix transpose. $\mathbf{I}$ is reserved for the identity matrix. All vectors are assumed to be columns. det stands for determinant of a square matrix.

9. Σ_i is used for summations, sampling interval is symbolized by T, and $\in$ denotes set inclusions.

10. Calligraphic script ($\mathscr{R}$, etc.) is reserved for sets which use capital letters. Elements of sets are then represented with the corresponding lowercase letter. Excepted are the well-known number sets which are rendered in doublestruck bold: $\mathbb{N}$, $\mathbb{Z}$, $\mathbb{Q}$, $\mathbb{R}$, and $\mathbb{C}$ for the naturals, integers, rationals, reals, and complex numbers, respectively. The natural numbers are taken to include 0. Restrictions to positive or negative subsets are indicated by a superscripted $+$ or $-$. The symbol $\triangleq$ is used for definitions. $\forall$ and $\exists$ have the usual meaning of "for all" and "there exists," respectively.

Conventions Used in the Book

Each chapter starts with a visual epigraph: the purpose is to evoke the intellectual curiosity of the reader. Chapters are divided into sections and subsections for clarity.

Figures, equations, and definitions are numbered consecutively in each chapter. The book has a variety of solved examples in light gray shade.

> Solved Examples

All references are placed at the end of each chapter for convenience. We use a number surrounded by square brackets for in-text references: [5]. We have strived to give credit to the original authors, keeping in mind Stigler's law of eponymy. If we incorrectly attributed an idea to the wrong original contributor, we sincerely apologize.

Important terminology and concepts are **boldfaced**. Important techniques are framed. In the electronic copy of this book, online URLs are colored and hyperlinked in midnight blue for ease of access. Also, we have hyperlinked any numbered definitions, equations, figures, etc. Hence, it would be prudent to purchase the ebook.

Figures were generated using a combination of xcircuit, xfig, tikZ, and PNG screen captures in UNIX that were converted to EPS.

We utilize Quite-Universal Circuit Simulator (QUCS), a functional open-source circuit simulator, which is introduced in lab component for Chap. 1. Installation instructions can be found in Appendix A. The mathematical plots were generated using SageMath. Computer code is in `verbatim font`. We have not given an extensive tutorial on SageMath because of the abundance of excellent tutorials online. An unintended consequence is: this book should help the reader learn a mathematical simulation tool (like SageMath) via the material (especially Chap. 5). QUCS and SageMath simulation files are available online at the companion website: http://www.harpgroup.org/muthuswamy/IntroToNonlinearCircuitsAndNetworks/.

On a concluding remark, when you find typos in the book please contact the authors with constructive comments: bharath.berkeley@gmail.com, santoban@gmail.com.

Contents

Acronyms

AC	Alternating Current
BJT	Bipolar Junction Transistor
CCCS	Current-Controlled Current Source
CCVS	Current-Controlled Voltage Source
CFOA	Current Feedback Operational Amplifier
DC	Direct Current
DIP	Dual Inline Package
DP	Driving-Point
FCAM	Flux Charge Analysis Method
KCL	Kirchhoff's Current Law
KVL	Kirchhoff's Voltage Law
MNA	Modified Nodal Analysis
NIC	Negative Impedance Converter
PWL	Piecewise-Linear
VCCS	Voltage-Controlled Current Source
VCVS	Voltage-Controlled Voltage Source
VTC	Voltage Transfer Characteristic

Chapter 1
Two-Terminal Network Elements

Chua, L. O. *Memristor - The Missing Circuit Element [2]*

Abstract This chapter will set the stage for the rest of this book. We will start by discussing what is the aim of circuit theory, what are the fundamental circuit variables, and when the techniques in this book are valid: the lumped circuit approximation holds (frequencies of interest are not too high). We will discuss the concepts of Kirchhoff's laws, basic circuit topology, Tellegen's theorem, and two-terminal circuit elements.

1.1 The Discipline of Circuit Theory

Circuit[1] theory is a fundamental engineering discipline that pervades all electrical engineering [5, 9]. For the present, by physical circuit, we mean any interconnection of electrical devices. Familiar examples of electrical devices include resistors, diodes, transistors, operational amplifiers (opamps), etc. The goal of circuit theory is to predict the electrical behavior of physical circuits. The purpose of these predictions is to improve their design: in particular, to decrease their cost and improve their performance under all conditions of operation (e.g., temperature effects, aging effects, possible fault conditions, etc.).

---

[1]In this book, we will use circuits and networks interchangeably, the justification will be discussed in Sect. 1.6. Also, at the outset, we encourage the reader to familiarize themselves with the companion website: http://www.harpgroup.org/muthuswamy/IntroToNonlinearCircuitsAndNetworks/.

© Springer International Publishing AG, part of Springer Nature 2019   1
B. Muthuswamy, S. Banerjee, *Introduction to Nonlinear Circuits and Networks*,
https://doi.org/10.1007/978-3-319-67325-7_1

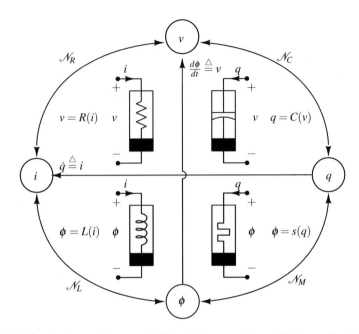

Fig. 1.1 The four fundamental two-terminal circuit elements along with the associated reference directions (Sect. 1.4) relate the fundamental circuit variables, through the **laws of elements**. The elements starting counterclockwise from the top-left are the **resistor, inductor, memristor,** and **capacitor**. Note that current is defined as the rate of flow of charge. There are **two mutually exclusive** definitions of voltage from electromagnetic field theory, refer to Sect. 1.2. The symbols used for the fundamental circuit elements are standard for nonlinear circuit elements, the reader may be familiar with the circuit symbol for the linear counterparts (excluding the memristor), see Sect. 1.9.4

Probably the most fascinating aspect is that lumped circuit theory uses only four fundamental circuit variables: current, charge, voltage and flux-linkage (flux). Moreover, current and voltage are related to charge and flux (Eqs. (1.3) and (1.4) respectively). Thus, fundamentally we have only four elements that are characterized by a mathematical relation between the abovementioned four circuit variables at the element's terminals [10], as shown in Fig. 1.1 [16].

Hence to start our study of circuit theory, we will first discuss the fundamental circuit variables, the topic of Sect. 1.2.

1.2 Fundamental Circuit Variables

We could say the advent of electricity [1] occurred with the discovery that dry substances such as amber tend to repel or attract each other upon being rubbed by different materials such as silk. This phenomenon was first explained by postulating the existence of a certain basic electrical quantity called the "electric charge" (charge), mathematical symbol q, which may be either positive or negative. Like charges exert a force of repulsion and unlike charges exert a force of attraction. The practical unit of charge is called the **coulomb** and has been defined to be equivalent to the total charge possessed by 6.24×10^{18} electrons. Charge can be measured by instruments such as the electroscope.

Since charged bodies exert forces on one another, energy or work is involved whenever one charged body is moved in the vicinity of another charged body. Hence if w is the work done by moving a charge q from point j to point k (assuming w is independent of the path taken),[2] the potential difference or voltage between these points is defined as the work per unit charge.

$$v_{jk} = \frac{w}{q} \tag{1.1}$$

Observe that the magnitude of the charge is arbitrary; only the ratio between work and charge is important. Hence, the incremental work dw required to move an incremental test charge dq from point j to point k must also satisfy Eq. (1.1). Thus:

$$v_{jk} = \frac{dw}{dq} \tag{1.2}$$

We will delete the subscripts j and k when there is no possibility of confusion and simply express voltage as v. The unit of voltage is called the **volt** and is measured using a voltmeter.

Charges can be caused to flow from one charged body to another by connecting a conducting wire between the two bodies. Hence, the quantity "rate of flow of charge" becomes very useful, and it has been given the name current with symbol i. By definition,

$$i = \frac{dq}{dt} \tag{1.3}$$

The unit of current is the **ampere**. One ampere represents a charge flowing at the rate of one coulomb per second. Current flow can be measured by an ammeter.

In 1819, Hans Christian Oersted discovered that current flowing through a wire produced a force on a compass needle in the vicinity of the wire. This indicates that the current (or moving charge) produces a magnetic field. This effect can be explained by the generation of a magnetic flux λ by the current. If the conductor is wound into a coil of n turns, then by defining $\phi = n\lambda$ to be the flux-linkage, Faraday discovered that the voltage between the two terminals of the coil is given by

$$v = \frac{d\phi}{dt} \tag{1.4}$$

The unit of flux-linkage is the **weber**. Flux-linkage can be measured by a fluxmeter.

---

[2]This assumption is valid only if the simultaneity postulate is satisfied, we will discuss more in Sect. 1.3.

1.3   The Simultaneity Postulate in Lumped Circuit Theory

Having discussed the fundamental circuit variables, the next question we need to address is: when are the techniques discussed in this book valid? The answer to this question is of paramount importance because the domain of application for circuit theory is extremely broad. For example, the size of circuits varies enormously: from very large-scale integrated circuits which include over a billion transistors on a chip the size of a fingernail to telecommunication circuits and power networks that span continents [9]. Throughout this book we shall consider only **lumped circuits** [9]. For a physical circuit to be considered lumped, its physical dimension must be small enough so that, for the problem at hand, **electromagnetic waves propagate across the circuit virtually instantaneously**. Consider the following example:

Example 1.3.1 Consider an audio circuit whose highest frequency of interest is $f = 20$ KHz. Discuss the lumped circuit approximation.

Solution For electromagnetic waves, $f = 20$ KHz corresponds to a wavelength of:

$$\lambda = \frac{c}{f}$$

$$= \frac{3 \times 10^8 \, \text{m/s}}{2.0 \times 10^4 \, \text{s}^{-1}}$$

$$= 15 \, \text{km}$$

Based on the calculations above, even if the circuit is spread across a tennis court, the size of the circuit is very small compared to the shortest wavelength of interest λ.

Definition 1.1 (Lumped Circuit Approximation) Lumped circuit approximation is valid if $d \ll c \cdot \Delta t$, where d is the largest dimension of the circuit, Δt the shortest time of interest, and c is the velocity of light.

When the conditions in Definition 1.1 are satisfied, electromagnetic theory proves [10] and experiments show that the lumped circuit approximation holds; namely, throughout the physical circuit the current $i(t)$ through any device terminal and the voltage difference $v(t)$ across any part of terminals, at any time t, are well-defined.[3]

---

[3]Unless otherwise stated, we will assume from now on throughout the book that analogous statements are true for $q(t)$ and $\phi(t)$. In this case, we can equivalently discuss $q(t)$ through any device terminal and $\phi(t)$ across any part of the terminals.

Example 1.3.2 Consider a circuit on a chip whose extent is 1 mm. Let the shortest signal time of interest be 0.1 ns. Discuss the lumped circuit approximation from Definition 1.1.

Solution Again, since electromagnetic waves travel at the speed of light, the time it would take for the electromagnetic wave to travel 1 mm is:

$$t = \frac{d}{c}$$

$$= \frac{1 \times 10^{-3}\,\text{m}}{3 \times 10^{8}\,\text{m/s}}$$

$$= 3.\bar{3}\,\text{ps}$$

Therefore the propagation time in comparison with the shortest signal time of interest is negligible and hence the lumped circuit approximation is valid.

Based on Examples 1.3.1 and 1.3.2, roughly speaking, the higher the frequency of operation, the smaller the device's physical dimension in order for the lumped circuit approximation to be satisfied. From an electromagnetic theory point of view, a lumped circuit reduces to a point since it is based on the approximation that electromagnetic waves propagate through the circuit instantaneously. For this reason, in lumped circuit theory, the respective locations of the elements of the circuit will not affect the behavior of the circuit. The approximation of a physical circuit by a lumped circuit is analogous to the modeling of a rigid body as a particle: in doing so, all data relating to the extent (shape, size, orientation, etc.) of the body are ignored by the theory.

In situations where lumped approximation is invalid, the physical dimensions of the circuit must be considered. To distinguish such circuits from lumped circuits we call them **distributed circuits**, typical examples are transmission lines and waveguides. In distributed circuits, the circuit variables depend not only on time, but also on space variables such as length and width. We need electromagnetic theory for predictions of the behavior of distributed circuits and hence they will not be discussed in this book.

1.4 Reference Directions

One of the most basic concepts in physical science is that any physical quantity is invariably measured with respect to some "assumed" frame of reference [10]. In electrical network theory, the frame of reference takes the form of an assumed reference direction of the current i and an assumed reference polarity of the

Fig. 1.2 An experiment
demonstrating that regardless
of which terminal of the black
box is chosen to be positive,
the actual voltage across
terminals a–b can be
unambiguously specified for
all time

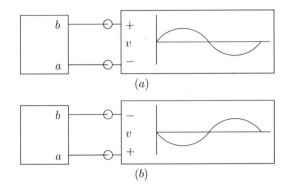

voltage v. A thorough understanding of the concept of reference current direction
and reference voltage polarity is absolutely essential in the study of (nonlinear)
network theory. It is a fact that a large percentage of the mistakes committed by
students of network theory can be traced to either the students' underestimation
of the full significance of reference current directions and voltage polarities or the
students' failure to maintain a consistent set of references.

The simplest way to understand the concept of assumed reference direction and
polarity is through the experiment illustrated in Fig. 1.2. We will discuss reference
voltage polarity. An analogous discussion holds for the reference current direction.
Suppose we are given a black box with a pair of **accessible ports** or **terminals** a–b,
as shown in Fig. 1.2, and we are required to measure the voltage across terminals
a–b. Let us measure the voltage by connecting a–b to the vertical input terminals
of an oscilloscope. Since one of the two vertical input terminals of an oscilloscope
is marked with a positive sign while the other is marked with a negative sign, the
question that immediately arises is which of the two terminals of the black box
should we connect to the positive terminal of the oscilloscope in order to obtain the
desired information?

The answer is that it does not matter. In order to see this, suppose we **arbitrarily
assume** terminal b is connected to the positive terminal as shown in Fig. 1.2a. The
assumption that terminal b is at the positive terminal does not mean that terminal
b is at a higher potential than terminal a. It does mean however that if at any time
$t = t_1$, $v(t_1) > 0$, then the potential at b is higher than the potential at a. On the
other hand, if $v(t_1) < 0$, then the potential at b at $t = t_1$ is actually lower than
the potential at a. For example, if the voltage $v(t)$ displayed on the oscilloscope (in
volts) is

$$v(t) = 10 \sin \pi t \tag{1.5}$$

then terminal b is at a higher potential than terminal a during the time interval
$0 < t < 1$ s. But during the time interval $1 < t < 2$ s, terminal b is at a lower
potential than terminal a.

Let us now consider what happens when we assume terminal a is connected to the positive terminal of the oscilloscope, instead of terminal b, as shown in Fig. 1.2b. Since this connection is opposite to the connection in Fig. 1.2a, it is clear that the voltage displayed on the oscilloscope (in volts) is given by:

$$v(t) = -10 \sin \pi t \tag{1.6}$$

Thus in either case, the final answers are identical. We can therefore conclude that in order to specify the voltage between any pair of terminals unambiguously, we may arbitrarily assume any one of the two possible terminals to be positive. By analogy, we can conclude that in order to specify the current through any wire unambiguously, we may arbitrarily assume any one of the possible two directions to be positive.

Let us consider next a two-terminal black box N and assume a reference direction for the terminal current i and a reference polarity for the terminal voltage v, see Fig. 1.3. Since the references for both i and v are arbitrary, there are four distinct sets of combinations of references. There is no reason to prefer any one combination over the others. However, in practice, it is usually convenient to choose the combination so that **positive** power represents power **entering** the black box.

From classical mechanics, we know that power is defined by Eq. (1.7).

$$p = \frac{dw}{dt} \tag{1.7}$$

But,

$$vi = \frac{dw}{dq}\frac{dq}{dt}$$
$$= \frac{dw}{dt} \tag{1.8}$$

Fig. 1.3 Two possible sets of reference direction for the passive sign convention from Definition 1.2

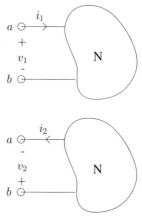

Thus we have:

$$p(t) = v(t)i(t) \tag{1.9}$$

From the simultaneity postulate, the same current must leave the negative terminal.[4] Hence, based on this observation, we have Definition 1.2.

Definition 1.2 (Associated Reference Direction or Passive Sign Convention) Whenever the reference direction for the current i in a two-terminal black box is in the direction of the reference voltage drop v across the black box ($v > 0, i > 0$), we use a positive sign in any expression that relates voltage to current. Otherwise, we use a negative sign.

Thus Definition 1.2 implies that the allowable reference combination must be either of the form shown in Fig. 1.3.

1.5 Kirchhoff's Laws

When circuit elements are interconnected to form a circuit, there are some governing laws that all elements in the network must obey. We shall refer to these laws as the **laws of interconnection**. Before we discuss these laws, we need the following definitions:

Definition 1.3 A **node** is a point in a circuit where two or more circuit elements are interconnected.

Definition 1.4 A **path** is a trace of adjoining elements, with no elements included more than once.

Definition 1.5 A **closed node sequence** is a path whose last node is the same as the starting node.

Definition 1.6 A **loop** is a closed node sequence that traverses only through two-terminal elements.

Definition 1.7 A **branch** is a path that connects two nodes.

Definition 1.8 A **connected circuit** is one in which any node can be reached from any other node, by traversing a path through the circuit elements.

Now, given any connected lumped circuit having n nodes, we may choose (arbitrarily) one of the nodes as a **ground** node, i.e., as a reference for measuring electric potentials. Note that a circuit does not have to be physically connected to ground for proper functionality, think about circuits inside our mobile phones.

---

[4]This is also a consequence of Kirchhoff's Current Law, see Sect. 1.5.1.

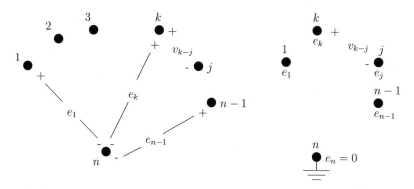

Fig. 1.4 Labeling node-to-ground voltages for a circuit with n nodes

With respect to the chosen ground node, we define $n - 1$ node-to-ground voltages as shown in Fig. 1.4. Since the circuit is a connected lumped circuit, these $n - 1$ voltages are well-defined and, in principle, physically measurable quantities. Henceforth, we shall label them $e_1, e_2, \ldots, e_{n-1}$ and dispense with the $+$ and $-$ signs indicating voltage reference direction. Note that $e_n = 0$ since node n is chosen as the ground node.

1.5.1 Kirchhoff's Current Law (KCL)

A fundamental law of physics asserts that electrical charge is conserved: There is no known experiment in which a net electric charge is either created or destroyed. KCL expresses this fundamental law in the context of lumped circuits. To state KCL, we first need the definition of a gaussian surface.

Definition 1.9 A **gaussian surface** $\mathscr{S}$ is a two-sided closed surface, that has an "inside" and an "outside."

To express the fact that the sum of the charges inside $\mathscr{S}$ is constant, we shall require that at all times, the algebraic sum of all the currents leaving the surface $\mathscr{S}$ is equal to zero.

Definition 1.10 KCL: For all lumped circuits, for all $\mathscr{S}$, for all times t, the algebraic sum of all the currents **leaving** $\mathscr{S}$ at time t is equal to zero.

We will choose $\mathscr{S}$ so that it cuts only the wires which connect the circuit elements, as discussed in Example 1.5.1.

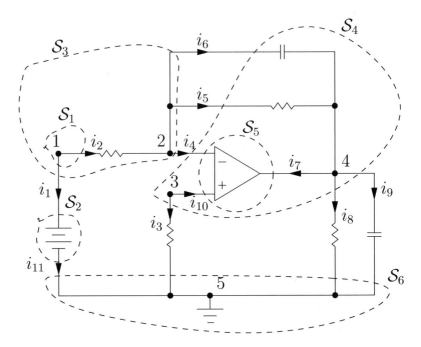

Fig. 1.5 An opamp circuit illustrating gaussian surfaces and KCL

Example 1.5.1 Write KCL expressions for the circuit in Fig. 1.5.

Solution In Fig. 1.5, we have used two-terminal elements and a three-terminal ideal operational amplifier (opamp) (that we will discuss in Sect. 2.5). In the figure, we have drawn six gaussian surfaces $\mathscr{S}_1, \mathscr{S}_2, \cdots, \mathscr{S}_6$. We will use these surfaces to illustrate KCL.

For $\mathscr{S}_1$, KCL states:

$$i_1(t) + i_2(t) = 0 \quad \forall t \tag{1.10}$$

Note that $\mathscr{S}_1$ contains only node 1 in its "inside." Thus a node may be considered as a special case of $\mathscr{S}$, i.e., the surface is shrunk to a point.

For $\mathscr{S}_2$, KCL states:

$$-i_1(t) + i_{11}(t) = 0 \tag{1.11}$$

Note that $\mathscr{S}_2$ encloses a two-terminal element. Thus we make the conclusion that for a **two-terminal element**, the current entering the element from one

(continued)

Example 1.5.1 (continued)
node at any time t is equal to the current leaving the element from the other node at t.

For $\mathscr{S}_3$, KCL states:

$$i_1(t) + i_4(t) + i_5(t) + i_6(t) = 0 \tag{1.12}$$

For $\mathscr{S}_4$, KCL states:

$$i_3(t) + i_8(t) + i_9(t) - i_4(t) - i_5(t) - i_6(t) = 0 \tag{1.13}$$

For $\mathscr{S}_5$, KCL states:

$$-i_4(t) - i_7(t) - i_{10}(t) = 0 \tag{1.14}$$

Note that these are the three currents pertaining to the opamp. Thus choosing an $\mathscr{S}$ that encloses any n-terminal element, we state that the algebraic sum of the currents leaving or entering the n-terminal element is equal to zero at all times t. n-terminal elements will be covered in more detail in Chap. 2.

For $\mathscr{S}_6$ (that encloses only the reference node), KCL states:

$$-i_3(t) - i_8(t) - i_9(t) - i_{11}(t) = 0 \tag{1.15}$$

We conclude this section by stating KCL for nodes:

Definition 1.11 KCL (Node Law): For all lumped circuits, for all $\mathscr{S}$, for all times t, the algebraic sum of currents **leaving** any node is equal to zero.

1.5.2 Kirchhoff's Voltage Law (KVL)

Let v_{k-j} denote the voltage difference between node k and node j as shown in Fig. 1.4. Kirchhoff's voltage law states:

Definition 1.12 KVL: For all lumped connected circuits, for all choices of ground node, for all times t, for all pairs of nodes k and j,

$$v_{k-j}(t) = e_k(t) - e_j(t) \tag{1.16}$$

Fig. 1.6 Circuit for
Example 1.5.2

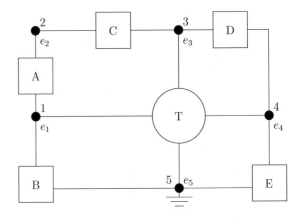

Example 1.5.2 Write KVL expressions for the circuit in Fig. 1.6.

Solution The connected circuit in Fig. 1.6 is made of 5 two-terminal
elements and 1 four-terminal element. There are five nodes. Choosing (arbitrarily) node 5 as the ground node, we define the four node-to-ground voltages
e_1, e_2, e_3, and e_4. Therefore by KVL, we may write the following seven
equations (for convenience, we drop the dependence on t):

$$v_{1-2} = e_1 - e_2$$

$$v_{2-3} = e_2 - e_3$$

$$v_{3-4} = e_3 - e_4$$

$$v_{1-4} = e_1 - e_4$$

$$v_{4-5} = e_4 - e_5 = e_4$$

$$v_{5-1} = e_5 - e_1 = -e_1 \tag{1.17}$$

Note that $v_{1-2}, v_{2-3}, v_{3-4}, v_{4-5}, v_{5-1}$ are the voltages across the two-
terminal elements A, C, D, E, B, respectively; v_{1-4}, v_{4-5} and v_{5-1} are the
voltages across the node pairs (1,4); (4,5) and (5,1) of the four-terminal
element T, respectively.

If we add the last three equations in Eq. (1.17), we find that:

$$v_{1-4} + v_{4-5} + v_{5-1} = 0 \tag{1.18}$$

Hence for this particular closed node sequence, the sum of the voltages is
equal to zero.

(continued)

Example 1.5.2 (continued)
Note also that if we add the first three and last two equations in Eq. (1.17), we find the sum of voltages around a loop is zero:

$$v_{1-2} + v_{2-3} + v_{3-4} + v_{4-5} + v_{5-1} = 0 \qquad (1.19)$$

Example 1.5.2 shows that we can state KVL in terms of closed node sequences:

Definition 1.13 KVL (Closed Node Sequences): For all lumped connected circuits, for all closed node sequences, for all times t, the algebraic sum of all node-to-node voltages around the chosen closed node sequence is equal to zero.

1.6 From Circuits to Graphs: The Definition of a Network

It should be clear from our discussions of KCL and KVL that the equations arising from laws of interconnection are independent of the type of elements in a network. We will now state the definition of a network.

Definition 1.14 A **network** is any interconnection of circuit elements.

Only the network connection diagram, the **topology**, needs to be specified in order to obtain the equations to the laws of interconnection. The topology of a circuit is best exhibited by way of a **graph**.

Definition 1.15 A **graph** $\mathscr{G}$ is specified by a set of nodes $\{1, 2, \cdots, n\}$ together with a set of branches $\{\beta_1, \beta_2, \cdots, \beta_n\}$.

If each branch is given an orientation, indicated by an arrow on the branch, we call the graph directed. For example, a two-terminal element and the associated **element graph** is shown in Fig. 1.7.

Notice that the **element graph** for a two-terminal element has two nodes and one branch. Also note that the directions of the current flow through and voltage drop across the two-terminal element are specified using the passive sign convention from Definition 1.2.

Fig. 1.7 A two-terminal element and its associated element graph representation

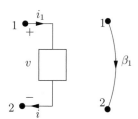

Fig. 1.8 Digraph associated
with the circuit in Fig. 1.5.
Detailed derivation of the
opamp digraph will be
covered in Sect. 2.5

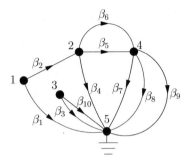

For a given circuit, if we replace each element by its associated element graph,
we obtain the **directed circuit graph** or **digraph** $\mathcal{G}$. In this book, whenever we
refer to a network, we mean the associated digraph of the circuit. We can use
either the digraph or the circuit for analysis. Hence, throughout this book, we
will use "circuits" and "networks" interchangeably. Note however that the laws of
interconnection in circuit theory, such as Tellegen's theorem from Sect. 1.6.1, are
essentially graph-theoretic concepts that arise from a network (associated with a
given circuit).

Example 1.6.1 Write KCL and KVL expressions for the digraph in Fig. 1.8.

Solution It is interesting to note that since the circuit contains a three-
terminal element, the digraph bears little resemblance to the circuit. In fact,
given the digraph, without specifying which nodes belong to the three-
terminal element, it is not possible to reconstruct the circuit. This observation
is false if the circuit contains only two-terminal elements.
  KCL gives:

$$i_1 + i_2 = 0$$

$$-i_2 + i_4 + i_5 + i_6 = 0$$

$$i_3 + i_{10} = 0$$

$$-i_5 - i_6 + i_7 + i_8 + i_9 = 0 \qquad\qquad (1.20)$$

(continued)

Example 1.6.1 (continued)

Let us rewrite Eq. (1.20) in matrix form:

$$
\begin{pmatrix}
1 & 1 & 0 & 0 & 0 & 0 & 0 & 0 & 0 & 0 \\
0 & -1 & 0 & 1 & 1 & 1 & 0 & 0 & 0 & 0 \\
0 & 0 & 1 & 0 & 0 & 0 & 0 & 0 & 0 & 1 \\
0 & 0 & 0 & 0 & -1 & -1 & 1 & 1 & 1 & 0
\end{pmatrix}
\begin{pmatrix}
i_1 \\ i_2 \\ i_3 \\ i_4 \\ i_5 \\ i_6 \\ i_7 \\ i_8 \\ i_9 \\ i_{10}
\end{pmatrix}
=
\begin{pmatrix}
0 \\ 0 \\ 0 \\ 0 \\ 0 \\ 0 \\ 0 \\ 0 \\ 0 \\ 0
\end{pmatrix}
\tag{1.21}
$$

Let:

$$
\mathbf{A} \triangleq
\begin{pmatrix}
1 & 1 & 0 & 0 & 0 & 0 & 0 & 0 & 0 & 0 \\
0 & -1 & 0 & 1 & 1 & 1 & 0 & 0 & 0 & 0 \\
0 & 0 & 1 & 0 & 0 & 0 & 0 & 0 & 0 & 1 \\
0 & 0 & 0 & 0 & -1 & -1 & 1 & 1 & 1 & 0
\end{pmatrix}
\tag{1.22}
$$

Thus, Eq. (1.21) can be written as:

$$
\mathbf{A}i = \mathbf{0}
\tag{1.23}
$$

Matrix **A** is called the **incidence matrix** (more details in Sect. 3.2.2).

We can express all ten branch voltages in terms of the reference node by using KVL:

$$v_1 = e_1$$

$$v_2 = e_1 - e_2$$

$$v_3 = e_3$$

$$v_4 = e_2$$

$$v_5 = e_2 - e_4$$

$$v_6 = e_2 - e_4$$

$$v_7 = e_4$$

$$v_8 = e_4$$

$$v_9 = e_4$$

$$v_{10} = e_3 \tag{1.24}$$

(continued)

Example 1.6.1 (continued)
Rewriting Eq. (1.24) in matrix form:

$$
\begin{pmatrix}
1 & 0 & 0 & 0 \\
1 & -1 & 0 & 0 \\
0 & 0 & 1 & 0 \\
0 & 1 & 0 & 0 \\
0 & 1 & 0 & -1 \\
0 & 1 & 0 & -1 \\
0 & 0 & 0 & 1 \\
0 & 0 & 0 & 1 \\
0 & 0 & 0 & 1 \\
0 & 0 & 1 & 0
\end{pmatrix}
\begin{pmatrix}
e_1 \\
e_2 \\
e_3 \\
e_4
\end{pmatrix}
=
\begin{pmatrix}
v_1 \\
v_2 \\
v_3 \\
v_4 \\
v_5 \\
v_6 \\
v_7 \\
v_8 \\
v_9 \\
v_{10}
\end{pmatrix}
\tag{1.25}
$$

Comparing Eqs. (1.21) and (1.25), we can see that the constant matrix on the LHS of Eq. (1.25) is A^T. Hence Eq. (1.25) can be written as:

$$\mathbf{A}^T \mathbf{e} = \mathbf{v} \tag{1.26}$$

Much more will be said about topological concepts in circuit theory throughout this book. Specifically, element graphs for multi-terminal elements will be discussed in Chap. 2. We will formalize the matrix formulation of Kirchhoff's laws in Chap. 3, before we discuss general circuit analysis techniques.

1.6.1 Generality of Digraphs: Tellegen's Theorem Example

To further illustrate the generality of the digraph approach, we will give an example of Tellegen's theorem [28]. We will formally state and prove the theorem in Chap. 3.
Consider the circuits in Fig. 1.9.
KVL and KCL in matrix form for the circuit in Fig. 1.9a are:

$$\mathbf{A}^T \mathbf{e} = \mathbf{v}$$
$$\mathbf{A} \mathbf{i} = \mathbf{0} \tag{1.27}$$

where:

$$
\mathbf{A} =
\begin{pmatrix}
1 & 1 & 0 & 0 & 1 & 0 \\
0 & -1 & 1 & 1 & 0 & 0 \\
0 & 0 & 0 & -1 & -1 & 1
\end{pmatrix}
\tag{1.28}
$$

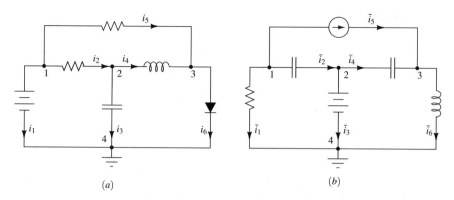

Fig. 1.9 Circuits for understanding Tellegen's theorem

Let the branch power $v_k i_k$ be summed for all N branches of the circuit. Then, by Eq. (1.27):

$$\sum_{k=1}^{N} v_k i_k = \mathbf{v}^T \mathbf{i}$$

$$= (\mathbf{A}^T \mathbf{e})^T \mathbf{i}$$

$$= \mathbf{e}^T (\mathbf{A}\mathbf{i})$$

$$= 0 \qquad\qquad\qquad (1.29)$$

In deriving Eq. (1.29), the familiar rules $(\mathbf{AB})^T = \mathbf{B}^T \mathbf{A}^T$, $(\mathbf{A}^T)^T = \mathbf{A}$, $\mathbf{e}^T \mathbf{0} = 0$ of vector algebra have been used.

The result in Eq. (1.29) should not be surprising since we have derived the conservation of power in a circuit from Kirchhoff's laws.

Consider however the circuit of Fig. 1.9b which has the same topological configuration, same reference directions and numbering, and hence the same $\mathbf{A}$ as the circuit in Fig. 1.9a. Hence, the incidence matrix for the circuit in Fig. 1.9b is also given by Eq. (1.28). Let the electrical quantities of the circuit be $\tilde{\mathbf{i}}, \tilde{\mathbf{v}}, \tilde{\mathbf{e}}$ in Fig. 1.9b. Then:

$$\mathbf{A}^T \tilde{\mathbf{e}} = \tilde{\mathbf{v}}$$

$$\mathbf{A}\tilde{\mathbf{i}} = \mathbf{0} \qquad\qquad\qquad (1.30)$$

Now consider:

$$\sum_{k=1}^{N} v_k \tilde{i}_k = \mathbf{v}^T \tilde{\mathbf{i}}$$

$$= (\mathbf{A}^T \mathbf{e})^T \tilde{\mathbf{i}}$$

$$= \mathbf{e}^T (\mathbf{A} \tilde{\mathbf{i}})$$

$$= 0 \qquad\qquad\qquad (1.31)$$

While the LHS of Eq. (1.31) has the dimensions of power, the quantity is physically meaningless since v_k and $\tilde{i}_k$ exist in two different circuits.

Similarly, we can show:

$$\tilde{\mathbf{v}}^T \mathbf{i} = 0 \qquad\qquad\qquad (1.32)$$

Equations (1.31) and (1.32) are general forms of Tellegen's theorem. We will apply Tellegen's theorem to derive some general properties of nonlinear resistive circuits in Chap. 3.

1.7 Circuit Theory from Electromagnetic Field Theory

Now that we have an understanding of the laws of interconnection, we will have a short discussion in this section on how to arrive at these laws, by using the fact that circuit theory is an approximation of electromagnetic field theory. Although we are only concerned with lumped circuits in this book, this (very short) section should be helpful because the approximation techniques used have roots in the very important concept of modeling [10].

1.7.1 The Art of Modeling

Engineers and scientists seldom analyze a physical system in its original form. Instead, they construct a model which approximates the behavior of the system. By analyzing the behavior of the model, they hope to predict the behavior of the actual system. The primary reason for constructing models is that physical systems are usually too complex to be amenable to a practical analysis. In most cases, the complexity of a system is due in part to the presence of many nonessential factors. **The basic principle of modeling consists, therefore, of extracting only the essential factors.**

As stated earlier, an electric circuit is an interconnection of electrical devices. We will encounter a plethora of electrical devices (diodes, transistors to name a few) throughout this book. But one has to again understand that we must extract only the essential factors of the device, based on the circuit in question. A classic example is frequency behavior. **No device behaves the same at all frequencies.** For example, at "low enough" frequencies, metallic parallel plates separated by a (small, compared to the width of the parallel plates) distance act like a capacitor. Increasing the frequency will lead to more complicated behavior, and the device does not behave like an ideal capacitor anymore. Hence, it is vital that the reader understands **device modeling is still more of an "art" than science**. Although **no general theory of device modeling** is presently available, there are **a variety of techniques available** that will help us model physical devices, in terms of the four fundamental circuit elements.

We will start discussing modeling of the fundamental circuit elements in Sect. 1.8 and continue the discussion throughout the book. With respect to circuit theory being an approximation of electromagnetic field theory, we will only discuss the laws of interconnection, namely KCL and KVL. A detailed discussion of how we get the terminal behavior for the four fundamental circuit elements from field theory is beyond the scope of this book. We hope to discuss this in our followup book, "Advanced Nonlinear Circuits and Networks."

1.7.2 KCL and KVL from Field Theory: A Very Brief Overview

From the node form of KCL in Definition 1.11, we know that the sum of the currents flowing out of a node must be equal to zero. From field theory (specifically Gauss' law), the surface integral of the current density over a closed surface must be equal to zero if no charge accumulates inside that surface. Definition 1.3 of a node implies that a node is a theoretical abstraction of a physical interconnection of wires: a node does not have any circuit elements such as capacitors associated with it. Hence, no charge can accumulate on a node, and the sum of currents leaving the node must be equal to zero.

KVL from Definition 1.13 is equivalent to Faraday's law of induction from Maxwell's theory of electromagnetism. This equivalence, however, is not directly evident as the relation between KCL and the law of conservation of charge. Indeed, KVL depends on how the branch voltages are defined in terms of the electromagnetic field. These details are also beyond the scope of this book, and will be discussed in our follow-up volume. But, we can get an intuitive idea by considering the fact that we defined branch voltage as the difference between node-to-ground voltages in Eq. (1.16). In fact, a practical device for measuring branch voltage—the voltmeter—is connected such that voltage is measured across a pair of nodes. Hence a voltmeter is designed to measure the line integral of the electric field along the path formed by the connecting leads. Thus, the sum of voltages around a closed loop in a circuit has the electromagnetic equivalent of the electric field around a closed path. The electric field involved in this integration is, by assumption, equal (or approximately so) to the negative gradient of a scalar potential [10]. Therefore, the line integral of the electric field should vanish, and this gives us KVL.

1.8   Characterization of a Two-Terminal Black Box

Now that we have discussed interconnection of circuit elements, it is time to discuss the circuit elements themselves. Although we will encounter many physical devices of varying complexity throughout this book, we will model them as black boxes [10]. These boxes may possess many terminals, but **only two of these are accessible to the external world in the sense that the device may be excited only through these terminals**. For our purpose, it is convenient to imagine that the device is enclosed in a box and that the two accessible terminals are brought out by two connecting wires, with the symbol shown in Fig. 1.10a.

It is important to emphasize that the content of the black box may be as simple as a light bulb, or as complicated as an arbitrary interconnection of black boxes as shown in Fig. 1.10b.

The choice of the term "black box" is quite appropriate here because the box is really black inside in the sense that we cannot see its contents. As a matter of fact, unless we open the box and peep inside, there is no way of determining its contents. However, as engineers, we are not so much interested in the contents of the box as in knowing what the box is capable of and how it behaves externally when it is connected with other black boxes into a network. In other words, we are primarily interested in predicting the external behavior of the black box. Our first step toward such an analytical approach is to "characterize" the black box. **To properly characterize a black box, it is paramount that we choose the correct set of terminal variables.** We will illustrate this idea in this section by modeling a "spring" from basic physics, refer to Fig. 1.11.

Suppose we did not know that in reality we had a spring inside the black box and we were asked to predict the behavior of the external terminals when an arbitrary force $f(t)$ is applied to one end (terminal) of the spring while the other

Fig. 1.10 Symbolic representation of two-terminal black boxes

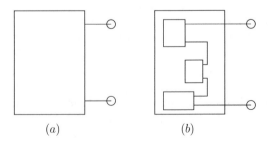

(a) (b)

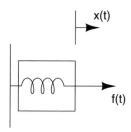

Fig. 1.11   An example illustrating the characterization of a mechanical black box

end (terminal) is fixed against a wall. The mechanical variables of interest here are the displacement x (displacement to the right of the initial 0 position is assumed positive, as shown in Fig. 1.11), velocity v of the terminal that is free to move and the force f (positive for tension, negative for compression).

Clearly the only way we can hope to characterize this black box (other than opening the box) is to start performing some experiments. Suppose we begin by applying a constant force $f = A$ and measure the corresponding velocity v. This would give us a point in the velocity-vs-force (v–f) plane.[5] By repeating the above experiment with several values of the force f, we obtain the data shown in Fig. 1.12.

We might be tempted to draw a smooth curve through these data points (which in this case happens to be the f axis) and claim to have characterized the black box in the sense that given any constant force f, we can analytically predict the associated velocity.

However a little thought will show that we have not really characterized the black box yet, for if, instead of applying a constant force we apply a slowly varying sinusoidal force, $f(t) = A\sin(t)$. The characteristics in Fig. 1.12 would predict that $v = 0$.

This is of course contrary to what we observe experimentally: namely, $v(t) = (A/k)\cos(t)$, where k is the spring constant. We might hope that this inconsistency can be resolved by plotting all points (v, f). Nevertheless we will again quickly conclude that the length of both axes of the resulting ellipse depends on the amplitude A of the applied force f. For each A we will obtain a unique ellipse and thus we will eventually fill the entire v–f plane. Even if we could draw an infinite set of ellipses, we would be able to predict the velocity only if f is sinusoidal. Using these ellipses to predict v due to non-sinusoidal f would again yield incorrect answers. We must now realize that the useful information we obtained from this experiment is that the black box cannot be characterized by a curve in the v–f plane.

---

[5]When we say x–y plane, we denote specifically x as the horizontal axis and y as the vertical axis of the plane. This is consistent with the conventional usage where the first variable denotes the abscissa and the second variable denotes the ordinate.

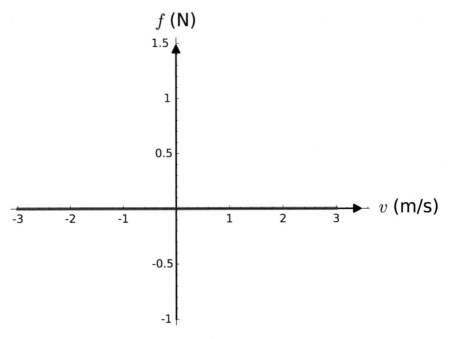

Fig. 1.12 Force-velocity plot for constant force $f = A$

Suppose we try another set of variables, say the force f and displacement x. Repeating the experiments, we will find that **provided $f(t)$ does not change rapidly, the black box can be characterized by a curve in the x–f plane.**

After experiencing the length of time needed to carry out the above experiments, we can now begin to appreciate the utility of such a conclusion; namely, the characterization of the black box permits an analytical solution and thereby eliminates the need to carry out any further experiments.

Observe however that our conclusion is based on the assumption that $f(t)$ does not change rapidly. If we were to repeat our experiment with higher-frequency sinusoidal waveforms, as well as non-sinusoidal waveforms which change rapidly, we will find deviations from our conclusions drawn using low frequency waveforms. This will suggest that our earlier assumption, that f does not change rapidly, is indeed necessary. In order to emphasize this restriction, it is a common practice to call a black box characterization as **static characterization**, in contrast to a **dynamic characterization** for higher frequencies. Hence for the black box in Fig. 1.11, the f–x curve is the static characteristic.

Since the deviation of the measured characteristic from the static characteristic increases slowly with frequency rather than abruptly, it is impossible to pick a definite frequency above which the static characteristic does not hold. Neither is it possible to find a single dynamic characteristic that would hold for all frequencies. Hence a certain amount of scientific judgment is involved in deciding whether a certain static characteristic curve can be used to satisfactorily solve a given problem.

It is encouraging, however, to know that a large percentage of practical networks can indeed be analyzed using only static characteristics. Moreover, even in cases where the static characteristic fails to give satisfactory solutions, we shall show in future chapters that we can often patch up the error by including "parasitic elements," namely, elements which are undesirable but which are invariably present in the black box in small quantities. Thus, in this book, **we will assume all characteristics are static and will utilize parasitic elements to model the necessary dynamic characteristics. We shall henceforth delete the adjective "static."**

For the example in Fig. 1.11, the parasitic element consists of the mass associated with the spring. At low frequencies, the mass being quite small, has relatively no effect on the $f-x$ curve. However as the frequency of the external force increases, the acceleration of the spring increases and the inertia force due to the mass becomes appreciable.

1.9 Two-Terminal Elements

From the previous section, we know that it is essential to choose the correct set of variables for characterizing a black box. For two-terminal elements, the circuit variables of interest are those that can be measured externally. Hence the terminal voltage v and terminal current i are of primary interest because they can be readily measured. The charge q and flux-linkage ϕ are also of interest because they can be indirectly measured by **integrating** the measured current $i(t)$ and voltage $v(t)$, respectively. From these measurements, we shall then try to establish a relationship, if any, between each pair of **independent** variables.

i and q are related by Eq. (1.3); v and ϕ are related by Eq. (1.4). Hence the only remaining combinations consist therefore of the relationship between the following variables.

1. Relationship between v and i, this is the two-terminal resistor shown in the top-left corner of Fig. 1.1.
2. Relationship between ϕ and i, this is the two-terminal inductor shown in the bottom-left corner of Fig. 1.1.
3. Relationship between ϕ and q, this is the two-terminal memristor shown in the bottom-right corner of Fig. 1.1.
4. Relationship between v and q, this is the two-terminal capacitor shown in the top-right corner of Fig. 1.1.

We will now discuss each of these elements in detail. But, before we begin our discussion of two-terminal elements, two important remarks:

- Time-varying elements: each of the four fundamental circuit elements we will discuss can be time-varying. For instance, a time-varying resistor is defined by the relation: $f_R(v, i, t) = 0$. A very simple example is a potentiometer (or variable resistor), whose arm is being rotated by say a motor. Nevertheless, the analysis of a nonlinear network containing time-varying elements is a very

Fig. 1.13 Symbol for a linear
resistor with resistance R

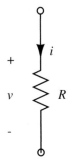

difficult mathematical problem requiring advanced mathematics. Hence we will
primarily discuss nonlinear time-invariant elements in this book and restrict our
discussion of time-varying elements to a few examples.

- Memristors: as we will see in Sect. 1.9.4, this device is the fourth fundamental
  circuit element. However, although a variety of circuit theoretic properties
  of the memristor can be obtained by studying the terminal behavior, a true
  understanding of a memristor's behavior requires us (unlike the other resistor,
  capacitor and inductor) to "peer inside" the black box (see Sect. 4.4.2).

1.9.1 Resistors

The **linear resistor** is probably the most familiar circuit element that one encounters
in basic physics. This device satisfies Ohm's law: that is, the voltage across such an
element is proportional to the current flowing through it. We represent it by the
symbol shown in Fig. 1.13 where the current i through the resistor and the voltage
v across it are measured using the passive sign convention from Definition 1.2.

Ohm's law states that at all times.

$$v(t) = Ri(t) \; or$$
$$i(t) = Gv(t) \tag{1.33}$$

where the constant R is the **resistance**[6] of the linear resistor (measured in the unit
of ohms (Ω)) and G is the **conductance** measured in units of siemens (S).

Equation (1.33) can be plotted on the i–v plane or v–i plane as shown in
Fig. 1.14.

---

[6]In nonlinear circuits, terms such as resistance (capacitance, inductance) become ambiguous and
only the terms "resistor, capacitor and inductor" should be used. Nevertheless, we will refer to
the appropriate small-signal quantity with the terms "resistance," "capacitance," and "inductance."
Note that there is no confusion with respect to "memristance," since the memristor is fundamentally
a nonlinear element.

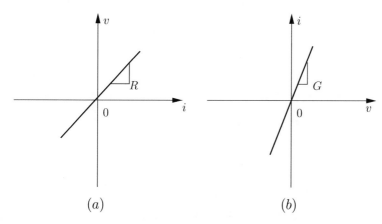

Fig. 1.14 Linear resistor characteristic plotted on the (**a**) i–v and (**b**) v–i plane

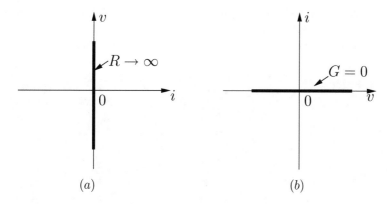

Fig. 1.15 Characteristic of an open circuit (**a**) i-v plane (**b**) v-i plane

There are two special cases of linear resistors which deserve special mention, namely, the **open circuit** and **short circuit**.

Definition 1.16 (Open Circuit) A two-terminal resistor is called an open circuit iff its current i is identically zero irrespective of the voltage v; i.e., $f(v, i) = i = 0$.

The characteristic of an open circuit is the v axis in the v–i plane, with zero slope ($G = 0$). In the i–v plane, it has an infinite slope, $R \rightarrow \infty$, refer to Fig. 1.15

Definition 1.17 (Short Circuit) A two-terminal resistor is called a short circuit iff its voltage v is identically zero irrespective of the current i; i.e., $f(v, i) = v = 0$.

The characteristic of a short circuit is the i axis in the v–i plane, with $G \rightarrow \infty$. In the i–v plane, the characteristic has zero slope, $G = 0$, refer to Fig. 1.16.

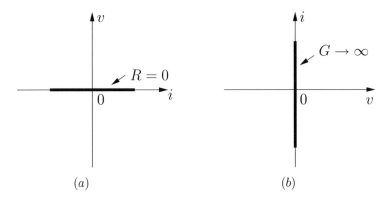

Fig. 1.16 Characteristic of a short circuit (**a**) i-v plane (**b**) v-i plane

Comparing Figs. 1.15 and 1.16, we see that the curve of the open circuit in one plane is identical to the curve of the short circuit in the other plane. For this reason, the open circuit is said to be the **dual** of the short circuit and vice versa.

Example 1.9.1 A linear resistor with resistance of $100\,\Omega$ is given. What is its dual?

Solution Consider a linear resistor with resistance $R = \frac{1}{100}\,\Omega$. The i–v and v–i characteristics are plotted in Fig. 1.17. Notice how the $i_1 - v_1$ characteristic of the resistor with $R = 100\,\Omega$ is equivalent to the $v_4 - i_4$ characteristic of the resistor with $R = \frac{1}{100}\,\Omega$. Similarly, the $v_2 - i_2$ characteristic for resistor with $R = 100\,\Omega$ is equivalent to the $i_3 - v_3$ characteristic for resistor with $R = \frac{1}{100}\,\Omega$. Hence the dual is a resistor with $R = \frac{1}{100}\,\Omega$.

From Eq. (1.9), the power delivered to a linear resistor at time t by the remainder of the circuit to which it is connected is:

$$p(t) = v(t)i(t)$$
$$= Ri^2(t)$$
$$= Gv^2(t) \tag{1.34}$$

Thus the power delivered to a linear resistor is always non-negative if $R \geq 0$. We say that a linear resistor is passive iff its resistance is non-negative. Thus a passive resistor always absorbs energy from the remainder of the circuit.

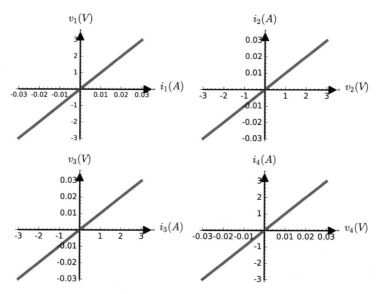

Fig. 1.17 $R = 100\,\Omega$ and $R = \frac{1}{100}\,\Omega$ are duals of each other. The dual of the top-left plot ($i_1 - v_1$ for $R = 100\,\Omega$) is in the bottom-right ($v_4 - i_4$ for $R = \frac{1}{100}\,\Omega$). Similarly, the dual of the top-right plot ($v_2 - i_2$ for $R = 100\,\Omega$) is in the bottom-left ($i_3 - v_3$ for $R = \frac{1}{100}\,\Omega$)

But from Eq. (1.34) we can see the power delivered to a linear resistor is negative if $R < 0$; i.e., as current flows through it, the resistor delivers energy to the remainder of the circuit. Therefore we call such a linear resistor with negative resistance an active resistor.

While linear passive resistors are familiar to everyone, linear active resistors are perhaps new to some readers. They are one of the basic circuit elements in the design of negative resistance oscillators. We will show how to synthesize piecewise-linear negative resistors using opamps in Sect. 2.5.3.2. We will discuss oscillator design in later parts of the book. For the present we only wish to mention that the linear active resistor is useful in modeling nonlinear devices and circuits over certain ranges of voltages, currents and frequencies.

While the linear resistor is perhaps the most prevalent circuit element in electrical engineering, nonlinear devices which can be modeled with nonlinear resistors have become increasingly important. Hence we will now define the concept of a nonlinear resistor in the most general way. Note that in keeping with the theme of the book, linear resistors (elements) will only be discussed as special cases of nonlinear resistors (elements).

Fig. 1.18 Nonlinear
resistor $\mathscr{R}$

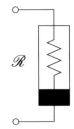

In general, a two-terminal element will be called a resistor if its voltage v and
current i satisfy the relation in Eq. (1.35):

$$\mathscr{R} = \{(v, i) : f_R(v, i) = 0\} \tag{1.35}$$

This relation is called the $v–i$ characteristic of the resistor and can be plotted
graphically in the $v–i$ (or $i–v$) plane. We have already done so for linear resistors.
The circuit symbol for the nonlinear resistor was shown in Fig. 1.1, reproduced in
Fig. 1.18.

Note that in view of the nonsymmetrical nature of the circuit symbol for the
nonlinear resistor (and nonlinear elements in general), we may avoid drawing the
associated voltage (flux) polarity and current (charge) direction signs beside the
symbol, **provided we agree to assume that the darkened edge is the negative
terminal and current (charge) enters the positive terminal.** This convention will
be followed in this book, when adding polarities and directions will clutter the
circuit diagram.

Now we can generalize the concept of duality to nonlinear resistors: we say that
the dual of a given resistor is another resistor whose $v–i$ characteristic in the $v–i$
plane is the same as that of the given resistor in the $i–v$ plane. We will revisit this
concept of duality throughout the book and study it in detail in Sect. 4.1.2, since it
helps us in understanding and analyzing circuits of great generality.

In order to be able to use nonlinear resistors effectively in a practical design, it
is necessary to understand some basic properties. We will illustrate these properties
by considering a prototypical example of a nonlinear resistor, the pn-junction diode
(henceforth referred to as diode).

Although we model diodes as nonlinear resistors, they are so important in
circuit theory that they have their own symbol, shown in Fig. 1.19. A typical $v–i$
characteristic is shown in Fig. 1.20.

In typical applications, the device is operated for values of diode current greater
than $-I_s$. The point where the current becomes equal to $-I_s$ is usually referred to
as the **knee** of the diode. For values of current greater than $-I_s$, the current obeys
the diode junction law in Eq. (1.36).

$$i(v) = I_s[e^{\frac{v}{V_T}} - 1] \tag{1.36}$$

Fig. 1.19 Circuit symbol for
a diode

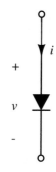

Fig. 1.20 Diode v–i

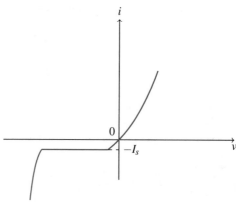

I_s is on the order of microamperes and it represents the reverse saturation current. The parameter $V_T = \frac{kT}{q}$ is called the thermal voltage, where q is the charge of an electron, k is Boltzmann's constant, and T is the temperature in K. At room temperature, V_T is approximately 0.026 V.

In Eq. (1.36), we have a nonlinear resistor whose current i is expressed as a function of its voltage v. This means that for any given voltage v, the current i is uniquely specified. A nonlinear resistor having this property is called a **voltage-controlled nonlinear resistor**. By contrast, if the voltage is a single-valued function of the current $v = v(i)$, we have a **current-controlled nonlinear resistor**. Another important property shared by some v–i curves is their symmetry with respect to the origin. Such elements are called **bilateral resistors** because in this case, the two terminals may be interchanged without effecting the v–i curve (see Exercise 1.2).

Finally, if for each pair of points (v_1, i_1) and (v_2, i_2) on the curve, we observe that whenever $v_2 > v_1$ then $i_2 > i_1$, then the corresponding element is said to be **strictly monotonically increasing resistor**. An example is a linear passive resistor.

Note that while Eq. (1.36) represents a good model for the diode at low frequencies (recall Sect. 1.8), we need to use additional circuit elements, capacitors, inductors, and linear resistors to model the device at higher frequencies. A very important physical property of the diode, namely charge-storage effects are modeled by memristors. Memristors will be discussed in Sect. 1.9.4.

Fig. 1.21 Circuit for
Example 1.9.2

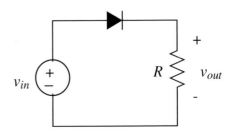

Many practical diode circuits can be analyzed by a very simply piecewise-linear
diode model, called the **ideal diode model**, described analytically by Eq. (1.37).

$$
\begin{aligned}
i &= 0 \quad \forall v < 0 \\
v &= 0 \quad \forall i > 0 \\
p &= vi = 0 \quad \forall v, i
\end{aligned}
\tag{1.37}
$$

Observe that the last constraint is introduced to eliminate any point in the fourth
quadrant from becoming a part of the v–i curve. It is also important to observe that
an ideal diode becomes an open circuit for $v < 0$ and a short circuit for $i > 0$.

Example 1.9.2 Consider the circuit shown in Fig. 1.21. Discuss what would
be the output voltage $v_{out}(t)$ if $v_{in}(t) = \sin(\pi t)$, assuming the ideal diode
model.

Solution The circuit in Fig. 1.21 is the first step in converting an AC
(alternating current or time-varying) voltage into a DC (direct current or
constant) voltage, a process called rectification. The terms AC and DC are
so named because in AC, the electric charge (and hence voltage) reverses (or
alternates) direction periodically. In DC, the electric charge flows in only one
direction.

The output voltage v_{out} for a sinusoidal v_{in} is shown in Fig. 1.22. When the
input voltage $v_{in}(t)$ is positive, the diode becomes a short circuit and $v_{out}(t) = v_{in}(t)$. When the input voltage is negative, the diode becomes an open circuit
and $v_{out}(t) = 0$. The result is that the output voltage becomes zero during
every other half cycle.

The circuit in Fig. 1.21 is called a half-wave rectifier, since the negative
half cycle is simply zeroed out, instead of being rectified.

Although the rectifier in Example 1.9.2 uses an ideal diode, the above example
illustrates a universal principle of creative design: first arrive at an idealized network
(which is usually much easier to come by) and then introduce physical non-idealities
as necessary.

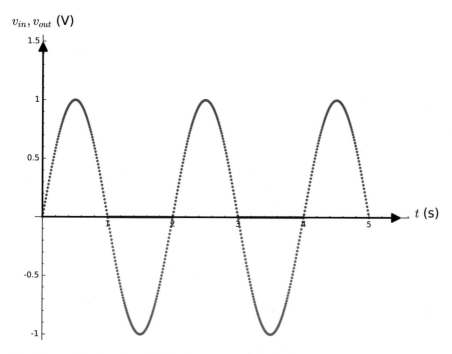

Fig. 1.22 $v_{in}(t)$ (red) and $v_{out}(t)$ (blue) for the circuit in Fig. 1.21

1.9.1.1 Concave and Convex Resistors

Consider the v–i curve shown in Fig. 1.23. The shape of the curve suggests the name: **concave resistor**. Its symbol is also shown in Fig. 1.23. The concave resistor shown in the figure is a piecewise-linear voltage-controlled resistor, which is uniquely specified by two parameters: G, the slope of the linear segment and E, the **breakpoint voltage**. In terms of a function representation, a concave resistor can be specified by:

$$i = \frac{1}{2}G\left[|v - E| + (v - E)\right] \qquad (1.38)$$

By using the definition of absolute value, one can easily understand that Eq. (1.38) represents Fig. 1.23.

Suppose $v < E$. Equation (1.38) becomes:

$$i = \frac{1}{2}G\left[-(v - E) + (v - E)\right]$$

$$= 0 \qquad (1.39)$$

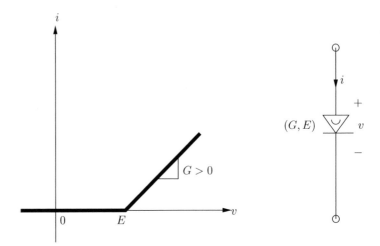

Fig. 1.23 Characteristic and symbol for a typical concave resistor

Similarly, if $v \geq E$, Eq. (1.38) becomes:

$$i = \frac{1}{2}G\left[(v - E) + (v - E)\right]$$

$$= G(v - E) \tag{1.40}$$

If $|E| = |I|$, we can apply the duality principle discussed earlier to define the **convex resistor** shown in Fig. 1.24. The convex resistor shown in Fig. 1.24 defines a piecewise-linear current-controlled resistor which is uniquely specified by two parameters: $G = \frac{1}{R}$, the slope of the linear segment and I, the breakpoint current. Being current-controlled, it can be represented by:

$$v = \frac{1}{2}R\left[|i - I| + (i - I)\right] \tag{1.41}$$

Using the same technique as the concave resistor, we can easily show that Eq. (1.41) is equivalent to the characteristic in Fig. 1.24. We leave that as an exercise for the reader.

1.9.1.2 Piecewise-Linear (PWL) Approximation

When we deal with complex circuits, we need to rely on computers for simulation. Therefore it is important to develop analytic methods to formulate problems precisely and to approximate nonlinear characteristics in mathematical form.

Sometimes, a mathematical characterization can be obtained from the physics of the device, as we did for the *pn*-junction diode. However, often we have to rely on

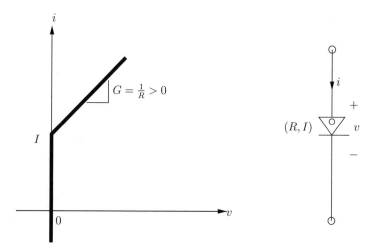

Fig. 1.24 Characteristic and symbol for a typical convex resistor

measurements or curves provided by the device manufacturer (as in the case of the bipolar junction transistor, to be discussed in Sect. 2.2.2.1). Therefore we need to introduce methods of approximation. For example, the tunnel-diode characteristic can be approximated by a polynomial:

$$i \approx \sum_{k=0}^{n} a_k v^k \qquad (1.42)$$

The subject of polynomial approximation or interpolation is well-developed, but it is beyond the scope of this book. On the other hand, piecewise-linear (PWL) approximation is useful in dealing with both simple and general circuits made up of nonlinear resistors. It is also straightforward and effective. We will devote this subsection to PWL approximation.

The PWL approximation of a tunnel-diode characteristic is shown in Fig. 1.25. The three linear segments have slopes:

$$G = \begin{cases} G_a & \text{for } v \le E_1 \\ G_b & \text{for } E_1 < v < E_2 \\ G_c & \text{for } v \ge E_2 \end{cases} \qquad (1.43)$$

Hence beginning from left to right in Fig. 1.25, we can decompose the PWL characteristic of the tunnel-diode into three parts:

1. A linear resistor with conductance G_0
2. A concave resistor characteristic which starts at E_1 with negative slope G_1
3. A concave resistor characteristic which starts at E_2 with positive slope G_2

Fig. 1.25 PWL approximation of the tunnel-diode characteristic

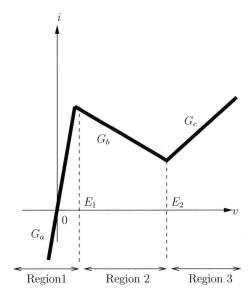

Fig. 1.26 Decomposition of the PWL tunnel diode characteristic into three components

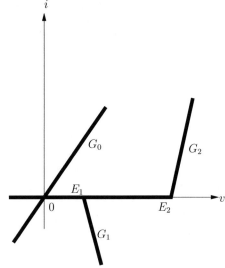

The corresponding characteristics are shown in Fig. 1.26. Comparing Figs. 1.26 and 1.25, we can see the following relations must be satisfied:

$$G_0 = G_a \tag{1.44}$$

$$G_0 + G_1 = G_b \tag{1.45}$$

$$G_0 + G_1 + G_2 = G_c \tag{1.46}$$

Thus: $G_0 = G_a$, $G_1 = -G_a + G_b$ and $G_2 = -G_b + G_c$. We can also obtain the current through the tunnel-diode in Fig. 1.26 by using the functional representation of the concave resistor in Eq. (1.38). Assuming the three component currents in Fig. 1.26 from left to right are i_0, i_1, i_2, we have:

$$i_0 = G_o v \tag{1.47}$$

$$i_1 = \frac{1}{2} G_1 \left[|v - E_1| + (v - E_1) \right] \tag{1.48}$$

$$i_2 = \frac{1}{2} G_2 \left[|v - E_2| + (v - E_2) \right] \tag{1.49}$$

Notice that i in Fig. 1.25 is simply: $i = i_0 + i_1 + i_2$. Hence we have:

$$i = -\frac{1}{2} (G_1 E_1 + G_2 E_2) + \left(G_0 + \frac{1}{2} G_1 + \frac{1}{2} G_2 \right) v + \frac{1}{2} G_1 |v - E_1| + \frac{1}{2} G_2 |v - E_2| \tag{1.50}$$

One can use the relationships between G_0, G_1, G_2 and G_a, G_b, G_c derived earlier to rewrite the equation in terms of G_a, G_b, G_c. We leave that as an exercise for the reader. Equation (1.50) may be written in the following general form:

$$i = a_0 + a_1 v + b_1 |v - E_1| + b_2 |v - E_2| \tag{1.51}$$

In fact, Eq. (1.51) can be fully generalized [8] (assuming no discontinuities at the breakpoints) as shown below.

$$i = a_0 + a_1 v + \sum_{k=1}^{n} b_k |v - E_k| \tag{1.52}$$

where $E_1 < E_2 < \cdots < E_n$ are the breakpoint voltages and

$$a_0 = i(0) - \sum_{k=1}^{n} b_k |E_k| \tag{1.53}$$

$$a_1 = \frac{1}{2} (m_0 + m_n) \tag{1.54}$$

$$b_k = \frac{1}{2} (m_k - m_{k-1}) \tag{1.55}$$

where m_0 is the slope of the first linear segment from the left and m_k is the slope of the $(k + 1)$th linear segment. Here, the segments are labeled consecutively from left to right, starting from zero. The interested reader is referred to [8] for further

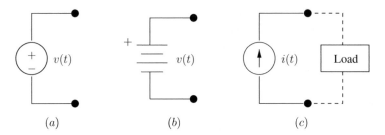

Fig. 1.27 Symbols for (**a**), (**b**) independent voltage source and (**c**) independent current source. An independent DC voltage source can also be indicated by the standard battery symbol shown in (**b**). Note that KCL from Sect. 1.5.1 implies that a current source must not be connected to an open circuit (unless $i(t) = 0$). Analogously, a voltage source must not be connected across a short-circuit (unless $v(t) = 0$)

details, where a very general representation (that includes discontinuities) is stated and proved.

1.9.2 Independent Sources

Note that in Example 1.9.2, we encountered a sinusoidal **voltage source**. Sources are a very important class of two-terminal devices because electrical energy must be supplied in order to move the charges which constitute current i. Of course energy cannot be created or destroyed, electrical sources simply transform some other form of energy into electrical energy. For instance, a battery transforms chemical energy into electrical energy. We will encounter two[7] types of sources[8]:

Definition 1.18 An **independent voltage source** is a two-terminal device whose terminal voltage v is always equal to some given function of time $v_s(t)$; regardless of the value of current flowing through it.

The dual of the independent voltage source is the independent current source.

Definition 1.19 An **independent current source** is a two-terminal device whose terminal current i is always equal to some given function of time $i_s(t)$; regardless of the value of voltage across its terminals.

The circuit symbol(s) for independent voltage and current sources are shown in Fig. 1.27.

On many occasions, we shall find it convenient to consider a DC voltage source and a DC current source as nonlinear resistors. This interpretation is valid because,

---

[7]We will not use charge and flux-linkage sources in this book.

[8]We will postpone discussion of the very important class of dependent sources till Sect. 2.2.1.2, after we have discussed two-port representation in Sect. 2.2.1.

Fig. 1.28 Nonlinear
inductor $\mathscr{L}$

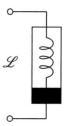

by definition, a DC voltage source with terminal voltage E can be represented by the vertical line $v = E$ in the v–i plane. Similarly, a DC current source with terminal current I can be represented by the horizontal line $i = I$ in the v–i plane.

1.9.3 Inductors and Capacitors

In this section, we introduce inductors and capacitors. To emphasize the "dual" character of these two elements, we will use a two-column format so that each statement on the left is the dual of the one on the right. Once the reader gets used to the idea of duality, they need only read one column while mentally reflecting on the dual statement in the other column.

An inductor is defined by

$$\mathscr{L} = \{(\phi, i) : f_L(\phi, i) = 0\} \tag{1.56}$$

The circuit symbol for an inductor is shown in Fig. 1.28, reproduced from Fig. 1.1.

If Eq. (1.56) can be solved for i as a single-valued function of ϕ, namely:

$$i = \hat{i}(\phi) \tag{1.57}$$

the inductor is said to be **flux-controlled**. If Eq. (1.56) can be solved as a single-valued function of i, namely:

$$\phi = \hat{\phi}(i) \tag{1.58}$$

then the inductor is said to be **current-controlled**. If the function $\hat{\phi}(i)$ is differentiable, we can apply the chain rule in Eq. (1.58) to obtain:

$$v = L(i)\frac{di}{dt} \tag{1.59}$$

with **small-signal inductance** $L(i)$:

$$L(i) \overset{\triangle}{=} \frac{d\hat{\phi}(i)}{di} \tag{1.60}$$

Fig. 1.29 Toroidal inductor

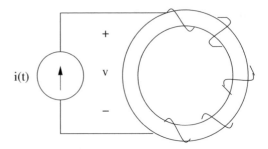

Example 1.9.3 Analyze the system shown in Fig. 1.29, where we have a conducting wire wound around a toroid made of a nonmetallic material.

Solution When $i(t)$ is applied, we recall from physics that a flux equal to $\phi(t) = Li(t)$ is induced at time t and circulates around the interior of the toroid. The constant of proportionality is given approximately by $L = \mu_0 \frac{N^2 A}{l}$ H where $\mu_0 = 4\pi \cdot 10^{-7}$ H/m is the permeability of the core, N is the number of turns of the coil, A is the cross-sectional area in m^2, and l is the midcircumference along the toroid in m. Hence, in Eq. (1.60), we will have $L(i) = L$ (a constant) and thus we have the classic linear time-invariant inductor from circuit theory, with the relation:

$$v = L\frac{di}{dt} \tag{1.61}$$

For the properties below, we will assume linear time-invariant inductors and address properties for the nonlinear counterparts in Chap. 4.

Memory Property

Suppose we apply a voltage source $v(t)$ across an inductor L. The inductor current can be obtained by integrating Eq. (1.61) (assuming $i(t \to -\infty) = 0$):

$$i(t) = \frac{1}{L} \int_{-\infty}^{t} v(\tau)d\tau \ t \geq t_0 \tag{1.62}$$

Hence the inductor current depends on the entire past history of $v(\tau)$. Therefore the inductor has **memory**.

Suppose however that current $i(t_0)$ at some time $t_0 < t$ is given, then we get:

$$i(t) = i(t_0) + \frac{1}{L} \int_{t_0}^{t} v(\tau)d\tau \ t \geq t_0 \tag{1.63}$$

Fig. 1.30 Initial condition transformation for L

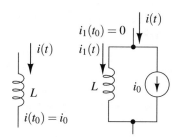

In other words, instead of specifying the entire past history, we need only specify $i(t)$ at some conveniently chosen initial time t_0. In effect, the initial condition $i(t_0)$ summarizes the effect of $v(\tau)$ from $\tau \rightarrow -\infty$ to $\tau = t_0$, on the present value of $i(t)$. We can draw an equivalent circuit symbolizing the memory effect as shown in Fig. 1.30.

Continuity Property Suppose we apply a voltage source described by a discontinuous square wave across an inductor, then the current through the inductor is given by Eq. (1.63). Assuming that $i(t_0) = 0$, we will obtain continuous inductor current waveform. This "smoothing" phenomenon turns out to be a general property.

If the voltage waveform $v_L(t)$ across a linear time-invariant inductor L remains bounded in a closed interval $[t_a, t_b]$, then the current waveform $i_L(t)$ through the inductor is a continuous function in the open interval (t_a, t_b). In particular, for any time T satisfying $t_a < T < t_b$, $i_L(T^-) = i_L(T^+)$.

A capacitor is defined by

$$\mathscr{C} = \{(q, v) : f_C(q, v) = 0\} \tag{1.64}$$

The circuit symbol for the capacitor is shown in Fig. 1.31, reproduced from Fig. 1.1.

If Eq. (1.64) can be solved for v as a single-valued function of q, namely:

$$v = \hat{v}(q) \tag{1.65}$$

the capacitor is said to be **charge-controlled**. If Eq. (1.64) can be solved as a single-valued function of v, namely:

$$q = \hat{q}(v) \tag{1.66}$$

Fig. 1.31 Nonlinear capacitor $\mathscr{C}$

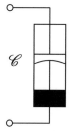

then the capacitor is said to be **voltage-controlled**. If the function $\hat{q}(v)$ is differentiable, we can apply the chain rule in Eq. (1.66) to obtain:

$$i = C(v)\frac{dv}{dt} \tag{1.67}$$

with **small-signal capacitance** $C(v)$:

$$C(v) \triangleq \frac{d\hat{q}(v)}{dv} \tag{1.68}$$

Example 1.9.4 Analyze the system shown in Fig. 1.32, where we have two flat parallel metal plates separated by a distance d.

Solution When $v(t)$ is applied, we recall from physics that a charge equal to $q(t) = Cv(t)$ is induced at time t on the upper plate, and an equal but opposite charge is induced on the lower plate at time t. The constant of proportionality is given approximately by $C = \epsilon_0 \frac{A}{d}$ F where $\epsilon_0 = 8.85 \cdot 10^{-12}$ F/m is the permittivity of free space, A is the plate area in m^2, and d is the separation of the plate in m. Hence, in Eq. (1.68), we have $C(v) = C$ (a constant) and thus we have the classic time-invariant capacitor from circuit theory:

$$i = C\frac{dv}{dt} \tag{1.69}$$

For the properties below, we will assume linear time-invariant capacitors and address properties for the nonlinear counterparts in Chap. 4.

Fig. 1.32 Parallel-plate capacitor

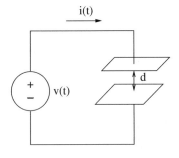

Memory Property Suppose we connect a current source $i(t)$ in series with capacitor C. The capacitor voltage can be obtained by integrating Eq. (1.69) (assuming $v(t \to -\infty) = 0$):

$$v(t) = \frac{1}{C} \int_{-\infty}^{t} i(\tau)d\tau \; t \geq t_0 \tag{1.70}$$

Hence the capacitor voltage depends on the entire past history of $i(\tau)$. Therefore the capacitor has **memory**.

Suppose however that voltage $v(t_0)$ at some time $t_0 < t$ is given, then we get:

$$v(t) = v(t_0) + \frac{1}{C} \int_{t_0}^{t} i(\tau)d\tau \; t \geq t_0 \tag{1.71}$$

In other words, instead of specifying the entire past history, we need only specify $v(t)$ at some conveniently chosen initial time t_0. In effect, the initial condition $v(t_0)$ summarizes the effect of $i(\tau)$ from $\tau \to -\infty$ to $\tau = t_0$, on the present value of $v(t)$. We can draw an equivalent circuit symbolizing the memory effect as shown in Fig. 1.33.

Continuity Property Suppose we apply a current source described by a discontinuous square wave through a capacitor, then the voltage across the capacitor is given by Eq. (1.71). Assuming $v(t_0) = 0$, we will obtain continuous capacitor voltage waveform. This "smoothing" phenomenon turns out to be a general property.

If the current waveform $i_C(t)$ in a linear time-invariant capacitor C remains bounded in a closed interval $[t_a, t_b]$, then the voltage waveform $v_C(t)$ across the capacitor is a continuous function in the open interval (t_a, t_b). In particular, for any time T satisfying $t_a < T < t_b$, $v_C(T^-) = v_C(T^+)$.

The continuity property for inductors and capacitors is so important that we will prove the continuity property for a capacitor (the inductor follows by duality).

Consider Eq. (1.71). Substituting $t = T$ and $t = T + dt$ into Eq. (1.71) where $t_a < T < t_b$ and $t_a < T + dt \leq t_b$, and subtracting, we get:

$$v_C(T + dt) - v_C(T) = \frac{1}{C} \int_{T}^{T+dt} i_C(\tau)d\tau \tag{1.72}$$

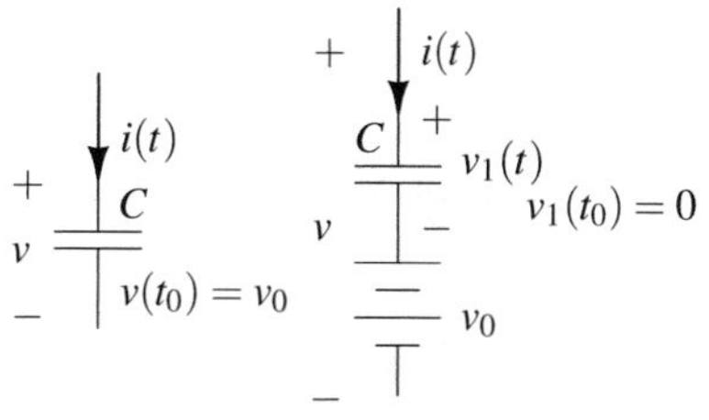

Fig. 1.33 Initial condition transformation for C

Fig. 1.34 Figure for
Example 1.9.5

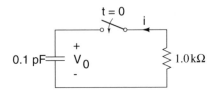

Since we have assumed $i_C(t)$ to be bounded in $[t_a, t_b]$, there is a finite constant M such that $|i_C(t)| < M, \forall t \in [t_a, t_b]$. It follows that the area under the curve $i_C(t)$ from T to $T + dt$ is at most $M dt$ (in absolute value), which tends to zero as $dt \to 0$. Hence Eq. (1.72) implies $v_C(T + dt) \to v_C(T)$. Therefore $v_C(t)$ is continuous at $t = T$.

Example 1.9.5 Find the value $i(0^+)$ in Fig. 1.34, assuming the capacitor is precharged to 0.5 V and the ideal switch instantaneously closes at $t = 0$.

Solution Since the capacitor is precharged to 0.5 V, $v_0(0^-) = 0.5$ V. By the continuity property for capacitors, $v_0(0^+) = 0.5$ V. Since we have a linear resistor, by Ohm's law and the passive sign convention:

$$i = -\frac{0.5\,\text{V}}{1.0\,\text{k}\Omega}$$

$$= -0.5\,\text{mA} \tag{1.73}$$

The continuity property will be further utilized in Chap. 4, where we apply it to solve[9] a variety of circuits that exhibit switching discontinuities.

1.9.4   Memristors

Looking at Fig. 1.1 and based on our discussions of the other fundamental circuit elements, it is only **natural** that, by **symmetry** arguments, there exists a fourth fundamental circuit element for establishing a ϕ–q relationship:

$$\mathcal{M} = \{(\phi, q) : f_M(\phi, q) = 0\} \tag{1.74}$$

Such an element was defined by Dr. Chua in 1971 [2]. Nevertheless investigations of this element began in earnest only after HP's announcement in 2008 [27]. HP's

---

[9]By "solve" a circuit, we mean to find the voltage across and current through every branch for all times t.

memristor is a very specific TiO$_2$ based device. We will instead study general characteristics of the memristor. Interestingly, literature survey gives a wealth of insight into the memristor, and hence we will begin our study with the very first citations of Chua's seminal work.

Dr. Penfield, in a MIT technical report [23], mentions the memristor in connection with the Josephson junctions. Throughout the late twentieth century, a plethora of research [13, 21, 22, 29] regarding the "phase-dependent conductance" in Josephson junctions were carried out. But a proper memristor approach to extracting the "phase-dependent conductance" occurred only with Peotta and Di Ventra's seminal paper in 2014 [24]. However, before we examine the ideal memristor in Josephson junctions, we will state some important properties of the memristor [2].

Consider a function of q based on Eq. (1.74):

$$\phi = s(q) \tag{1.75}$$

Differentiating both sides of Eq. (1.75) with respect to time and applying the chain rule, we get:

$$\begin{aligned}
\frac{d\phi}{dt} &= \frac{ds(q)}{dt} \\
&= \frac{ds}{dq}\frac{dq}{dt} \tag{1.76}
\end{aligned}$$

From Eq. (1.3), $i = \frac{dq}{dt}$ and from Eq. (1.4), $v = \frac{d\phi}{dt}$. Hence we have the memristor v–i relation in Eq. (1.77).

$$v(t) = M(q(t))i(t) \tag{1.77}$$

$M(q(t))$ in Eq. (1.78) is defined as the **incremental memristance**, we can analogously define a $W(\phi(t))$ as **incremental menductance**.[10]

$$i(t) = W(\phi(t))v(t) \tag{1.78}$$

We can make the following observations from Eq. (1.77) (analogous observations hold for Eq. (1.78)):

1. $M(q(t)) = M\left(\displaystyle\int_{-\infty}^{t} i(\tau)d\tau\right)$. Hence the fact that memristor stands for "memory-resistor" can be justified: the value of the memristance at any time t depends on the time integral of the memristor current from $-\infty$ to t. Therefore while the memristor behaves like an ordinary resistor at a **given instant** of time,

---

[10]To avoid clutter, we will use the terms memristance and menductance from now on. We will reserve the use of "incremental" for clarity purposes.

Fig. 1.35 A Josephson
junction formed by using a
superconductor–insulator–
superconductor setup. The
barrier I is thin enough (on
the order of a few angstroms)
that superconducting Cooper
pairs can tunnel across the
junction when $v = 0$ [12]

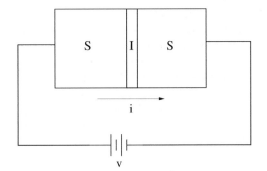

its **resistance depends on the complete past history (or memory)** of memristor
current.
2. In the very special case where the memristor ϕ–q curve is a straight line, we
obtain $M(q) = R$, the memristor reduces to a linear time-invariant resistor.

Point 2. above illustrates why the memristor is not relevant in linear circuit
theory: unlike the other three fundamental circuit elements (resistor, inductor,
capacitor), **a memristor is a fundamentally nonlinear device, a linear memristor
is simply a resistor**. In other words, memristors are not used to model linear circuits,
such circuits do not exhibit "memristive" effects. Nevertheless, since a memristor **is**
the fourth fundamental circuit element, it is only logical that we discuss memristors
in a book on circuit theory. However, understand that research into circuit theoretic
properties of the memristor is still[11] in its infancy. Hence the motivated reader
will probably enhance (and even prove) several fundamental properties. As an
analogy, Maxwell completed Ampere's law for non-static situations with a time-
derivative of electric flux, and published his "Treatise on Electricity and Magnetism"
in 1873. At his death 6 years later, his theory was unfortunately neither well
understood nor widely accepted [11]. It took the Herculean efforts of primarily four
individuals—John Francis Fitzgerald, Oliver Lodge, Oliver Heaviside, along with
key contributions from Heinrich Hertz—to transform Maxwell's fertile ideas from
his treatise into the theory now known today as "Maxwell's Equations". Similarly,
a memristor correctly completes the four fundamental circuit element graph in
Fig. 1.1. A few of the properties predicted by Chua [2], Chua and Kang [7] for a
memristor have been validated. Many of the fertile hypotheses set forth in those
classic works still need to be investigated.

We can now discuss the phase-dependent conductance in Josephson junction as
an ideal memristor. Before doing so we will derive the Josephson relation from first
principles, since not only does this relation utilize fundamental physical principles,
but it also helps us practice our definitions of the fundamental circuit variables.

Consider the Josephson junction (JJ) shown in Fig. 1.35.

---

[11] As of the 2017 first edition of this book.

From basic physics, we know that energy E is quantized from the Planck–Einstein relation:

$$E = h\nu \tag{1.79}$$

Rewriting Eq. (1.79), we get:

$$E = h\frac{\omega}{2\pi} \tag{1.80}$$

In Eq. (1.79), ν is frequency in Hz and in Eq. (1.80), ω is angular frequency in rad/s.

In a physical JJ in the superconducting state, a quantum mechanical phase difference Φ is established between the two superconductors. Therefore, we have:

$$E = \frac{h}{2\pi}\frac{d\Phi}{dt} \tag{1.81}$$

Based on Eq. (1.1), we have:

$$2e^- v = \frac{h}{2\pi}\frac{d\Phi}{dt} \tag{1.82}$$

We have used $2e^-$ because Cooper pairs carry the charge in the superconducting state. Using Eq. (1.4), defining $\hbar \overset{\Delta}{=} \frac{h}{2\pi}$ and simplifying, we get:

$$\frac{d\phi}{dt}\left(\frac{2e^-}{\hbar}\right) = \frac{d\Phi}{dt} \tag{1.83}$$

In Eq. (1.83), we can define $\phi_0 \overset{\Delta}{=} \phi/\frac{\hbar}{2e^-}$. In other words, the quantum mechanical phase difference across the junction is quantized as a function of the magnetic flux through the loop ($\frac{h}{2e^-}$ is the magnetic flux quantum): $\Phi = \phi_0$. Equation (1.83) is the **fundamental Josephson relation**.

The current i through the junction can be written [12] as:

$$i(v) = I_c \sin(\phi_0) + \sigma_0(v)v + \epsilon \cos(\phi_0)v + \cdots \tag{1.84}$$

In Eq. (1.84), I_c, ϵ are constants based on the physical superconducting materials and $\sigma_0(v)$ is the nonlinear conductance for the particular JJ. An interesting observation is that the current i will be non-zero even if the voltage across the junction is zero! This is due to the Josephson current $I_j = I_c \sin(\phi_0)$. The mindful reader should have noticed that we can model a JJ as a nonlinear inductor, when $v = 0$.

But notice the third term in Eq. (1.84) can be written as:

$$i_3(v) = W(\phi_0)v \tag{1.85}$$

This equation is precisely the equation of an ideal memristor (the device is technically a menductor). But there are two issues in trying to design an ideal memristor:

1. The Josephson current I_j is usually much larger when compared to the memristance term i_3.
2. The memristance term is non-zero only when the voltage across the junction is non-zero. And in this case, it is oscillating at a very high frequency.[12]

Nevertheless, Peotta and Di Ventra propose an elegant approach [24] to isolate the memristance term: utilize two Josephson junctions of different material, connected in parallel, to cancel the Josephson current.

But how do we identify a memristive two-terminal black box? The answer lies in our definition: since $v(t)$ $(i(t))$ has to be zero whenever $i(t)$ $(v(t))$ is zero for a memristor (menductor), under periodic excitation, a memristor distinctly displays a Lissajous figure in the v–i plane.

However, in the case of the ideal memristor in the Josephson junction, we still have the issue of the $\cos\phi_0$ term oscillating at very high frequencies for practical measurements. Thus, are there other devices that can be modeled as memristors and can be easily studied experimentally?

Fortunately, the answer is yes. For studying such devices, we will use the generalization of an ideal memristor[13] to a general memristive device, as defined by Chua and Kang [7].

An nth-order current-controlled memristive one-port is represented by:

$$\dot{\mathbf{x}} = f(\mathbf{x}, i, t)$$
$$v = R(\mathbf{x}, i, t)i \tag{1.86}$$

An nth-order voltage-controlled memristive one-port is represented by:

$$\dot{\mathbf{x}} = f(\mathbf{x}, v, t)$$
$$i = G(\mathbf{x}, v, t)v \tag{1.87}$$

In both equations above, $\mathbf{x}$ denotes the **state** of the memristive device. $f : \Re^n \times \Re \times \Re \to \Re^n$ is a continuous n-dimensional vector function, $R, G : \Re^n \times \Re \times \Re \to \Re$ are continuous scalar functions. It is assumed that the state equation(s) have a unique solution for any initial state $x_0 \in \Re^n$.

A variety of physical devices exhibit memristive effects. We will now examine one of these devices: a discharge tube, whose resistance can be modeled as a

---

[12]Private email communication from Dr. Brian Josephson to Dr. Muthuswamy on March 14th 2014.

[13]Unfortunately, as of the writing of this book, the Josephson junction is the only (that we know of) physical device that serves as a model for an ideal memristor.

function of the number of conduction electrons n_e [16]. Consider Eq. (1.88):

$$v_M = R(n_e)i_M$$
$$\dot{n}_e = -\beta n + \alpha R(n_e)i_M^2 \tag{1.88}$$

v_M is the voltage across the discharge tube, i_M is the current flowing through it and n_e is the number of conduction electrons. $R(n_e) = \frac{F}{n_e}$. α, β, F are parameters depending on the dimensions of the tube and the gas fillings. Comparing Eqs. (1.86) and (1.88), we can clearly see that a discharge tube can be modeled as a current-controlled memristor.

Figure 1.36 shows a simulated $v_M - i_M$ curve and Fig. 1.37 shows an oscilloscope screenshot of a measured discharge tube characteristic.

We will have more to discuss about other devices with memristive behavior (such as *pn*-junctions) in Sect. 4.4.2.

Fig. 1.36 Simulated Lissajous figure for $i_M(t) = \sin(\omega t)$ in Eq. (1.88), with $\alpha = 0.1, \beta = 0.1, F = 1, \omega = 0.063$ [16]

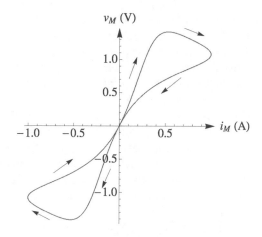

Fig. 1.37 Measured discharge tube characteristics, showing the classic pinched hysteresis loop [6]. We have plotted v_M on the Y-axis (2 V/div) and have scaled i_M to voltage for ease of plotting on the X-axis (5 V/div)

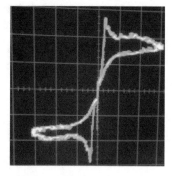

1.9.5   Higher-Order Circuit Elements

Sections 1.9.1 through 1.9.4 have helped us discuss the four fundamental circuit elements in Fig. 1.1. The elements are fundamental in the sense that no element from this basic set can be derived from the other three elements [3].

In fact, we can generalize Fig. 1.1 to **higher-order** circuit elements. As a motivating example, consider Eq. (1.89) (a, b, c are constant real numbers) of the Duffing oscillator. This oscillator is used to model a variety of phenomena in science. We will discuss a circuit implementation of this oscillator in Sect. 5.5.

$$\ddot{v} + c\dot{v} + v(b + a \cdot v^2) = i(t) \tag{1.89}$$

Since we are forcing a current input on the RHS, each expression on the LHS of Eq. (1.89) represents current. Thus, by KCL, we simply have three elements in parallel to an ideal current source.

From Eq. (1.69), we know that the current through a capacitor is proportional to the first derivative of the voltage across it. Hence, $c\dot{v}$ in Eq. (1.69) can be modeled by a linear time-invariant capacitor. We will see in Sect. 5.5 that a tunnel diode has a cubic $i(v)$ and hence can be used to synthesize the $v(b + a \cdot v^2)$ term. But, does there exist a two-terminal element whose current through the terminals is proportional to the **second derivative** of the voltage across it?

Although the current answer to the question is "we do not know," Eq. (1.89) shows the **necessity** of defining such a circuit element. Of course, one could also ask: why not simply build an analog computer that solves Eq. (1.89)? The answer is: the analog computer will not help us study the underlying physical phenomenon. As an analogy, consider a mass-spring-damper model of a second-order system. The equivalent analog computer implementation is simply a "signal flow" graph and cannot yield insightful information, say, the energy transfer between the mass and the spring.

Hence, we need to expand our repertoire of fundamental circuit elements from Fig. 1.1, by introducing a sufficiently rich family of elementary circuit elements [3, 4] which play the same role as the set of basis vectors to define a vector space. The key concept is the following definition.

Definition 1.20 (α–β **Element**)   A two-terminal or one-port black box characterized by a constitutive relation in the $v^{(\alpha)}$-versus-$i^{(\beta)}$ plane is called an (α, β)

element, where $v^{(\alpha)}$ and $i^{(\beta)}$ are defined below.

$$v^{(\alpha)}(t) \overset{\triangle}{=} \begin{cases} \frac{d^\alpha}{dt^\alpha}v(t) & \text{if } \alpha > 0 \\ v(t) & \text{if } \alpha = 0 \\ \underbrace{\int_\infty^t \cdots \int_\infty^t v(\tau_1)d\tau_2 \cdots d\tau_{|\alpha|}}_{|\alpha|} & \text{if } \alpha < 0 \end{cases} \qquad (1.90)$$

or

$$i^{(\beta)}(t) \overset{\triangle}{=} \begin{cases} \frac{d^\beta}{dt^\beta}i(t) & \text{if } \beta > 0 \\ i(t) & \text{if } \beta = 0 \\ \underbrace{\int_\infty^t \cdots \int_\infty^t i(\tau_1)d\tau_2 \cdots d\tau_{|\beta|}}_{|\beta|} & \text{if } \beta < 0 \end{cases} \qquad (1.91)$$

The circuit symbol for $v^{(\alpha)}$-$i^{(\beta)}$ element is shown in Fig. 1.38.

In other words, based on Eqs. (1.90) and (1.91), we can make the following observations:

1. Every $(0,0)$ element is a resistor $\mathscr{R}$
2. Every $(-1,0)$ element is an inductor $\mathscr{L}$
3. Every $(0,-1)$ element is a capacitor $\mathscr{C}$
4. Every $(-1,-1)$ element is a memristor $\mathscr{M}$

Thus, based on our discussion so far and KCL, a circuit equivalent of Eq. (1.89) is shown in Fig. 1.39.

We will revisit and synthesize (α, β) elements later in the text, once we understand concepts behind resistive and dynamic nonlinear networks.

It is instructive to visualize the (α, β) elements in the form of a "periodic table" in Fig. 1.40, that expands the basic four element quadrangle from Fig. 1.1. In Fig. 1.40,

Fig. 1.38 Two terminal or one-port representation of (α, β) element

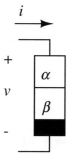

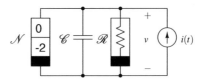

Fig. 1.39 A circuit realization of the Duffing oscillator

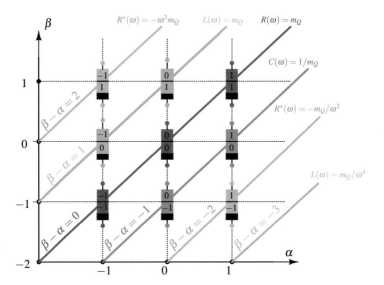

Fig. 1.40 The "periodic table" of all two-terminal (α, β) elements. All elements printed in the same color belong to the same circuit element species, namely, frequency-dependent resistors (in red), frequency-dependent inductors (in blue), frequency-dependent capacitors (in green), and frequency-dependent negative resistors (in orange). m_Q is defined as a small-signal slope about an operating point and will be discussed in Sect. 4.6.2

all elements printed in the same color belong to the same element "species" and, notice that there are only four colors. The justification for the "periodic table" label and an analysis of Fig. 1.40 will be done in Sect. 4.6.2.

1.10   Series and Parallel Connections of Resistors

We are now in a position to consider a special but very important class of circuits: circuits formed by series and parallel connections of two-terminal resistors. First, we wish to generalize the concept of the v–i characteristic of a resistor to that of a two-terminal circuit made of two-terminal resistors, or more succinctly, a **resistive one-port**. We will demonstrate that the series and parallel connections of two-terminal resistors will yield a one-port whose v–i characteristic is again that of a resistor.

Fig. 1.41 Two nonlinear
resistors connected in series
together with the rest of the
circuit $\mathcal{N}$

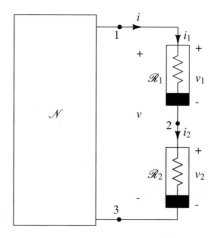

We say that two resistive one-ports are equivalent iff their v–i characteristics are the same.

When we talk about resistive one-ports, we naturally use **port voltage** and **port current** as the pertinent variables. The v–i characteristic of a one-port in terms of its port voltage and port current is often referred to as the **driving-point or DP characteristic** of the one-port. The reason we call it the DP characteristic is that we may consider the one-port as being driven by an independent voltage source v_s or an independent current source i_s. In the former, the input is $v_s = v$ and the response is the current i. In the latter, the input is $i_s = i$ and the response is v. We will next illustrate how to determine DP plots graphically, a technique that we will return to throughout the book.

Consider the circuit shown in Fig. 1.41 where two nonlinear resistors $\mathcal{R}_1$ and $\mathcal{R}_2$ are connected at node 2. Nodes 1 and 3 are connected to the rest of the circuit, which is designated by $\mathcal{N}$. Looking towards the right from nodes 1 and 3, we have a one-port which is formed by the **series connection of two resistors** $\mathcal{R}_1$ and $\mathcal{R}_2$. For our present purposes, the nature of $\mathcal{N}$ is irrelevant. We are interested in obtaining the DP characteristic of the one-port with port voltage v and port current i.

Let us assume that both resistors are current-controlled, i.e.,

$$v_1 = \hat{v}_1(i_1)$$

$$v_2 = \hat{v}_2(i_2) \tag{1.92}$$

Notice that these are the laws of elements. Next, applying KVL for the node sequence 1–2–3–1 gives:

$$v = v_1 + v_2 \tag{1.93}$$

Applying KCL to nodes 1 and 2 gives:

$$i = i_1 = i_2 \tag{1.94}$$

Combining Eqs. (1.92)–(1.94), we obtain:

$$v = \hat{v}(i)$$

$$\text{where } \hat{v}(i) \stackrel{\triangle}{=} \hat{v}_1(i) + \hat{v}_2(i) \tag{1.95}$$

Note that Eq. (1.95) can be extended to n nonlinear resistors $\mathscr{R}_n$ in series. Thus, we can conclude that:

1. KVL requires the port voltage v to be equal to the sum of the branch voltages of the resistors.
2. KCL forces all branch currents to be equal to the port current.
3. If each resistor is current-controlled, the resulting DP characteristic of the one-port is also a current-controlled resistor.

Example 1.10.1 Determine $\hat{v}(i)$ if the two terminals of $\mathscr{R}_1$ in Fig. 1.41 are turned around.

Solution The new circuit is redrawn in Fig. 1.42. Hence, the v–i characteristic for $\mathscr{R}_1$ is now:

$$v_1 = \hat{v}_1(-i_1) \tag{1.96}$$

KVL gives:

$$v = -v_1 + v_2 \tag{1.97}$$

Thus, we have:

$$v = -\hat{v}_1(-i) + \hat{v}_2(i) \tag{1.98}$$

Consider the circuit shown in Fig. 1.43 where two nonlinear resistors $\mathscr{R}_1$ and $\mathscr{R}_2$ are connected across nodes 1 and 2 to the rest of the circuit, which is designated by $\mathscr{N}$. Looking towards the right from nodes 1 and 2, we have a one-port which is formed by the **parallel connection of two resistors** $\mathscr{R}_1$ and $\mathscr{R}_2$. For our present purposes, the nature of $\mathscr{N}$ is irrelevant. We are interested in obtaining the DP characteristic of the one-port with port voltage v and port current i.

Let us assume that both resistors are voltage-controlled, i.e.,

$$i_1 = \hat{i}_1(v_1)$$

$$i_2 = \hat{i}_2(v_2) \tag{1.99}$$

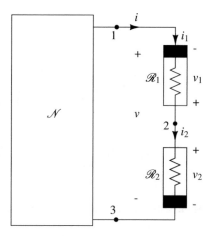

Fig. 1.42 Circuit for Example 1.10.1

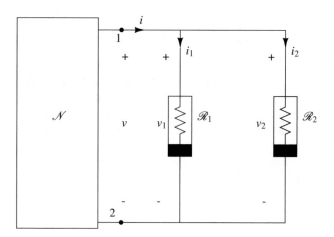

Fig. 1.43 Two nonlinear resistors connected in parallel together with the rest of the circuit $\mathcal{N}$

Notice that these are the laws of elements. Next, applying KVL gives:

$$v = v_1 = v_2 \tag{1.100}$$

Applying KCL at node 1 gives:

$$i = i_1 + i_2 \tag{1.101}$$

Combining Eqs. (1.99)–(1.101), we obtain:

$$i = \hat{i}(v)$$

$$\text{where } \hat{i}(v) \overset{\triangle}{=} \hat{i}_1(v) + \hat{i}_2(v) \tag{1.102}$$

Table 1.1 Dual terms

S	S*
Branch voltage	Branch current
Resistance	Conductance
Current-controlled resistor	Voltage-controlled resistor
Open circuit	Short circuit
Independent voltage source	Independent current source
Inductor	Capacitor
KVL	KCL
Port voltage	Port current
Series connection	Parallel connection

Note that Eq. (1.102) can be extended to n nonlinear resistors $\mathscr{R}_n$ in parallel. Thus, we can conclude that:

1. KVL forces all branch voltages to be equal.
2. KCL requires the port current i to be equal to the sum of the branch currents of the resistors.
3. If each resistor is voltage-controlled, the resulting DP characteristic of the one-port is also a voltage-controlled resistor.

The careful reader would have noticed that Eqs. (1.92)–(1.94) and Eqs. (1.99)–(1.101) are duals of each other! In other words, if we make the substitutions for all the v's with i's and for all the i's with v's in one set of equations, we obtain precisely the other set. For this reason, we can extend and generalize the concept of duality introduced earlier for resistors to circuits.

In Table 1.1, we list two sets of terms S and S^* which we have encountered and which are said to be dual to one another.

Before we end this section, we would like to solve an example that illustrates a variety of concepts from this chapter.

Example 1.10.2 In Fig. 1.44, determine the value of I.

Solution Before attempting to solve any problem, it is a good idea to **understand** the problem and **devise a plan of action**. We then **carry out the plan** and **check our answer** [25].

In this case, a quick examination of the problem will indicate that we need to determine the current through a linear resistor and hence if we know the voltage across it, we can apply Ohm's law.

Starting at node B and applying KVL, we get:

$$v_{AB} + 3 - 2 = 0 \tag{1.103}$$

(continued)

Example 1.10.2 (continued)

Notice our judicious choice of voltage polarity as v_{AB} and not v_{BA}. This choice is no accident: in this problem, the current direction has been clearly specified, so we must choose v_{AB} to comply with the passive sign convention definition from Definition 1.2.

We can in fact now carry out the plan and get the result as:

$$I = \frac{v_{AB}}{20\,\text{k}}$$

$$= -0.05\,\text{mA}. \tag{1.104}$$

Notice a negative I implies the voltage drop across the resistor is opposite to the direction we picked.

How do we check our answer? One approach would be to make sure that the power delivered is equal to power absorbed. This is essential because our circuit is a closed system. We need to first find the current through the branch B–D–A. This can be done by finding the voltage across the $10\,\text{k}$ resistor which in turn can be found by using KVL around D–A–C–B–D:

$$4 + 3 - 2 - v_{BD} = 0 \tag{1.105}$$

Thus $v_{BD} = 5\,\text{V}$. Hence the current through the $10\,\text{k}$ is $I_{BD} = 0.05\,\text{mA}$. KCL at node B gives $I_{CB} = 0.1\,\text{mA}$. We now have all the necessary variables to find the power associated with each element, keeping in mind the passive sign convention form Definition 1.2.

$$P_{3\,\text{V}} = -0.3\,\text{mW}$$

$$P_{2\,\text{V}} = +0.2\,\text{mW}$$

$$P_{4\,\text{V}} = -0.2\,\text{mW}$$

$$P_{10\,\text{k}} = +0.25\,\text{mW}$$

$$P_{20\,\text{k}} = +0.05\,\text{mW}$$

$$\Sigma = 0\,\text{mW} \tag{1.106}$$

Hence we should have good confidence that our answer is correct. We will discuss more circuit analysis techniques based on energy (including expressions for energy stored in an inductor, etc.) and power in Sect. 4.5.

Fig. 1.44 Circuit with only
linear elements

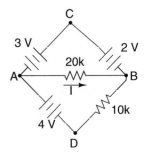

1.11 Conclusion

In this chapter, we discussed the fundamental circuit variables, elements, and Kirchhoff's laws. To summarize:

1. We will assume the lumped circuit approximation.
2. The four fundamental circuit variables are: charge $(q(t))$, flux-linkage $(\phi(t))$, voltage $(v(t))$, and current $(i(t))$.
3. There are four fundamental circuit elements: resistors establish a $v–i$ relationship, capacitors a $q–v$ relationship, inductors a $\phi–i$ relationship, and memristors establish a $\phi–q$ relationship.
4. We will follow a black box approach and model static (or "low frequency") characteristics. We will use parasitic components as necessary to model "high frequency" effects. We emphasize that "low frequency" and "high frequency" depend on the particular device being modeled.
5. Kirchhoff's laws relate the voltages and currents as defined by the topology of the network.
6. The laws of interconnection (KVL, KCL, Tellegen's theorem) are independent of the laws of elements.
7. Elements are said to be in series when they have the same current flowing through them.
8. Elements are said to be in parallel when they have the same voltage across them.

In the next chapter, as a natural follow-up, we will study multi-terminal elements such as the operational amplifier.

Exercises

1.1 Given the $v–i$ characteristic Γ of a resistor $\mathscr{R}$ on the $v–i$ plane, show that the dual characteristic is obtained by reflecting Γ about the 45° line through the origin.

1.2 Find a necessary and sufficient condition for a nonlinear two-terminal element (resistor, inductor, capacitor, and memristor) to be bilateral.

Fig. 1.45 Circuit for
Exercise 1.4 (Part 1)

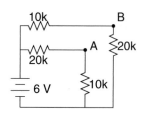

Fig. 1.46 Circuit for
Exercise 1.4 (Part 2)

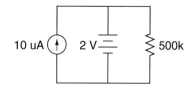

Fig. 1.47 Circuit for
Exercise 1.4 (Part 3)

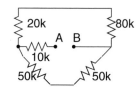

Fig. 1.48 Circuit for
Exercise 1.4 (Part 4)

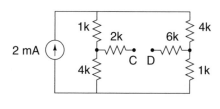

1.3 Discuss mechanical analogies to the four fundamental circuit elements. For the memristor, a good starting point is the classic paper by Oster and Auslander [19].

1.4 This exercise (courtesy of Dr. Oldham from UC Berkeley [18]) is designed to test the reader's fundamental understanding of the conceptual material from this chapter, and is very similar to Example 1.10.2. As a result, the reader should strive to find the correct solution mentally, without the use of pen and paper.

Find the values of the indicated variables below.

1. V_{AB} in Fig. 1.45.
2. V_{CD} in Fig. 1.46.
3. Power associated with the $500\,k\Omega$ resistor in Fig. 1.47.
4. Equivalent resistance at AB in Fig. 1.48.

1.5 Consider the circuit in Fig. 1.49. Assuming $\mathscr{R}_1$ is current-controlled ($v_1 = \hat{v}_1(i_1)$), $\mathscr{R}_2$ and $\mathscr{R}_3$ are voltage-controlled ($i_2 = \hat{i}_2(v_2), i_3 = \hat{i}_3(v_3)$), determine the characteristic for $\mathscr{R}$.

Fig. 1.49 A ladder circuit
with nonlinear resistors

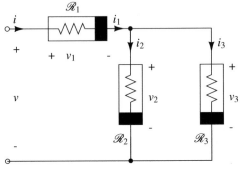

Fig. 1.50 Circuit for
problem 1.7

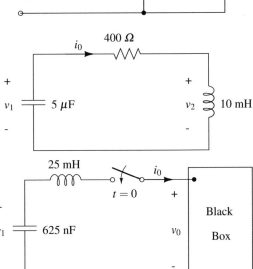

Fig. 1.51 Circuit for
problem 1.8

1.6 Based on the ideas from Sect. 1.10, discuss:

1. $\mathscr{L}_n$ inductors in series and parallel
2. $\mathscr{C}_n$ capacitors in series and parallel
3. $\mathscr{M}_n$ memristors in series and parallel

1.7 The current in the circuit in Fig. 1.50 [17] is known to be $i_0 = 5e^{-2000t}(2\cos 4000t + \sin 4000t)$ mA to $t \geq 0^+$. Find the values of $v_1(0^+)$ and $v_2(0^+)$.

1.8 At $t = 0$, a series-connected capacitor and inductor are placed across the terminals of a black box, as shown in Fig. 1.51 [17]. For $t > 0$, it is known that:

$$i_0 = 1.5e^{-16000t} - 0.5e^{-4000t} \text{ A} \tag{1.107}$$

If $v_1(0) = -50$ V, find and sketch v_0 for $t \geq 0$.

1.9 Although in this book we will deal with nonlinear circuits, understanding the distinction between linearity and nonlinearity is vital. The goal of this exercise is to explore this distinction.

Simply stated, a **linear system** satisfies the **principle of superposition**, while a nonlinear **system** does not. The principle of superposition can be phrased from basic physics courses as: "response of the sum is equal to the sum of the responses." Mathematically [14], a **system** $H : [\mathscr{R} \to \mathscr{R}] \to [\mathscr{R} \to \mathscr{R}]$ is linear iff $\forall x, y \in [\mathscr{R} \to \mathscr{R}]$ and $\alpha, \beta \in \mathscr{R}$:

$$H(\alpha x + \beta y) = \alpha H(x) + \beta H(y) \tag{1.108}$$

Notice that H is a function of a **function space**. x, y are **signals**, for our purposes, a time-domain signal. In other words $x(t)$ is a real number. But, the input to H is not a real number, but rather an entire signal! The distinction between function spaces and signals may be subtle but it is very important.

To understand this distinction and hence the difference between linearity and nonlinearity, answer the following: Is the system: $S(x) = \alpha x + \beta$ (given the definitions of x, α, β above) linear? Prove or disprove using Eq. (1.108).

Note that you may be tempted to conclude S is linear because "it looks like the equation of a straight-line." Do not jump to conclusions! Remember that S is a **system**!

One of the (many) "nice" properties of a linear system (as opposed to a nonlinear system) is a linear system's response to sinusoidal signals. We will see in Sect. 4.3 that if the input to the linear system is a sinusoid, the output signal is also a **sinusoid at the same frequency** (with amplitude and phase changed as per the frequency response). Exercise 4.13 will explore this idea further in the context of S above.

Lab 1: Introduction to Quite Universal Circuit Simulator (QUCS)

Objective: To successfully install QUCS
Theory:
The goal of circuit simulation is to predict the behavior of a circuit before we physically construct the circuit. A **very important** point about simulation in general: simulation is a **necessary but not sufficient** step.[14] In other words, if a circuit gives us the correct result in simulation, it may have the same physical

---

[14]This of course assumes that the simulator has been setup correctly **and** can simulate the problem at hand! You will see, especially in Chap. 5, that quite a few nonlinear circuit simulations "converge" to incorrect results in "traditional" circuit simulators. But a discussion of numerical simulation techniques for chaotic systems is beyond the scope of this book, the interested reader is referred to [20].

Fig. 1.52 QUCS startup screen in OS X Sierra

behavior. However, **the physical circuit may not display the same behavior as simulation** because again, the idea of device modeling: we may not have taken into account all the physical characteristics in simulation. But, if a circuit does not "work" in simulation, it will definitely not "work" in reality. This concept should become clear once the reader understands one of the main ideas in this book, namely, device modeling.

A variety of circuit simulators exist. In keeping with the introductory nature of this text, we would like to use a circuit simulator that is easy to use, has a robust graphical user interface (GUI) and is supported across multiple platforms (Windows, OS X and Linux based computers). Moreover, as stated in the online QUCS FAQ [26], classic SPICE based simulators have a variety of limitations that QUCS aims to overcome.

In this lab component, we simply install QUCS and make sure that the program is functional.

Lab Exercise:

1. Download and install the correct version of QUCS from [26] for your platform. Detailed instructions are in Appendix A.
2. Start QUCS. If successful, you should see Fig. 1.52.
3. Once you start QUCS, we encourage you to read the associated documentation [26] and try some of the sample simulations. You should also go through the tutorial video on QUCS in the youtube channel for this book. More will be explained about the different simulation (transient, etc.) throughout the book.
4. A very important resource is the online QUCS workbook. We will refer to this workbook throughout this text, so please make sure you start reading through it.

References

1. Chua, L.O.: Introduction to Nonlinear Network Theory. McGraw-Hill, New York (1969, out of print)
2. Chua, L.O.: Memristor – the missing circuit element. IEEE Trans. Circuit Theory **18**(5), 507–519 (1971)
3. Chua, L.O.: Device modeling via basic nonlinear circuit elements. IEEE Trans. Circuits Syst. **CAS-27**(11), 1014–1044 (1980)
4. Chua, L.O.: Nonlinear circuit foundations for nanodevices, part I: the four-element torus (invited paper). Proc. IEEE **91**(11), 1830–1859 (2003)
5. Chua, L.O.: University of California, Berkeley EE100 Fall 2008 Supplementary Lecture Notes (2008). Available online: Two-Terminal Elements: http://inst.eecs.berkeley.edu/~ee100/fa08/lectures/EE100supplementary_notes_1.pdf; Kirchhoff's Laws: http://inst.eecs.berkeley.edu/~ee100/fa08/lectures/EE100supplementary_notes_2.pdf; Two-Terminal Resistors: http://inst.eecs.berkeley.edu/~ee100/fa08/lectures/EE100supplementary_notes_3.pdf. Last accessed 10 Feb 2017
6. Chua, L.O.: If it's pinched it's a memristor. Semicond. Sci. Technol. **29**(10), 104001–104043 (2014)
7. Chua, L.O., Kang, S.M.: Memristive devices and systems. Proc. IEEE **64**(2), 209–223 (1976)
8. Chua, L.O., Kang, S.M.: Section-wise piecewise-linear functions: canonical representation, properties, and applications. Proc. IEEE **65**(6), 915–929 (1977)
9. Chua, L.O., Desoer, C.A., Kuh, E.S.: Linear and Nonlinear Circuits. McGraw-Hill, New York (1987, out of print)
10. Fano, R.M., Chu, L.J., Adler, R.B.: Electromagnetic Fields, Energy and Forces, Fifth reprinting. MIT Press, Cambridge (1968)
11. Hunt, B.J.: The Maxwellians. Cornell University Press, Ithaca (1991)
12. Josephson, B.: Supercurrents through barriers. Adv. Phys. **14**(56), 419–451 (1965)
13. Langenberg, D.N.: Physical interpretation of the $\cos \phi$ term and implications for detectors. Rev. Phys. Appl. (Paris) **9**, 35–40 (1974)
14. Lee, E.A., Varaiya, P.P.: Structure and Interpretation of Signals and Systems, 2nd edn. (2011). LeeVaraiya.org. ISBN 978-0-578-07719-2
15. Makin, J.G.: A Computational Model of Human Blood Clotting: Simulation, Analysis, Control and Validation. Tech. Rep. UCB/EECS-2008-165. Electrical Engineering and Computer Sciences Department, University of California, Berkeley (2008)
16. Muthuswamy, B., et al.: Memristor Modelling. In: Proceedings of the 2014 IEEE International Symposium on Circuits and Systems, pp. 490–493 (2014)
17. Nilsson, J.W., Riedel, S.A.: Electric Circuits, 9th edn. Prentice-Hall, Upper Saddle River (2011)
18. Oldham, W.: EECS 40 Midterm 1 (1999). Available online: https://hkn.eecs.berkeley.edu/examfiles/ee40_fa99_mt1.pdf Last accessed February 20th 2017
19. Oster, G.F., Auslander, D.M.: The memristor: a new bond graph element. J. Dyn. Sys. Meas. Control **94**(3), 249–252 (1972)
20. Parker, T.S., Chua, L.O.: Practical Numerical Algorithms for Chaotic Systems. Springer, Berlin (1989)
21. Pedersen, N.F., et al.: Magnetic field dependence and Q of the Josephson plasma resonance. Phys. Rev. B **11**(6), 4151–4159 (1972)
22. Pedersen, N.F., et al.: Evidence for the existence of the Josephson quasiparticle-pair interference current. Low Temp. Phys. **13**, 268–271 (1974)
23. Penfield, P.L.: Frequency-Power Formulas for Josephson Junctions. MIT Technical Report on Microwave and Millimeter Wave Techniques, QPR No. 113, pp. 31–32 (1974). Available online: https://dspace.mit.edu/bitstream/handle/1721.1/56460/RLE_QPR_113_V.pdf. Last accessed 20 Feb 2017

24. Peotta, A., Di Ventra, M.: Superconducting memristors. Phys. Rev. Applied **2**, 034011-1–034011-10 (2014)
25. Polya, G.: How to Solve It. Princeton University Press, Princeton (1945)
26. QUCS: Quite Universal Circuit Simulator. Available online: http://qucs.sourceforge.net/index.html. Last accessed 20 Feb 2017
27. Strukov, D.B., et al.: The missing memristor found. Nature **453**, 80–83 (2008)
28. Temes, G.C., LaPatra, J.W.: Introduction to Circuit Synthesis and Design. McGraw-Hill, New York (1977)
29. Thompson, E.D.: Power flow for Josephson elements. IEEE Trans. Electron Devices **20**(8), 680–683 (1973)

Chapter 2
Multi-Terminal Network Elements

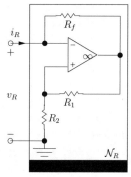

Two-terminal piecewise linear negative resistor, synthesized using a multi-terminal opamp

Abstract This chapter will naturally expand upon the ideas in Chap. 1 and discuss black boxes that have more than two terminals. We will first discuss characterization of a multi-terminal black box, followed by a discussion of the two-port representation technique. We will then talk about resistive, inductive (including transformers), and capacitive three-terminal elements. Circulators and opamps are next discussed. After this, we discuss the family of two-port scalors, rotators, reflectors, and gyrators. A current feedback opamp-based implementation approach is used for studying mutators.

2.1 Characterization of a Multi-Terminal Black Box

While the conventional resistor is probably the most familiar circuit element [6] [4], the transistor is certainly the electronic device that heralded the computer revolution. A transistor is a three-terminal device which behaves like a two-terminal nonlinear resistor when viewed from any pair of terminals, at low enough frequencies. This is

B. Muthuswamy, S. Banerjee, *Introduction to Nonlinear Circuits and Networks*,
https://doi.org/10.1007/978-3-319-67325-7_2

why its inventors (Nobel laureates Bardeen, Brattain, and Shockley) christened it as a **transfer resistor**, or transistor in brief.

The transistor is not the only multi-terminal device, many devices have more than two terminals. Our objective in this section is to learn how these multi-terminal devices may be characterized so that we shall be in a position to use them more effectively [3]. The basic principles discussed in the preceding chapter for characterizing two-terminal devices are still applicable. A set of measurable **independent** variables is selected and a series of external measurements are taken with the objective of deriving a consistent relationship among the variables. Once this relationship is found, we have characterized the black box because from then on, any design using this device can be undertaken on the basis of this relationship alone, thereby obviating the need for further measurements.

To discuss the selection of an independent set of variables, let us consider first the three-terminal black box shown in Fig. 2.1a. The most obvious variables are the currents i_1, i_2, and i_3 entering the terminals, and the voltages v_{12}, v_{23}, and v_{31} across the terminals. However, the black box in Fig. 2.1a can be enclosed by a Gaussian surface and hence the currents i_1, i_2, and i_3 entering this surface must satisfy KCL, namely, $i_1 + i_2 + i_3 = 0$.

Thus, if we know the value of any two of these currents, we can calculate the third, and therefore there is no need to measure all three currents. This observation is equivalent to saying that the three variables i_1, i_2, and i_3 are not independent. Similarly, from KVL we have $v_{12} + v_{23} + v_{31} = 0$ and hence the three variables v_{12}, v_{23}, and v_{31} are not independent.

Consequently, among the six variables shown in Fig. 2.1a, only two currents and two voltages are independent. For this reason, we may select any terminal to be ground and define the two currents i_1, i_2 and voltages v_1, v_2 as shown in Fig. 2.1b. In theory, there is no reason for preferring one terminal over another as the ground terminal. In practice, however, such a preference may be desirable

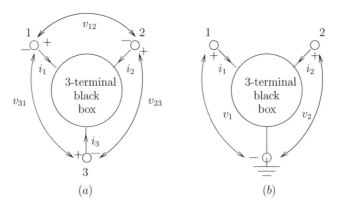

(a) (b)

Fig. 2.1 In the process of characterizing a three-terminal black box, one terminal is arbitrarily chosen as the ground terminal. The voltages of the remaining terminals are measured with respect to the common terminal

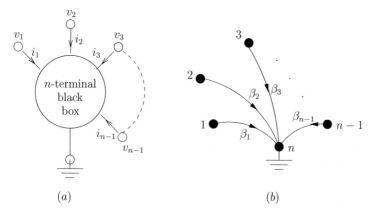

Fig. 2.2 For an n-terminal element, we can arbitrarily choose one terminal as ground. With n chosen as the ground terminal, we have the associated element graph

because the measurements may be easier and more accurately obtained.[1] To avoid ambiguity, it is of the utmost importance to specify the common terminal associated with the measured characteristics of a particular device. A ground terminal can also be arbitrarily chosen for a generic multi-terminal (or henceforth, n-terminal) element, as shown in Fig. 2.2a. Based on our choice of the ground terminal, we can also easily draw the associated element graph of the n-terminal device, as shown in Fig. 2.2b. Notice that we will have n possible element graphs for an n-terminal element, depending on our choice of the ground node.

With the abovementioned precaution of choosing the common terminal associated with the measured characteristic, let us investigate the type of measurements that may be taken. Just as in the two-terminal case, it is necessary to excite the black box by a voltage source or a current source. However, the response to these excitations need not restricted to currents and voltages. Recall from Eq. (1.3) that it is possible to measure the charge q_j entering terminal j by integrating the current i_j, namely:

$$q_j(t) = \int_{-\infty}^{t} i_j(\tau)d\tau \ \ j = 1, 2, \cdots, n-1 \qquad (2.1)$$

Similarly, from Eq. (1.4), we can measure the flux-linkage ϕ_j associated with each voltage v_j between terminal j and ground by integrating the voltage v_j:

$$\phi_j(t) = \int_{-\infty}^{t} v_j(\tau)d\tau \ \ j = 1, 2, \cdots, n-1 \qquad (2.2)$$

---

[1]This is especially true for the transistor, where the characteristic curves can be more accurately, and more easily, measured if a particular terminal (called emitter for *npn* junction transistors) is chosen to be the ground terminal.

Hence, among the variables of interest to us are q_j, i_j, ϕ_j, and v_j ($j = 1, 2, \cdots, n-1$). Any **independent** combination of these variables constitute a valid set of measurements. Observe that the combination q_j and i_j (ϕ_j and v_j) is not valid because these variables are already related by Eqs. (2.1) and (2.2). If a certain set of measurements leads to some consistent relationship, then the device is said to be characterized by that relationship.

The corresponding element classifications now take the following forms:

1. n-terminal resistors, involving only $v_1, v_2, \cdots, v_{n-1}; i_1, i_2, \cdots, i_{n-1}$.
2. n-terminal inductors, involving only $i_1, i_2, \cdots, i_{n-1}; \phi_1, \phi_2, \cdots, \phi_{n-1}$.
3. n-terminal capacitors, involving only $v_1, v_2, \cdots, v_{n-1}; q_1, q_2, \cdots, q_{n-1}$.
4. n-terminal memristors, involving only $\phi_1, \phi_2, \cdots, \phi_{n-1}; q_1, q_2, \cdots, q_{n-1}$.

But, in general, in order to completely characterize an n-terminal black box, $n - 1$ distinct laboratory setups are required. For example, Fig. 2.3 shows the setups necessary to characterize a four-terminal device.

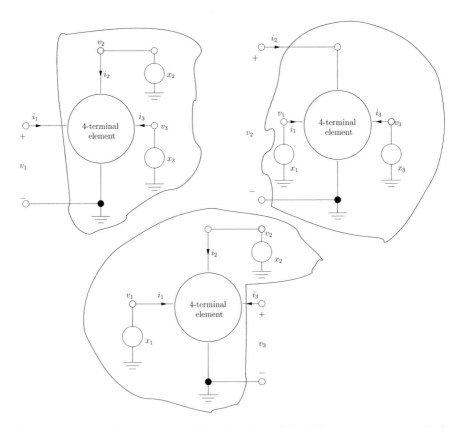

Fig. 2.3 To characterize a four-terminal black box, three distinct laboratory setups are required. Each setup involves as many sets of measurements as necessary to include all desired combinations of parameter values of the controlling variables

Thus, it is in general impractical to completely characterize an n-terminal black box when $n >> 3$. Fortunately, most practical devices have single digit values for n, and those devices will be discussed in the remainder of this chapter, starting with $n = 3$.

2.2 Three-Terminal Resistors, Inductors, and Capacitors

2.2.1 Two-Port Representation

The concept of a port was first introduced in Sect. 1.4. Recall that a port can be created from a circuit by connecting two leads to a pair of nodes of the circuit. Thus, a one-port can be viewed as a black box which has one pair of terminals from the outside. In the case of a multi-port such as the four-terminal black box from Fig. 2.3, we see that the box can be completely characterized with three sets of measurements, using three pairs of terminals.

As discussed previously, because of the complexity involved in practically characterizing a multi-terminal device for $n >> 3$ (n is the number of terminals), we will primarily discuss three-terminal elements or two-ports, with $n = 3$. For details on multi-ports, refer to section 4 from [6].

The generalization from a two-terminal to a three-terminal element amounts to extending from scalar port variables to $n - 1$-dimensional vector variables. For the purposes of clarity, we will discuss resistive two-ports in detail. For inductors, capacitors, and memristors, the principles are identical and hence for those elements, we will only discuss one form of representation.

A three-terminal element, or a two-port, will be called a (time-invariant) resistor if its port voltages and port currents satisfy the following relation:

$$\mathscr{R}_R = \{(v_1, v_2, i_1, i_2); f_1(v_1, v_2, i_1, i_2) = 0 \text{ and } f_2(v_1, v_2, i_1, i_2) = 0\} \qquad (2.3)$$

This relation, similar to the two-terminal resistor given by Eq. (1.35) in Chap. 1, will be called **the $v - i$ characteristic of a three-terminal resistor or a resistive two-port**. The difference with respect to Eq. (1.35) is that we now need two scalar functions $f_1(\cdot)$ and $f_2(\cdot)$ to characterize a two-port and there are four scalar variables v_1, v_2, i_1, i_2. The characteristic is in general a two-dimensional surface in a four-dimensional space.

When we deal with two-ports, we often need to distinguish the ports, so one of them is marked as port 1 and the other is marked as port 2, as shown in Fig. 2.4. As a tradition, port 1 is often referred to as the **input port** and port 2 is often referred to as the **output port**.

We will now first consider linear resistors and use them to bring out pertinent concepts in the generalization from a two-terminal (one-port) to multi-terminal (two-port) element. Nonlinear two-ports such as transistors will be discussed in Sect. 2.2.2.

Fig. 2.4 A two-port with its port voltages v_1, v_2 and port currents i_1, i_2

Table 2.1 Six representations of a two-port

Representations	Dependent variables	Independent variables
Current-controlled	v_1, v_2	i_1, i_2
Voltage-controlled	i_1, i_2	v_1, v_2
Hybrid 1	v_1, i_2	i_1, v_2
Hybrid 2	i_1, v_2	v_1, i_2
Transmission 1	v_1, i_1	v_2, i_2
Transmission 2	v_2, i_2	v_1, i_1

Table 2.2 Equations for the six representations of a linear resistive two-port

Representations	Scalar equations	Vector equations
Current-controlled	$v_1 = r_{11}i_1 + r_{12}i_2$ $v_2 = r_{21}i_1 + r_{22}i_2$	$\mathbf{v} = \mathbf{R}\mathbf{i}$
Voltage-controlled	$i_1 = g_{11}v_1 + g_{12}v_2$ $i_2 = g_{21}v_1 + g_{22}v_2$	$\mathbf{i} = \mathbf{G}\mathbf{v}$
Hybrid 1	$v_1 = h_{11}i_1 + h_{12}v_2$ $i_2 = h_{21}i_1 + h_{22}v_2$	$\begin{bmatrix} v_1 \\ i_2 \end{bmatrix} = \mathbf{H} \begin{bmatrix} i_1 \\ v_2 \end{bmatrix}$
Hybrid 2	$i_1 = h'_{11}v_1 + h'_{12}i_2$ $v_2 = h'_{21}v_1 + h'_{22}i_2$	$\begin{bmatrix} i_1 \\ v_2 \end{bmatrix} = \mathbf{H}' \begin{bmatrix} v_1 \\ i_2 \end{bmatrix}$
Transmission 1	$v_1 = t_{11}v_2 - t_{12}i_2$ $i_1 = t_{21}v_2 - t_{22}i_2$	$\begin{bmatrix} v_1 \\ i_1 \end{bmatrix} = \mathbf{T} \begin{bmatrix} v_2 \\ -i_2 \end{bmatrix}$
Transmission 2	$v_2 = t'_{11}v_1 + t'_{12}i_1$ $-i_2 = t'_{21}v_1 + t'_{22}i_1$	$\begin{bmatrix} v_2 \\ i_2 \end{bmatrix} = \mathbf{T}' \begin{bmatrix} v_1 \\ i_1 \end{bmatrix}$

For the transmission representations, for historical reasons, a minus sign is used in conjunction with i_2. Because of the reference direction chosen for i_2, $-i_2$ gives the current leaving the output port

   With four scalar variables v_1, v_2, i_1, i_2 and two equations to characterize a resistive two-port, there are $C_2^4 = 6$ possible two-port representations, since we may choose any two of the four variables as independent variables (the remaining two are then the dependent variables). Table 2.1 gives the classification of the six representations according to dependent and independent variables.

   Table 2.2 gives the equations of the six possible representations of a linear resistive two-port.

In Table 2.2, $\mathbf{G}$ is the inverse matrix of $\mathbf{R}$. Similarly, we also have $\mathbf{H}' = \mathbf{H}^{-1}$ and $\mathbf{T}' = \mathbf{T}^{-1}$. We call $\mathbf{H}$ and $\mathbf{H}'$ hybrid matrices because both the dependent and independent variables are mixtures of a voltage and current. We call $\mathbf{T}$ and $\mathbf{T}'$ the transmission matrices because they relate the variables pertaining to one port to that pertaining to the other and the two-port serves as a transmission media. Hence, transmission matrices are important in the study of communication networks. A discussion of these networks is the beyond the scope of this book, but the interested reader is referred to Chapter 13 in [6].

Example 2.2.1 Consider a resistive two-port made up of three linear resistors as shown in Fig. 2.5. Determine the current-controlled and voltage-controlled representations.

Solution Let us apply two independent current sources to the two-port as shown in Fig. 2.6. KCL applied to nodes 1, 2, and 3 yields:

$$i_{s1} = i_1$$

$$i_{s2} = i_2$$

$$i_3 = i_1 + i_2 \tag{2.4}$$

Using Ohm's law and KVL for node sequences $1 - 3 - 4 - 1$ and $2 - 3 - 4 - 2$, we get:

$$v_1 = i_1 R_1 + R_3(i_1 + i_2) = (R_1 + R_3)i_1 + R_3 i_2$$

$$v_2 = i_2 R_2 + R_3(i_1 + i_2) = R_3 i_1 + (R_2 + R_3)i_2 \tag{2.5}$$

We will rewrite Eq. (2.5) in matrix form, to obtain the current-controlled representation from Table 2.2.

$$\begin{pmatrix} v_1 \\ v_2 \end{pmatrix} = \begin{pmatrix} R_1 + R_3 & R_3 \\ R_3 & R_2 + R_3 \end{pmatrix} \begin{pmatrix} i_1 \\ i_2 \end{pmatrix} \tag{2.6}$$

Hence, we have the resistance matrix $\mathbf{R}$ as defined in Eq. (2.7).

$$\mathbf{R} \triangleq \begin{pmatrix} R_1 + R_3 & R_3 \\ R_3 & R_2 + R_3 \end{pmatrix} \tag{2.7}$$

Notice that $\mathbf{R}$ is symmetrical: $\mathbf{R}^{\mathrm{T}} = \mathbf{R}$. Such symmetries will be exploited when we discuss resistive nonlinear networks in Chap. 3.

(continued)

Example 2.2.1 (continued)

Now, that we have **R**, **G** (conductance matrix) for the voltage-controlled representation is simply $\mathbf{R}^{-1}$:

$$\mathbf{G} \triangleq \mathbf{R}^{-1} = \frac{1}{R_1 R_2 + R_2 R_3 + R_3 R_1} \begin{pmatrix} R_2 + R_3 & -R_3 \\ -R_3 & R_1 + R_3 \end{pmatrix} \qquad (2.8)$$

In Example 2.2.1, we could have derived the voltage-controlled representation first by using independent voltage sources v_{s1} and v_{s2}, then used the fact that $\mathbf{R} \triangleq \mathbf{G}^{-1}$. In other words, it is quite simple to transform one two-port representation to another, as shown in Example 2.2.2.

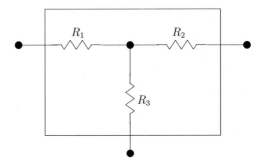

Fig. 2.5 The resistive T-network

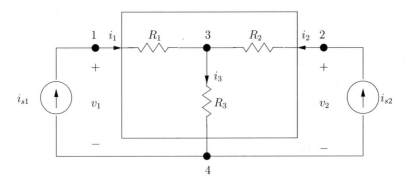

Fig. 2.6 For Example 2.2.1, we will use two independent current sources in Fig. 2.5 for obtaining the current-controlled representation

Example 2.2.2 In Example 2.2.1, let $R_1 = 1\ \Omega, R_2 = 2\ \Omega, R_3 = 3\ \Omega$. Determine the numerical current-controlled representation and the other representations from Table 2.2.

Solution The numerical current-controlled representation is given by Eq. (2.6):

$$\begin{pmatrix} v_1 \\ v_2 \end{pmatrix} = \begin{pmatrix} 4 & 3 \\ 3 & 5 \end{pmatrix} \begin{pmatrix} i_1 \\ i_2 \end{pmatrix} \tag{2.9}$$

The voltage-controlled representation can be found using $\mathbf{G}$ in Eq. (2.8) or by inverting the numerical square matrix in Eq. (2.9):

$$\begin{pmatrix} i_1 \\ i_2 \end{pmatrix} = \begin{pmatrix} \dfrac{5}{11} & \dfrac{-3}{11} \\ \dfrac{-3}{11} & \dfrac{4}{11} \end{pmatrix} \begin{pmatrix} v_1 \\ v_2 \end{pmatrix} \tag{2.10}$$

It is straightforward to derive the other four representations from the equations above. The general treatment is beyond the scope of this book but can be found in classic references such as [6]. However, it is easy to obtain, for example, the hybrid representations.

For the Hybrid 2 representation, we first solve for i_1 in terms of v_1 and i_2 by using the first row from Eq. (2.9). Next, we solve for v_2 in terms of v_1 and i_2 by using the second row from Eq. (2.10). Thus:

$$\begin{pmatrix} i_1 \\ v_2 \end{pmatrix} = \begin{pmatrix} \dfrac{1}{4} & \dfrac{-3}{4} \\ \dfrac{3}{4} & \dfrac{11}{4} \end{pmatrix} \begin{pmatrix} v_1 \\ i_2 \end{pmatrix} \tag{2.11}$$

The hybrid 1 representation can be found by inverting $\mathbf{H}'$ from Eq. (2.11):

$$\mathbf{H} = \begin{pmatrix} \dfrac{11}{5} & \dfrac{3}{5} \\ \dfrac{-3}{5} & \dfrac{1}{5} \end{pmatrix} \tag{2.12}$$

The transmission matrices can be obtained in a similar manner and is left as an exercise for the reader.

2.2.1.1 Physical Interpretations

In the examples from Sect. 2.2.1, we derived various two-port representations. In particular, we derived the current-controlled representation by using two current sources at the two-ports and determining the two-port voltages (as shown in Fig. 2.6).

For a physical interpretation of two-ports, recall from Chap. 1 that we defined a linear two-terminal resistor as one having a straight line characteristic passing through the origin in the $v - i$ plane. For two-ports, we have four variables and two equations, e.g., the current-controlled representation is:

$$v_1 = r_{11}i_1 + r_{12}i_2$$

$$v_2 = r_{21}i_1 + r_{22}i_2 \tag{2.13}$$

These two equations impose two linear constraints on the port voltages and the port currents and hence the point representing the four variables; namely, (v_1, v_2, i_1, i_2) is constrained to a two-dimensional subspace in the four-dimensional space spanned by v_1, v_2, i_1, i_2. Of course, this is difficult to visualize. However, if we take one equation at a time, we can represent it by a family of curves in the appropriate $i - v$ planes, as shown in Fig. 2.7.

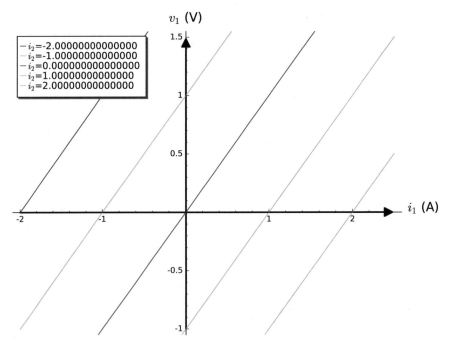

Fig. 2.7 Two-port characteristics plotted on the $i_1 - v_1$ plane, with i_2 as parameter. $r_{11} = 1, r_{12} = -1$. A similar plot can be generated for $i_2 - v_2$ plane

From the first equation in Eq. (2.13), we can give the following interpretations for r_{11} and r_{12}:

$$r_{11} = \frac{v_1}{i_1}\bigg|_{i_2=0} \tag{2.14}$$

Thus, r_{11} is called the **driving-point resistance at port 1** when $i_2 = 0$, i.e., port 2 is kept open circuited. Similarly, r_{12} can be interpreted by:

$$r_{12} = \frac{v_1}{i_2}\bigg|_{i_1=0} \tag{2.15}$$

Hence, r_{12} is called the **transfer resistance when $i_1 = 0$**, i.e., port 1 is kept open circuited.

Analogously, we can derive the following relationships from the second equation in Eq. (2.13):

$$r_{21} = \frac{v_2}{i_1}\bigg|_{i_2=0} \tag{2.16}$$

$$r_{22} = \frac{v_2}{i_2}\bigg|_{i_1=0} \tag{2.17}$$

r_{21} is the **transfer resistance when $i_2 = 0$** and r_{22} is the **driving-point resistance at port 2**. Figure 2.8 gives the physical interpretations of Eq. (2.14) through (2.17).

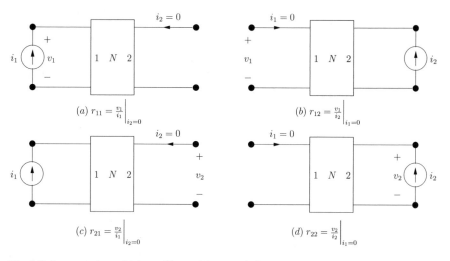

Fig. 2.8 Interpretations of (**a**) r_{11}, (**b**) r_{12}, (**c**) r_{21}, and (**d**) r_{22}

Example 2.2.3 Give the physical interpretation of the hybrid 1 linear resistive two-port representation from Table 2.2.

Solution The two equations for the hybrid 1 representation read:

$$v_1 = h_{11}i_1 + h_{12}v_2 \tag{2.18}$$

$$i_2 = h_{21}i_1 + h_{22}i_2 \tag{2.19}$$

Following the same treatment as the current-controlled representation, we write:

$$h_{11} = \left.\frac{v_1}{i_1}\right|_{v_2=0} \tag{2.20}$$

$$h_{12} = \left.\frac{v_1}{v_2}\right|_{i_1=0} \tag{2.21}$$

$$h_{21} = \left.\frac{i_2}{i_1}\right|_{v_2=0} \tag{2.22}$$

$$h_{22} = \left.\frac{i_2}{v_2}\right|_{i_1=0} \tag{2.23}$$

The physical interpretations of the sources, responses, and external connections for the four hybrid representations are shown in Fig. 2.9.

Note that the four hybrid parameters h_{11}, h_{12}, h_{21}, h_{22} represent a **driving-point resistance, a reverse voltage transfer ratio, a forward current transfer ratio**, and a **driving-point conductance**, respectively. As we will see in Sect. 3.1, the hybrid representation is obtained when we derive the small-signal model for the common-emitter configuration of the bipolar junction transistor.

Analogous interpretations can be given for other two-port representations such as the current-controlled representation.

2.2.1.2 Dependent Sources

Up to this point, we have encountered independent voltage and current sources. Independent sources are used as inputs to a circuit. In this section, we will introduce another type of source, called **controlled sources** or **dependent sources**.

A controlled source is a resistive two-port element consisting of two branches: a primary branch which is either an open circuit or a short circuit and a secondary branch which is either a voltage source or a current source. The voltage or current waveform in the secondary branch is **controlled** by (or **dependent** upon) the

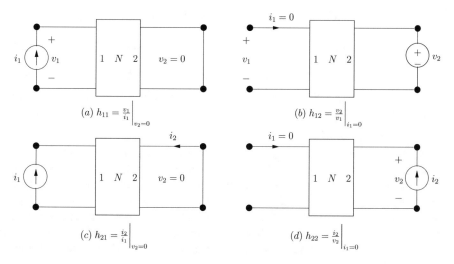

$$(a)\ h_{11} = \frac{v_1}{i_1}\Big|_{v_2=0} \qquad (b)\ h_{12} = \frac{v_2}{v_1}\Big|_{i_1=0}$$

$$(c)\ h_{21} = \frac{i_2}{i_1}\Big|_{v_2=0} \qquad (d)\ h_{22} = \frac{i_2}{v_2}\Big|_{i_1=0}$$

Fig. 2.9 Interpretations of (**a**) h_{11}, (**b**) h_{12}, (**c**) h_{21}, and (**d**) h_{22}

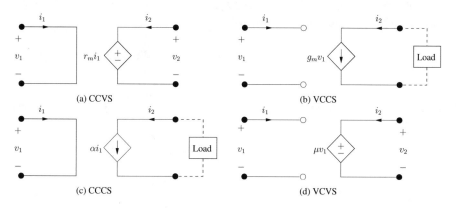

(a) CCVS   (b) VCCS

(c) CCCS   (d) VCVS

Fig. 2.10 Four types of linear controlled sources

voltage or current of the primary branch. Therefore, there exist four types of controlled sources depending on whether the primary branch is an open circuit or a short circuit and whether the secondary branch is a voltage source or a current source. The four types of controlled sources are shown in Fig. 2.10. They are the **current-controlled voltage source** (CCVS), **voltage-controlled current source** (VCCS), **current-controlled current source** (CCCS), and **voltage-controlled current source** (VCCS). Note that we use a **diamond-shaped**[2] symbol to denote controlled sources. This is to differentiate them from the independent sources.

---

[2]Diamond-shaped symbol for controlled sources was used for the first time in [3].

Each linear controlled source is characterized by two linear equations:

$$\text{CCVS:} \qquad\qquad v_1 = 0 \qquad v_2 = r_m i_1 \qquad\qquad (2.24)$$

$$\text{VCCS:} \qquad\qquad i_1 = 0 \qquad i_2 = g_m v_1 \qquad\qquad (2.25)$$

$$\text{CCCS:} \qquad\qquad v_1 = 0 \qquad i_2 = \alpha i_1 \qquad\qquad (2.26)$$

$$\text{VCVS:} \qquad\qquad i_1 = 0 \qquad v_2 = \mu v_1 \qquad\qquad (2.27)$$

r_m is the **transresistance**, g_m is the **transconductance**, α is called the **current transfer ratio**, and μ is called the **voltage transfer ratio**. They are all constants, thus the four controlled sources are linear time-invariant two-port resistors. More generally, if a CCVS is characterized by the two equations: $v_1 = 0$, $v_2 = f(i_1)$, where $f(\cdot)$ is a given **nonlinear** function, then that CCVS is a **nonlinear controlled source**. Similarly, if a CCCS is characterized by the two equations $v_1 = 0$, $i_2 = \alpha(t)i_1$, where $\alpha(\cdot)$ is a given function of time, then this CCCS is a **linear time-varying controlled source**.

Recall from Table 2.2, a linear resistive two-port has six representations. In the case of linear controlled sources, Eq. (2.24) to (2.27) can be put in matrix form for each corresponding to one representation:

$$\text{CCVS:} \qquad\qquad \begin{pmatrix} v_1 \\ v_2 \end{pmatrix} = \begin{pmatrix} 0 & 0 \\ r_m & 0 \end{pmatrix} \begin{pmatrix} i_1 \\ i_2 \end{pmatrix} \qquad\qquad (2.28)$$

$$\text{VCCS:} \qquad\qquad \begin{pmatrix} i_1 \\ i_2 \end{pmatrix} = \begin{pmatrix} 0 & 0 \\ g_m & 0 \end{pmatrix} \begin{pmatrix} v_1 \\ v_2 \end{pmatrix} \qquad\qquad (2.29)$$

$$\text{CCCS:} \qquad\qquad \begin{pmatrix} v_1 \\ i_2 \end{pmatrix} = \begin{pmatrix} 0 & 0 \\ \alpha & 0 \end{pmatrix} \begin{pmatrix} i_1 \\ v_2 \end{pmatrix} \qquad\qquad (2.30)$$

$$\text{VCVS:} \qquad\qquad \begin{pmatrix} i_1 \\ v_2 \end{pmatrix} = \begin{pmatrix} 0 & 0 \\ \mu & 0 \end{pmatrix} \begin{pmatrix} v_1 \\ i_2 \end{pmatrix} \qquad\qquad (2.31)$$

In Eq. (2.28), we have the current-controlled representation for the CCVS. Since the resistance matrix is singular, its inverse does not exist. Therefore, there is no voltage-controlled representation for a CCVS. In fact, it is easy to see that neither of the hybrid representations exists as well. We can make similar statements for the other three controlled sources, i.e., only one of the representations in the first four rows of Table 2.2 exists.

Linear controlled sources are extremely useful in modeling electronic devices and circuits, as we will see in Sect. 2.2.2. In Sect. 2.5.2.1, we will see that all four controlled sources can be realized physically (to a good approximation) by using operational amplifiers.

Fig. 2.11 Figure for
Example 2.2.4

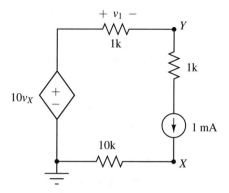

Example 2.2.4 In Fig. 2.11, determine the values of v_X and v_Y.

Solution We have a VCVS, whose input depends on v_X (voltage at node X with respect to ground). To avoid clutter, we have not explicitly drawn the two-port form for the VCVS. But, the reader **must** understand that all dependent sources are two-ports.

Since all elements in the circuit are in series, the current flowing through all elements is 1 mA, due to the constant current source. Since all resistors are also linear, by Ohm's law and the passive sign convention, we have:

$$v_X = 1 \cdot 10 \text{ V}$$
$$= 10 \text{ V} \qquad (2.32)$$

From KVL:

$$v_Y + v_1 - 10v_X = 0 \qquad (2.33)$$

Hence:

$$v_Y = 10v_X - v_1$$
$$= 99 \text{ V} \qquad (2.34)$$

2.2.1.3 Transformers

The **ideal transformer** is an ideal two-port resistive circuit element which is characterized by the following two equations:

$$v_1 = nv_2 \tag{2.35}$$

$$i_2 = -ni_1 \tag{2.36}$$

where n is a real number called the **turns ratio**. The symbol for the ideal transformer is shown in Fig. 2.12.

The ideal transformer is a **linear** resistive two-port, since its equations impose **linear** constraints on its port voltages and port currents. Note that neither the current-controlled representation nor the voltage-controlled representation exists for the ideal transformer. Eqs. (2.35) and (2.35) can be written in matrix form in terms of the hybrid matrix representation:

$$\begin{pmatrix} v_1 \\ i_2 \end{pmatrix} = \mathbf{H} \begin{pmatrix} i_1 \\ v_2 \end{pmatrix} = \begin{pmatrix} 0 & n \\ -n & 0 \end{pmatrix} \begin{pmatrix} i_1 \\ v_2 \end{pmatrix} \tag{2.37}$$

The ideal transformer is an idealization of a physical transformer, constructed using coupled inductors, that is used in many applications. The properties of the physical transformer will be discussed in Sect. 2.2.3.

We wish to stress that because the ideal transformer is an ideal element defined by Eq. (2.37), the relation between port voltages and port currents holds for all waveforms and for all frequencies, including DC.

Two fundamental properties of the ideal transformer are:

1. The ideal transformer neither dissipates nor stores energy. Indeed, the power entering the two-port at time t from Eq. (2.37) is:

$$p(t) = v_1(t)i_1(t) + v_2(t)i_2(t) = 0 \tag{2.38}$$

Thus, the ideal transformer is a **non-energic** element (another non-energic element is the ideal diode).

Fig. 2.12 An ideal transformer defined by the single parameter n, the turns ratio. Notice that the sign of i_2 is negative in the expression for n, confirming to the passive sign convention

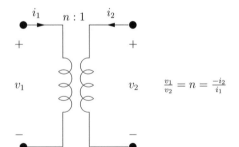

2. When the ideal transformer is terminated at the output port with an $R - \Omega$ resistor, the input port behaves as a linear resistor with resistance $n^2 R$. In other words:

$$v_2 = -i_2 R \qquad (2.39)$$

Therefore, $\frac{v_1}{i_1} = \frac{n v_2}{-i_2/n} = n^2 R$.

2.2.2 Three-Terminal Resistors

In the previous sections, we discussed linear resistive two-ports and their various characterizations and properties. In the real world, we need to deal with **nonlinear** resistive two-ports and three-terminal devices, such as transistors. Much of the material given in the previous two sections can be extended and generalized to the nonlinear case. For brevity, we will simply summarize the six nonlinear representations in Table 2.3.

2.2.2.1 The npn Bipolar Transistor

Perhaps, the most commonly used three-terminal nonlinear resistor is a transistor. These devices come in mainly two variants—the bipolar junction transistor (BJT) and the metal-oxide-semiconductor field-effect transistor (MOSFET). We will discuss the low-frequency characteristics of the *npn* BJT here, together with some aspects of modeling. A discussion of MOSFETs can be found in excellent texts such as [7].

Consider the common-base *npn* transistor as shown in Fig. 2.13. The nodes are labeled e, b, and c corresponding to the emitter, base, and collector, respectively.

Table 2.3 Equations for the six representations of a nonlinear resistive two-port

Representations	Scalar equations
Current-controlled	$v_1 = \hat{v}_1(i_1, i_2)$
	$v_2 = \hat{v}_2(i_1, i_2)$
Voltage-controlled	$i_1 = \hat{i}_1(v_1, v_2)$
	$i_2 = \hat{i}_2(v_1, v_2)$
Hybrid 1	$v_1 = \hat{v}_1(i_1, v_2)$
	$i_2 = \hat{i}_2(i_1, v_2)$
Hybrid 2	$i_1 = \hat{i}_1(v_1, i_2)$
	$v_2 = \hat{v}_2(v_1, i_2)$
Transmission 1	$v_1 = \hat{v}_1(v_2, -i_2)$
	$i_1 = \hat{i}_1(v_2, -i_2)$
Transmission 2	$v_2 = \hat{v}_2(v_1, i_1)$
	$-i_2 = \hat{i}_2(v_1, i_1)$

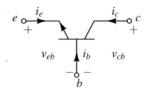

Fig. 2.13 The common-base *npn* transistor

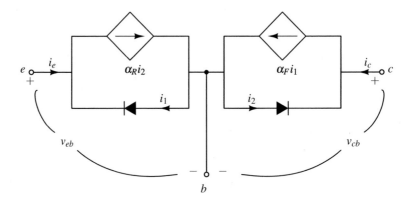

Fig. 2.14 Ebers–Moll circuit model of *npn* transistor

A good low-frequency characterization is given by the one-dimensional diffusion model which yields the **Ebers–Moll** equations:

$$i_e = -I_{ES}\left(e^{\frac{-v_{eb}}{V_T}} - 1\right) + \alpha_R I_{CS}\left(e^{\frac{-v_{cb}}{V_T}} - 1\right) \tag{2.40}$$

$$i_c = \alpha_F I_{ES}\left(e^{\frac{-v_{eb}}{V_T}} - 1\right) - I_{CS}\left(e^{\frac{-v_{cb}}{V_T}} - 1\right) \tag{2.41}$$

I_{ES}, I_{CS}, α_R, and α_F are device parameters. V_T is the thermal voltage defined earlier in Sect. 1.9.1, where we discussed the diode. Typically, $\alpha_R = 0.5$–0.8, $\alpha_F = 0.99$; I_{ES}, I_{CS} are on the order of 10^{-12} to 10^{-10} at 25°C. $V_T \approx 26$ mV at 25°C. Note that an *npn* BJT is in essence two **interacting** *pn*-junction diodes connected back to back to form a three-terminal device. Thus, with the base terminal as the ground node, the currents i_e and i_c entering the device at the emitter and the collector, respectively, are functions of two node-to-ground voltages v_{eb} and v_{cb}. From Eqs. (2.40) and (2.41), we see that the transistor is a **three-terminal voltage-controlled nonlinear resistor**. It can be represented by the equivalent circuit in Fig. 2.14, where the two *pn*-junctions are connected at the base node b to model the terms $-I_{ES}\left(e^{\frac{-v_{eb}}{V_T}} - 1\right)$ and $-I_{CS}\left(e^{\frac{-v_{cb}}{V_T}} - 1\right)$. The two CCCS are used to

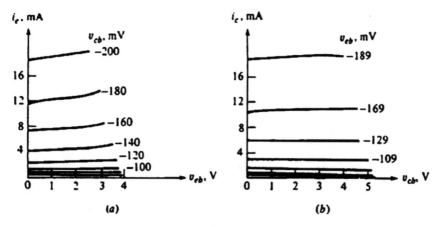

Fig. 2.15 Characteristics of an *npn* BJT in the common-base configuration [6]

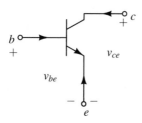

Fig. 2.16 The common-emitter *npn* transistor

model the terms $\alpha_R I_{CS}\left(e^{\frac{-v_{cb}}{V_T}} - 1\right)$ and $\alpha_F I_{ES}\left(e^{\frac{-v_{eb}}{V_T}} - 1\right)$ which represent the interaction between the two diodes.

The characteristics of Eqs. (2.40) and (2.41) are shown in Fig. 2.15 in the $v_{eb} - i_e$ plane and the $v_{cb} - i_c$ plane, respectively. Note that v_{cb} serves as a parameter in the family of curves in the $v_{eb} - i_e$ plane. Similarly, v_{eb} serves as a parameter in the family of curves in the $v_{cb} - i_c$ plane.

In most amplifier circuits, the common-emitter configuration shown in Fig. 2.16 is used. It is possible to derive equations for the common-emitter configuration directly from those of the common-base configuration of Eqs. (2.40) and (2.41). For the common-emitter configuration, the two-port voltages are v_{be} and v_{ce}. The two-port currents are i_b and i_c. These can be related to the variables of the common-base configuration by simply using Kirchhoff's laws:

$$v_{be} = -v_{eb} \tag{2.42}$$

$$v_{ce} = v_{cb} - v_{eb} \tag{2.43}$$

$$i_b = -(i_e + i_c) \tag{2.44}$$

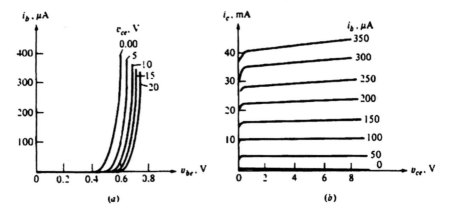

Fig. 2.17 Characteristics of an *npn* BJT in the common-emitter configuration [6]

Substituting the above equations into Eqs. (2.40) and (2.41), we can express the port currents i_b and i_c for the common-emitter configuration in terms of the port voltages v_{be} and v_{ce}. They are:

$$i_b = (1 - \alpha_F)I_{ES}\left(e^{\frac{v_{be}}{V_T}} - 1\right) + (1 - \alpha_R)I_{CS}\left(e^{\frac{v_{be}-v_{ce}}{V_T}} - 1\right) \qquad (2.45)$$

$$i_c = \alpha_F I_{ES}\left(e^{\frac{v_{be}}{V_T}} - 1\right) - I_{CS}\left(e^{\frac{v_{be}-v_{ce}}{V_T}} - 1\right) \qquad (2.46)$$

Thus, we again have a voltage-controlled representation for the common-emitter configuration. This set of equations is not particularly useful, because in practice, the measured data are usually expressed in terms of the hybrid 1 representation, i.e.,

$$v_{be} = \hat{v}_{be}(i_b, v_{ce}) \qquad (2.47)$$

$$i_c = \hat{i}_c(i_b, v_{ce}) \qquad (2.48)$$

Furthermore, as a tradition, we usually plot i_b vs v_{be} with v_{ce} as a parameter, and i_c vs v_{ce} with i_b as a parameter, as shown in Fig. 2.17. This is because we get a smoothly varying family of collector-to-emitter $v - i$ curves.

2.2.2.2   BJT Piecewise-Linear Approximation

As stated earlier, we often rely on measured data for characterizing physical (particularly nonlinear) electronic devices, such as the transistor. We will use the PWL approximation from Sect. 1.9.1.2, which will help us obtain circuit models, given the characteristic curves provided by device manufacturers in respective **datasheets** of their devices. The PWL characteristics of an *npn* BJT are shown in

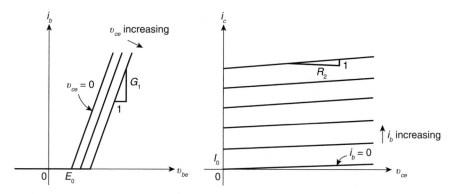

Fig. 2.18 PWL approximation of common-emitter characteristics [6]

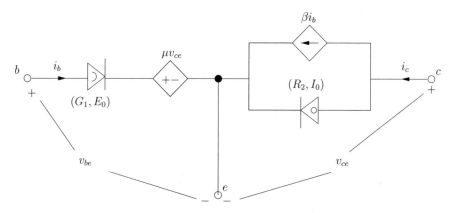

Fig. 2.19 PWL model of common-emitter BJT configuration

Fig. 2.18. The equivalent circuit for this representation is shown in Fig. 2.19. Note that with $v_{ce} = 0$, the $v_{be} - i_b$ characteristic in Fig. 2.18 is precisely that of a concave resistor specified by E_0 and slope G_1. In Fig. 2.19, we can see that if $v_{ce} = 0$, we simply have a concave resistor across the base-emitter terminal. In Fig. 2.18, the $v_{be} - i_b$ characteristic shifts to the right as v_{ce} increases. This is modeled in Fig. 2.19 by a VCVS with transfer voltage ratio μ.

Similarly, in the $v_{ce} - i_c$ characteristic in Fig. 2.18, with $i_b = 0$ the characteristic is of a convex resistor with i_c-axis intercept equal to I_0 and the slope equal to $\frac{1}{R}$. As i_b increases, the current i_c increases. These behaviors are modeled using a convex resistor and CCCS in Fig. 2.19, respectively.

For many large-signal applications, example H-bridges that simply run DC motors forward or backwards, these models are unnecessarily complicated and further simplifications are possible. But, it is important to again (recall Sect. 1.7) bear in mind that models are developed with specific applications in mind. Obviously, the simpler the model, the easier the circuit analysis. Thus, for applications where only an approximate solution is called for, we should use the simplest but valid model to get an idea of how the circuit functions. In other situations, example in determining the precise operating points using a computer, we need to use a more precise model for the transistor than that of the Ebers–Moll model. A variety of such models exist and are implemented by programs such as QUCS and SPICE. We will not cover such models in this book.

2.2.3 Three-Terminal Inductors

A three-terminal element is called a **three-terminal inductor** if it can be characterized by two sets of curves, or relationships, involving the variables i_1, i_2, ϕ_1, ϕ_2. Just as for three-terminal resistors, there are several possible forms of representation. Since the principles are identical, only one form will be discussed here, namely:

$$\phi_1 = \phi_1(i_1, i_2) \tag{2.49}$$

$$\phi_2 = \phi_2(i_1, i_2) \tag{2.50}$$

To find the voltages v_1 and v_2 corresponding to any current waveforms i_1 and i_2, we apply the chain rule, thereby obtaining:

$$v_1(t) = \frac{\partial \phi_1}{\partial i_1} \frac{di_1}{dt} + \frac{\partial \phi_1}{\partial i_2} \frac{di_2}{dt} \tag{2.51}$$

$$v_2(t) = \frac{\partial \phi_2}{\partial i_1} \frac{di_1}{dt} + \frac{\partial \phi_2}{\partial i_2} \frac{di_2}{dt} \tag{2.52}$$

Practically speaking, we will only discuss linear three-terminal inductors. Hence, Eqs. (2.51) and (2.52) reduce to:

$$v_1(t) = L_{11} \frac{di_1}{dt} + L_{12} \frac{di_2}{dt} \tag{2.53}$$

$$v_2(t) = L_{21} \frac{di_1}{dt} + L_{22} \frac{di_2}{dt} \tag{2.54}$$

The reason for discussing only linear three-terminal inductors is that the most common type of commercially available three-terminal inductor is that of a toroidal coil with a center tap, which is precisely a **physical transformer** (or transformers). These devices are of crucial importance in power circuitry and are hence discussed in a separate subsection below.

2.2.3.1 Physical Transformers

A transformer that implements Eqs. (2.53) and (2.54) is shown in Fig. 2.20. The ferromagnetic material used for the torus in Fig. 2.20 is typically ferrite or thin sheets of special steel. As shown in Fig. 2.20, we have wound on this torus two coils; we thus obtain a two-port. If we drive the first port with a generator so that the current i_1 is positive and have the second port open (hence $i_2 = 0$), there will be a strong magnetic field setup in the torus, $\mathbf{H}$ as indicated in the figure. Note if i_1 varies with time, since the magnetic field links the second coil, there will be a time-varying flux through that second coil. Hence, by Faraday's law, a voltage will be induced and $v_2(t) \neq 0$. Thus, electrical energy is transferred between the two-ports via electromagnetic induction.

Referring back to Eqs. (2.53) and (2.54), from fundamental energy considerations in physics, $L_{12} = L_{21} = M$, where M is the **mutual inductance** of inductor 1 and inductor 2. We know from our discussion of the two-terminal inductor in Sect. 1.9.3, L_{11} is the **self-inductance** of inductor 1 and L_{22} is the self-inductance of inductor 2. The schematic symbol for coupled coils is shown in Fig. 2.21. Note that we can rewrite Eqs. (2.53) and (2.54) in matrix form:

$$\begin{pmatrix} v_1 \\ v_2 \end{pmatrix} = \begin{pmatrix} L_{11} & M \\ M & L_{22} \end{pmatrix} \begin{pmatrix} \dot{i}_1 \\ \dot{i}_2 \end{pmatrix} \tag{2.55}$$

The square matrix L in Eq. (2.55) is called the **inductance matrix**. There is a very important relationship between a physical transformer and the ideal transformer discussed in Sect. 2.2.1.3, as Example 2.2.5 shows.

Fig. 2.20 Two coupled coils wound on a torus of ferromagnetic material

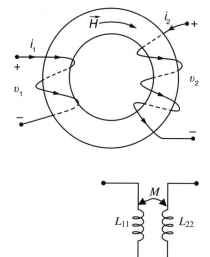

Fig. 2.21 Schematic symbol used for coupled coils with mutual inductance M, with self-inductances L_{11}, L_{22}

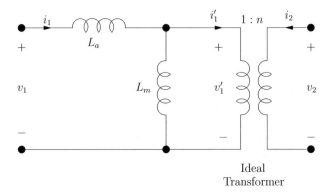

Fig. 2.22 A two-port equivalent to a pair of coupled inductors

Example 2.2.5 Show that Fig. 2.22, a two-port made up of an ideal transformer and two (uncoupled) inductors L_a and L_m, is equivalent to a pair of linear time-invariant coupled inductors modeled by Eq. (2.55).

Solution We will need to derive a form of Eq. (2.55) from Fig. 2.22. First, notice that for the ideal transformer, we have the following:

$$v_1' = \frac{1}{n} v_2 \tag{2.56}$$

$$i_2 = \frac{-1}{n} i_1' \tag{2.57}$$

Using the $v - i$ relationship for a two-terminal inductor and applying KCL to the node between L_a and L_m, we get:

$$v_1(t) = L_a \frac{di_1}{dt} + L_m \frac{d(i_1 - i_1')}{dt} \tag{2.58}$$

Hence, we have:

$$v_1(t) = (L_a + L_m) \frac{di_1}{dt} - L_m \frac{di_1'}{dt} \tag{2.59}$$

Substituting for i_1' from Eq. (2.57), we get:

$$v_1(t) = (L_a + L_m) \frac{di_1}{dt} + n L_m \frac{di_2}{dt} \tag{2.60}$$

(continued)

Example 2.2.5 (continued)
Using Eq. (2.56), we get:

$$v_2 = nv_1' \tag{2.61}$$

But, from Fig. 2.22, we get:

$$v_1' = L_m \frac{di_1 - i_1'}{dt} \tag{2.62}$$

Thus:

$$v_2 = nL_m \frac{di_1}{dt} - nL_m \frac{di_1'}{dt} \tag{2.63}$$

Again using Eq. (2.57), we get:

$$v_2 = nL_m \frac{di_1}{dt} + n^2 L_m \frac{di_2}{dt} \tag{2.64}$$

Rewriting Eqs. (2.60) and (2.64), we get the following matrix form:

$$\begin{pmatrix} v_1 \\ v_2 \end{pmatrix} = \begin{pmatrix} L_a + L_m & nL_m \\ nL_m & n^2 L_m \end{pmatrix} \begin{pmatrix} \dot{i}_1 \\ \dot{i}_2 \end{pmatrix} \tag{2.65}$$

The equations above do indeed model a pair of linear time-invariant coupled inductors.

The physical interpretations of L_a and L_m are as follows: L_a is the **leakage inductance**, that is, the inductance seen at the first port due to the leakage flux, i.e., the lines of magnetic field that do not link both coils. Indeed, from Exercise 2.2, as $n^2 \to 1$, $M^2 \to L_{11}L_{22}$ and thus $L_a \to 0$. L_m is called the **magnetizing inductance**: its role is to model the magnetic flux common to both coils.

Suppose we wish to build a high-quality transformer. We choose a torus of magnetic material with a very high permeability μ (e.g., ferrite, etc.). We then wind tightly on the torus the two coils forming a two-part, as in Fig. 2.20. Suppose that we are able to find magnetic materials with increasingly high μ: As μ becomes larger and larger, the leakage flux would get smaller and hence $L_a \to 0$. Also, the common flux would keep increasing, hence $L_m \to \infty$. Referring to Fig. 2.22, we see that we are left with an ideal-transformer!

2.2.4   Three-Terminal Capacitors

Analogous to a three-terminal inductor, we will define a three-terminal capacitor using the form:

$$q_1 = q_1(v_1, v_2) \tag{2.66}$$

$$q_2 = q_2(v_1, v_2) \tag{2.67}$$

Considering q_1 and q_2 to be linear functions of v_1, v_2 and using the fact that a capacitor is a dual of the inductor, we get:

$$i_1(t) = C_{11}\frac{dv_1}{dt} + C_{12}\frac{dv_2}{dt} \tag{2.68}$$

$$i_2(t) = C_{21}\frac{dv_1}{dt} + C_{22}\frac{dv_2}{dt} \tag{2.69}$$

Physical three-terminal capacitors are beyond the scope of this book. Nevertheless, nonlinear three-terminal capacitors find a variety of applications such as parametric amplification in solid-state circuits [9].

2.3   Three-Terminal Memristors

Finally, we have the three-terminal memristor:

$$\phi_1 = \phi_1(q_1, q_2) \tag{2.70}$$

$$\phi_2 = \phi_2(q_1, q_2) \tag{2.71}$$

We will not discuss three-terminal memristors as they are no physical examples yet. However, the possibility of their future availability cannot be dismissed.

2.4   The Three-Port Circulator

Circulators[3] are very useful microwave devices, used in communication systems and measurements. An **ideal three-port circulator** is a linear circuit element specified

---

[3]This section was added after a discussion on June 6th 2017, with Dr. Yuping Huang from the Stevens Institute of Technology. His group uses circulators in optical quantum computing applications.

by the following three equations:

$$f_1(v_1, v_2, v_3, i_1, i_2, i_3) \triangleq v_1 - Ri_2 + Ri_3 = 0 \tag{2.72}$$

$$f_2(v_1, v_2, v_3, i_1, i_2, i_3) \triangleq v_2 + Ri_1 - Ri_3 = 0 \tag{2.73}$$

$$f_3(v_1, v_2, v_3, i_1, i_2, i_3) \triangleq v_3 - Ri_1 + Ri_2 = 0 \tag{2.74}$$

where R is a real constant called the **reference resistance**. We can recast the equations above in an elegant matrix form:

$$\begin{pmatrix} v_1 \\ v_2 \\ v_3 \end{pmatrix} = \begin{pmatrix} 0 & R & -R \\ -R & 0 & R \\ R & -R & 0 \end{pmatrix} \begin{pmatrix} i_1 \\ i_2 \\ i_3 \end{pmatrix} \tag{2.75}$$

The circuit symbol for a circulator is shown in Fig. 2.23a. Observe that a three-port circulator is **non-energic** because the instantaneous power **entering** the three-port is identically zero, from Fig. 2.23a:

$$p_{\text{circulator}} = v_1 i_1 + v_2 i_2 + v_3 i_3$$

$$= (Ri_2 - Ri_3)i_1 + (-Ri_1 + Ri_3)i_2 + (Ri_1 - Ri_2)i_3$$

$$= 0 \tag{2.76}$$

Hence, energy is neither stored nor dissipated in the circulator. To demonstrate how energy is being **redistributed**, suppose we connect three identical resistors whose values are chosen equal to R in the setup shown in Fig. 2.23b. Since $v_2 = -Ri_2$ and $v_3 = -Ri_3$, it follows from Eq. (2.75) that

$$v_1 = Ri_2 - Ri_3$$

$$-Ri_2 = -Ri_1 + Ri_3$$

$$-Ri_3 = Ri_1 - Ri_2 \tag{2.77}$$

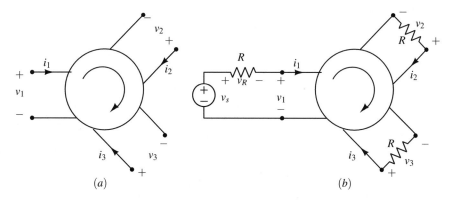

(a)   (b)

Fig. 2.23 A three-port circulator and a typical application

Solving these equations, we obtain:

$$v_1 = Ri_1$$
$$i_1 = i_2$$
$$i_3 = 0 \tag{2.78}$$

Now, conservation of energy[4] in the circuit in Fig. 2.23b implies:

$$v_s(-i_1) + v_Ri_1 + p_{\text{circulator}} + v_2(-i_2) + v_3(-i_3) = 0 \tag{2.79}$$

But, we have shown that $p_{\text{circulator}} = 0$ and $i_3 = 0$. If we let $p_s = v_s i_1$ be the power supplied by the voltage source, we get:

$$p_s = v_Ri_1 + v_2(-i_2)$$
$$= Ri_1^2 + Ri_2^2$$
$$= 2(Ri_1^2) \tag{2.80}$$

We conclude that half of the power supplied by the voltage source is dissipated in its associated series resistor, while the other half is dissipated in the resistor across port 2. In other words, all power entering port 1 is **redirected** to port 2 (to be dissipated in the terminating resistor), with nothing left for port 3 (recall we obtained $i_3 = 0$).

If we repeat the preceding analysis but with the voltage source inserted in port 2, instead of port 1, we will find that all power entering port 2 gets delivered to port 3 with nothing left for port 1. Similarly, inserting the voltage source in port 3, we find that all the power entering port 3 gets delivered to port 1 with nothing left for port 2. Hence, the circulator functions by "circulating" the energy entering one port into the next port **whenever all ports are terminated by resistors equal to the reference resistor** R.

This property is widely exploited in many communication systems for diverting power into various desired channels. For example, the setup in Fig. 2.23b can be used to model the following situation: Let the voltage-source resistor combination model a portable radio transmitter. Let the resistor R across port 2 model an antenna, and let the resistor R across port 3 model a receiver. Because of the circulator, no outgoing signal transmitted from port 1 will reach the receiver. Conversely, any incoming signal from elsewhere that is received by the antenna (when port 1 is not transmitting) will be delivered to the receiver in port 3. Without the circulator, two separate antennas will be needed, one to keep the receiver from receiving its own transmitted signal and the other to keep the transmitter from receiving unwanted signals intended for the receiver.

---

[4]We could also apply Tellegen's theorem.

2.5 Operational Amplifier (Opamp)

The opamp is an extremely versatile and inexpensive semiconductor device. It has been the workhorse of the electronics hobbyist and students for nearly six decades and hence is of paramount importance.

For **low-frequency** applications, the opamp behaves like a **multi-terminal nonlinear resistor**, which can often be represented by an ideal opamp model. This model greatly simplifies the analysis and design of opamp circuits. In fact, one of the reasons why opamps are so popular is that, at low frequencies,[5] they behave almost like the ideal model! Exercise 2.5 helps the reader understand this justification: the exercise instructs the reader to analyze a typical opamp circuit using the more complicated **finite gain** model and then compare results with those predicted by the ideal opamp model.

Depending on the **dynamic range** of the input signals, the opamp may operate in the **linear** or **nonlinear** region. Section 2.5.2 is devoted to those circuits where the opamp is operating only in the linear region. This restriction allows us to simplify the (nonlinear) ideal opamp model into a **linear** model, called the **virtual short-circuit model**. This model is used extensively in Sect. 2.5.2 for analyzing both simple circuits by **inspection** as well as complicated circuits via a **systematic** method.

In Sect. 2.5.3, we use the nonlinear ideal opamp model to analyze opamps operating in the nonlinear region. We will primarily discuss voltage feedback opamps, but Sect. 2.5.5 will discuss current feedback opamps.

Note that we use a variety of examples. We encourage the reader to simulate these examples using QUCS[6] and also have access to the necessary electronics equipment ("breadboard," etc.) so they can construct the discussed circuits and see opamps "in action."

2.5.1 Device Description, Characteristics, and Model

Opamps are multi-terminal devices, shown in Fig. 2.24, and are sold in several standard packages. For the "breadboard," the most convenient is the DIP (Dual Inline Package) versions of the integrated circuit (IC). Figure 2.25 gives the schematic of the μA741, an opamp introduced by Fairchild Semiconductor in 1968, and still in use today. The seven terminals brought out through the package leads (Fig. 2.24) are labeled **inverting input** IN$-$, **noninverting input** IN$+$, **output**

---

[5]Unless otherwise stated, we will assume that all opamp circuits operate at low enough frequencies so the ideal opamp model is valid.

[6]Although we cover circuit simulation in QUCS in Chap. 3 lab, the reader should be able to use their "native intelligence" to easily simulate the circuits in this chapter, using the QUCS online workbook as a guide.

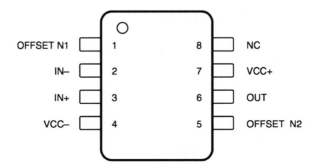

Fig. 2.24 μA741 opamp, in 8-pin SOIC, DIP, and SO versions. Opamp is not to scale

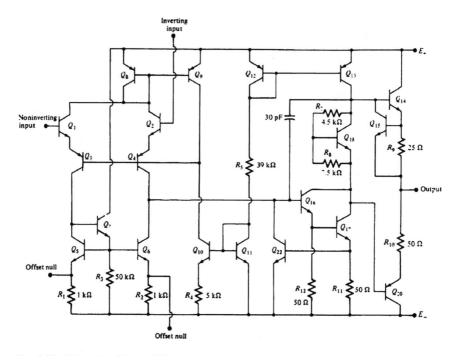

Fig. 2.25 Schematic of the μA741

OUT, **positive power supply** (VCC+), **negative power supply** (VCC−), and **offset nulls** (OFFSET N1, OFFSET N2). The remaining terminals of the package are not connected to the IC and are labeled NC (no connection). The additional terminals such as OFFSET N1 are usually connected to some external nulling or compensation circuit for improving the performance of the opamp. We will not use such external circuits in this book.

Some opamps have more than seven terminals; others have less. For most applications, however, only the five terminals indicated in the standard opamp symbol in Fig. 2.26a are essential. Note that the opamp can be considered a 4-

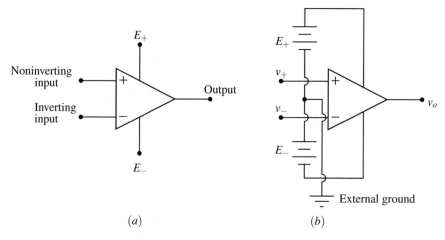

Fig. 2.26 Standard opamp symbol and a typical biasing scheme. (**a**) The $+$ and $-$ signs inside the triangle denote the noninverting and inverting input terminals, respectively. (**b**) A "biased" opamp

terminal device for circuit analysis and design purposes, in the sense that both E_+ and E_- (Fig. 2.26) are referenced to a common external ground. All voltages are also measured with respect to this ground. So, from a circuit theoretic standpoint, we only need v_+, v_-, v_o, and the external ground (four terminals). However, to be consistent with most electronics literature, we will not show the opamp as a four-terminal device. Rather, we will **implicitly assume** that the opamp is connected properly, as in Fig. 2.26b.

In order for the opamp to function properly, its internal transistors must be biased at appropriate operating points (we will discuss small-signal analysis in Sect. 3.1.1, the concept of biasing should become clear then). The power supply terminals are provided for this purpose. In Fig. 2.24, the supplies are labeled as VCC+ and VCC−. The justification for using the CC label is that the μA741 is a BJT opamp, CC is an acronym for "collector." Since other transistor (for example, FET)-based opamps exist, we will use E_+ and E_- for generality.

In general, E_+ and E_- are connected to a split power supply as shown in Fig. 2.26b, with respect to an **external ground**. Typically, $E_+ = 15$ V and $E_- = 15$ V (they do not have to be symmetrical with respect to ground). For clarity purposes, we will henceforth use the symbol shown in Fig. 2.27 (assuming that we have a symmetrical external power supply, $E = \pm 15$ V). In Fig. 2.27a, i_- and i_+ denote the current entering the opamp inverting and noninverting terminals, respectively. Similarly, v_-, v_+, and v_o denote, respectively, the voltage from the inverting terminal, noninverting terminal, and output terminal to ground. The variable v_d in Fig. 2.27b is called the **differential input voltage** and will play an important role in opamp circuit analysis.

To derive an exact characterization of an opamp would require analyzing the entire integrated circuit, such as the one shown in Fig. 2.25. Fortunately, for many

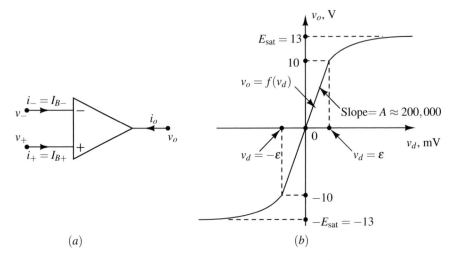

Fig. 2.27 Experimental characterization of a typical opamp. In (**b**), $v_d \triangleq v_+ - v_-$

low-frequency applications, the opamp terminal currents and voltages have been found **experimentally** to obey the following **approximate** relationships:

$$i_- = I_{B-} \tag{2.81}$$

$$i_+ = I_{B+} \tag{2.82}$$

$$v_o = f(v_d) \tag{2.83}$$

where I_{B-} and I_{B+} are called the **input bias currents** and $f(v_d)$ denotes the v_o-vs-v_d **voltage transfer characteristic (VTC)**, since the plot shows how one voltage v_{in} is "transferred" to another voltage v_o.[7] Apart from a scaling factor which depends on the power supply voltage, $f(v_d)$ follows approximately an odd-symmetric function as shown in Fig. 2.27b (drawn for a ±15 V supply voltage). Moreover, this function has been found to be rather insensitive to changes in the output current i_o.

The transfer characteristic in Fig. 2.27b displays three remarkable properties:

1. v_o and v_d have different scales: one is in volts, the other in millivolts.
2. In a small interval $-\epsilon < v_d < \epsilon$ of the origin, $f(v_d) \approx A v_d$ is nearly **linear** with a very steep slope A—called the **open-loop voltage gain**. It is called "open loop" because there is no feedback in the circuit (that is, the output is not connected back to any of the inputs). This is a terminology from control systems. If there is feedback, we say the loop is "closed" (or "closed loop").

---

[7]The word "transfer" means that the response variable does not appear at the same port as the source serving as input. There are four types of TCs possible: v_o-vs-v_{in}, v_o-vs-i_{in}, i_o-vs-v_{in}, and i_o-vs-i_{in}.

3. $f(v_d)$ **saturates** at $v_o = \pm E_{sat}$, where E_{sat} is typically 2 V less than the power
   supply voltage, if the opamp in question is realized using BJTs. In that case, we
   say the opamp is not "rail-to-rail." On the other hand, FET opamps usually have
   rail-to-rail behavior and E_{sat} for such FET opamps usually range from E_- to E_+.

Also, the bias currents for opamps using BJTs as inputs are much larger than
opamps that use FET input transistors. For example, the **average** input bias current
$I_B \triangleq \frac{1}{2}(|I_{B+}| + |I_{B-}|)$ is equal to 0.1 mA for the μA741 but only 0.1 nA for the
μA740 (which uses a part of FET input transistors).

The open-loop voltage gain A is typically equal to at least 100,000 (200,000 for
the μA741). On the other hand, the voltage ϵ at the end of the **linear** region in
Fig. 2.27b is typically less than 0.1 mV.

In view of the typical magnitudes of I_{B-}, I_{B+}, A, and ϵ, little accuracy is lost
by assuming $I_{B-} = I_{B+} = \epsilon = 0$, $A \to \infty$. This simplifying assumption leads
to the **ideal** opamp model shown in Fig. 2.28. To emphasize that $A \to \infty$ in the
linear region, we added ∞ inside the triangle to distinguish the **ideal** opamp symbol
from other models. Unless otherwise stated, the ideal opamp model will be used
throughout this book. Note that the VTC of the ideal opamp model reduces to the
three-segment PWL characteristic shown in Fig. 2.28a. The ideal opamp model can
be described analytically as follows:

$$i_- = 0 \tag{2.84}$$

$$i_+ = 0 \tag{2.85}$$

$$v_o = E_{sat}\frac{|v_d|}{v_d}, \quad v_d \neq 0 \tag{2.86}$$

$$v_d = 0, \quad -E_{sat} < v_o < E_{sat} \tag{2.87}$$

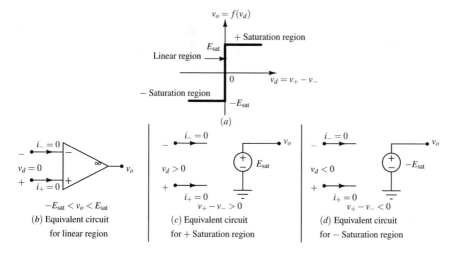

Fig. 2.28 Ideal opamp model

Because these equations are rather cumbersome and difficult to manipulate analytically, it is more practical to represent each region by the simple equivalent circuits
shown in Fig. 2.28b, c, and d. Note that these three equivalent circuits contain
exactly the same information as Eq. (2.84) through (2.87). In particular, when the
opamp is operating in the linear region, the ideal opamp model reduces to that shown
in Fig. 2.28b. Note that in the linear region, v_d is constrained to be zero at all times
while $|v_o|$ is constrained to be less than the saturation voltage E_{sat}. Hence, the circuit
is described by Eqs. (2.84), (2.85), and (2.87).

The circuit in Fig. 2.28c is described by Eqs. (2.84), (2.85), and (2.86) with $v_d >$
0. Likewise, the circuit in Fig. 2.28d is described by Eqs. (2.84), (2.85), and (2.86)
with $v_d < 0$.

Opamp circuits designed to operate exclusively in the linear region are analyzed
in Sect. 2.5.2. Note that although the opamp is operating in the linear region, the
circuit itself may contain nonlinear elements. Opamp circuits operating in both
linear and nonlinear regions will be analyzed in Sect. 2.5.3.

Example 2.5.1 The datasheet for a μA741 shows a typical open-loop voltage
gain of 200,000. Calculate the value of ϵ for a power supply voltage of
± 20 V. Assume E_{sat} = magnitude of power supply voltage ± 2 V (use $\pm$
as appropriate).

Solution Given the information above, we have $E_{\text{sat}} = \pm 18$ V. Hence:

$$\epsilon = \frac{\pm 18 \text{ V}}{200000}$$

$$= \pm 0.09 \text{ mV} \tag{2.88}$$

Example 2.5.2 An opamp manufacturer's datasheet usually specifies the
typical value of the average input bias current I_B (defined earlier as
$\frac{1}{2}(|I_{B+}| + |I_{B-}|)$) and the offset current $I_{\text{os}} \triangleq |I_{B+}| - |I_{B-}|$. Express $|I_{B+}|$
and $|I_{B-}|$ in terms of I_B and I_{os}.

Solution From the definition of I_B, we get:

$$2I_B = |I_{B+}| + |I_{B-}| \tag{2.89}$$

Adding the equation above to the definition of I_{os}, we get:

$$2I_B + I_{\text{os}} = 2|I_{B+}| \tag{2.90}$$

(continued)

Example 2.5.2 (continued)
Subtracting the definition of I_{os} from the equation for $2I_B$, we get:

$$2I_B - I_{\text{os}} = 2|I_{B-}| \tag{2.91}$$

Hence, we have:

$$|I_{B+}| = \frac{1}{2}(2I_B + I_{\text{os}}) \tag{2.92}$$

$$|I_{B-}| = \frac{1}{2}(2I_B - I_{\text{os}}) \tag{2.93}$$

2.5.2 Linear Opamp Circuits

The methods to be developed in this section are valid **only** if the opamp output voltage satisfies

$$-E_{\text{sat}} < v_o(t) < E_{\text{sat}} \tag{2.94}$$

for all times t. We will henceforth refer to the expression in Eq. (2.94) as the **validating inequality** for the **linear** region. If this inequality is violated in any time interval $[t_1, t_2]$, the solution in this interval is incorrect and must be recalculated using the nonlinear model from Sect. 2.5.3.

Recall from Fig. 2.1 in Sect. 2.1 that a three-port is characterized by three relationships among the associated voltage and current variables. Notice that in the linear region, the ideal opamp in Fig. 2.28b can be described analytically by the three equations[8]:

$$i_- = 0 \tag{2.95}$$

$$i_+ = 0 \tag{2.96}$$

$$v_+ - v_- = 0 \tag{2.97}$$

---

[8]These correspond to Eqs. (2.84), (2.85), and (2.87).

Fig. 2.29 The voltage
follower, or unity-gain buffer

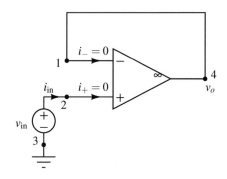

Consequently, we can think of the ideal opamp model in Fig. 2.28b as a three-port.[9]
For purposes of analysis, Eqs. (2.95) through (2.97) are equivalent to:

1. Connecting a **short circuit** across the opamp input terminals.
2. Stipulating that the **currents** through the **input terminals are zero at all times**.

To emphasize the special nature of this short circuit, we will henceforth refer to the
model from Eqs. (2.95) through (2.97) as the **virtual short-circuit model**. Notice
that the word "virtual" is very important, $v_+ = v_-$ because of the opamp, not
because v_+ is physically connected to v_-.

Using the virtual short-circuit model, many opamp circuits can be analyzed by
inspection. This method usually requires no more than three calculations and is often
implemented by invoking KCL and Eqs. (2.95) through (2.97) mentally, perhaps
with an occasional scribble on the "back of an envelope." It is best illustrated via
some useful opamp circuits as examples.

Example 2.5.3 Determine the v_o-vs-v_{in} VTC for the circuit in Fig. 2.29.

Solution First, let us apply KCL at node 2 and obtain:

$$i_{in} = i_+ = 0 \tag{2.98}$$

Applying next KVL around the closed node sequence $4 - 3 - 2 - 1 - 4$:

$$-v_o + v_{in} - v_d = 0 \tag{2.99}$$

(continued)

---

[9]Recall that an opamp always has an external reference terminal, hence an ideal opamp can also
be considered as a four-terminal resistor.

Example 2.5.3 (continued)
where we have used the usual definition: $v_d = v_+ - v_-$. But, because of the
virtual short-circuit model $v_d = 0$, so:

$$v_o = v_{in} \tag{2.100}$$

To complete the analysis, we apply the validating inequality from Eq. (2.94)
and obtain:

$$-E_{sat} < v_{in} < E_{sat} \tag{2.101}$$

This gives the dynamic range of input voltages beyond which the opamp no
longer operates in the linear region.

Note that the voltage follower in Example 2.5.3 defines a unity-gain VCVS. This
circuit has an infinite input resistance because $i_{in} = 0$ and its output "duplicates"
the input voltage, regardless of the external load. Consequently, it is also called an
isolation amplifier. It is widely used between 2 two-ports to prevent one two-port
from "loading down" the other two-port. This isolation technique is one of the most
useful tools in a designer's "toolbox."

Example 2.5.4 Determine the v_o-vs-v_{in} VTC for the inverting amplifier
circuit in Fig. 2.30. Note that this circuit contains linear resistors, as opposed
to the voltage follower.

Solution Since $v_d \triangleq v_+ - v_- = 0$, we have:

$$v_1 = v_{in} \tag{2.102}$$

By Ohm's law:

$$i_1 = \frac{v_1}{R_1} \tag{2.103}$$

Since $i_- = 0$, we have $i_2 = i_1$. Hence:

$$v_2 = R_f i_1 = R_f \left(\frac{v_{in}}{R_1} \right) \tag{2.104}$$

(continued)

Example 2.5.4 (continued)
Applying KVL around the closed node sequence $4 - 2 - 1 - 4$:

$$v_o = \left(\frac{-R_f}{R_1}\right) v_{in} \qquad (2.105)$$

To complete the analysis, we apply the validating inequality from Eq. (2.94) and obtain:

$$\left(-\frac{R_1}{R_f}\right) E_{sat} < v_{in} < \left(\frac{R_1}{R_f}\right) E_{sat} \qquad (2.106)$$

Hence, so long as the input signal satisfies Eq. (2.106), the circuit functions as a voltage amplifier with voltage gain equal to $-R_f/R_1$ (assuming $R_f > R_1$).

Exercise 2.3 gives an example of a noninverting amplifier configuration. Of course, we can have nonlinear elements in conjunction with the ideal opamp model, exercise 2.4 shows one such circuit that functions as a "clipper." Note again the versatility of the ideal opamp model comes from the fact that even if we did assume that the open-loop gain A is finite, the answers obtained using a **finite gain model** are nearly identical to the results from the ideal opamp model. Exercise 2.5 explores this further.

The inspection method often fails whenever it is necessary to solve two or more simultaneous equations. In such cases, it is desirable to develop a systematic method for writing a system of linearly independent equations involving as few variables as possible. The following example illustrates the basic steps involved.

Fig. 2.30 The inverting amplifier

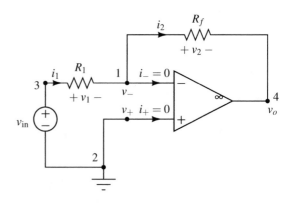

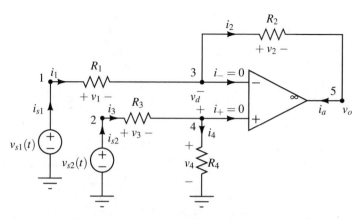

Fig. 2.31 An opamp summing amplifier for illustrating the systematic method

Example 2.5.5 Consider the opamp circuit in Fig. 2.31. Determine the VTC for this circuit using the systematic method.

Solution Although this circuit can be solved by inspection, we will leave that to the reader as an exercise. Given below are the steps for the systematic method approach:

1. Label the nodes consecutively and let e_j denote as usual the voltage from node j to the ground node. In our case, $j = 1, 2, \cdots, 5$. Express **all** resistor voltages and the differential opamp voltage v_d in terms of the node-to-ground voltages via KVL:

$$v_1 = e_1 - e_3 \tag{2.107}$$

$$v_2 = e_3 - e_5 \tag{2.108}$$

$$v_3 = e_2 - e_4 \tag{2.109}$$

$$v_4 = e_4 \tag{2.110}$$

$$v_d = e_4 - e_3 \tag{2.111}$$

2. Express the branch current in each linear resistor in terms of node-to-ground voltages via Ohm's law:

$$i_1 = \frac{e_1 - e_3}{R_1} \tag{2.112}$$

$$i_2 = \frac{e_3 - e_5}{R_2} \tag{2.113}$$

(continued)

Example 2.5.5 (continued)

$$i_3 = \frac{e_2 - e_4}{R_3} \qquad (2.114)$$

$$i_4 = \frac{e_4}{R_4} \qquad (2.115)$$

3. Identify all other branch **current** variables which **cannot** be expressed in terms of node-to-ground voltages, namely, the currents i_{s1}, i_{s2}, and i_a. Note that the opamp input currents i_- and i_+ are not variables in an ideal opamp model, because they are equal to zero. Our objective is to write a system of linearly independent equations in terms of node-to-ground voltages and the **identified** current variables $\{i_{s1}, i_{s2}, i_a\}$.
4. Write KCL at each node (except the ground node) in terms of $\{e_1, e_2, e_3, e_4, e_5, i_{s1}, i_{s2}, i_a\}$:

$$\text{Node 1:} \quad \frac{e_1 - e_3}{R_1} = i_{s1} \qquad (2.116)$$

$$\text{Node 2:} \quad \frac{e_2 - e_4}{R_3} = i_{s2} \qquad (2.117)$$

$$\text{Node 3:} \quad \frac{e_3 - e_5}{R_2} = \frac{e_1 - e_3}{R_1} \qquad (2.118)$$

$$\text{Node 4:} \quad \frac{e_4}{R_4} = \frac{e_2 - e_4}{R_3} \qquad (2.119)$$

$$\text{Node 5:} \quad i_a = \frac{e_3 - e_5}{R_2} \qquad (2.120)$$

5. Eqs. (2.116) through (2.120) consists of five equations with eight variables. Hence, we need to write three more independent equations. Since we have already made use of KVL (Step 1), KCL (Step 4), and the resistor characteristics (Step 2), these three equations must come from the characteristics of the voltage sources and the opamp:

$$e_1 = v_{s1} \qquad (2.121)$$

$$e_2 = v_{s2} \qquad (2.122)$$

$$e_4 - e_3 = 0 \qquad (2.123)$$

(continued)

Example 2.5.5 (continued)

6. Together, the equations in the previous two steps constitute a system of eight linearly independent equations in terms of eight variables. Solving these equations for the desired opamp output voltage $v_o = e_5$ by any elimination or any other method, we obtain:

$$v_o = \left[\frac{R_4(1 + R_2/R_1)}{R_3 + R_4} \right] v_{s2}(t) - \left(\frac{R_2}{R_1} \right) v_{s1}(t) \tag{2.124}$$

7. Determine the dynamic range of the input voltages for which Eq. (2.124) holds, i.e., where the opamp is operating in the linear region:

$$-E_{\text{sat}} < \left[\frac{R_4(1 + R_2/R_1)}{R_3 + R_4} \right] v_{s2}(t) - \left(\frac{R_2}{R_1} \right) v_{s1}(t) < E_{\text{sat}} \tag{2.125}$$

We should of course perform some sanity checks for Example 2.5.5. For example, if $v_{s2} = 0$, we obtain an inverting amplifier. The expression and dynamic range correctly reduce to the corresponding expressions for an inverting amplifier.

We could have also quite easily derived Eqs. (2.124) and (2.125) by using the inspection method: since R_3 and R_4 are in series, we can quickly obtain an expression for e_4 and simply write a KCL expression at e_3 (since $e_3 = e_4$ by the virtual short-circuit model). The point of the example was to illustrate the systematic method.

The preceding systematic method is applicable to any opamp circuit containing linear resistors, independent voltage, and current sources, and opamps modeled by virtual short circuits. This method will be generalized in Sect. 4.2.2.1, called **modified nodal analysis (MNA)**, for arbitrary circuits.

2.5.2.1 Implementation of Dependent Sources

A very elegant opamp application is implementation of dependent sources. In fact, the inverting amplifier from Exercise 2.3 is an example of a VCVS. We will consider the implementation of the other dependent sources in the examples below.

Fig. 2.32 CCVS using opamp

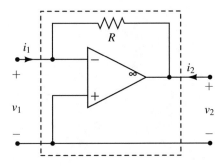

Example 2.5.6 The circuit in Fig. 2.32 (boxed to highlight the two-port variables) implements a linear CCVS. Determine the transresistance r_m and the dynamic range for the opamp.

Solution We can easily derive the two-port CCVS relationship in Eq. (2.24) using the inspection method. Since $v_1 = v_- - v_+$, we have:

$$v_1 = 0 \tag{2.126}$$

Applying Ohm's law:

$$v_1 - v_2 = i_1 R \tag{2.127}$$

Thus:

$$v_2 = -Ri_1 \tag{2.128}$$

Hence, the transresistance $r_m = -R$. Applying the validating inequality and using the relationship between v_2 and i_1 derived above, we get the dynamic range:

$$-\frac{E_{\text{sat}}}{R} < i_1 < \frac{E_{\text{sat}}}{R} \tag{2.129}$$

Fig. 2.33 VCCS using
opamp

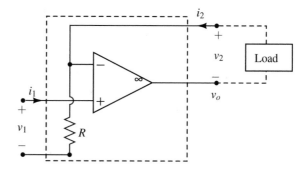

Example 2.5.7 The circuit in Fig. 2.33 (boxed to highlight the two-port
variables) implements a linear VCCS. Determine the transconductance g_m
and the dynamic range for the opamp.

Solution We can easily derive the two-port VCCS relationship in Eq. (2.25)
using the inspection method. Since $i_1 = i_+ = 0$, we have:

$$i_1 = 0 \tag{2.130}$$

Applying Ohm's law and using the virtual short-circuit model, we get:

$$i_2 = \frac{v_1}{R} \tag{2.131}$$

Hence, the transconductance $g_m = \frac{1}{R}$. From KVL: $v_1 - v_2 = v_o$ and applying
the validating inequality, we get the dynamic range:

$$v_2 - E_{\text{sat}} < v_1 < v_2 + E_{\text{sat}} \tag{2.132}$$

Example 2.5.8 The circuit in Fig. 2.34 (boxed to highlight the two-port
variables) implements a linear CCCS. Determine the current gain α and the
dynamic range for the opamp.

Solution This circuit illustrates the importance of understanding that an
opamp is biased via external power supplies. Our goal is to derive Eq. (2.26)
using the inspection method. Applying KVL around loop $1 - 2 - 3 - 4 - 1$

(continued)

Example 2.5.8 (continued)
and using node-to-ground voltages, we get:

$$(e_1 - e_2) + (e_2 - e_3) + (e_3 - e_4) = 0 \tag{2.133}$$

We will simplify the KVL equation by first noting that the opamp virtual short-circuit model implies $e_1 = e_4$. We will then apply Ohm's law to resistors R_1, R_2 and use KCL at node 2. Thus, we can simplify the KVL equation to:

$$i_1 R_1 + (i_1 - i_2) R_2 = 0 \tag{2.134}$$

We thus have:

$$i_2 = \left(1 + \frac{R_1}{R_2}\right) i_1 \tag{2.135}$$

Therefore, the current gain for the CCCS is $\alpha = 1 + \frac{R_1}{R_2}$. Since $e_1 = e_4 = 0$, we get from KVL:

$$-i_1 R_1 + v_2 = v_o \tag{2.136}$$

Now, we can apply the validating inequality:

$$-E_{\text{sat}} < -i_1 R_1 + v_2 < E_{\text{sat}} \tag{2.137}$$

Simplifying:

$$E_{\text{sat}} > i_1 R_1 - v_2 > -E_{\text{sat}} \tag{2.138}$$

Hence, the dynamic range is:

$$\frac{v_2 - E_{\text{sat}}}{R_1} < i_1 < \frac{v_2 + E_{\text{sat}}}{R_1} \tag{2.139}$$

2.5.3 Nonlinear Opamp Circuits

There are many applications where the opamp operates in all three regions of the ideal opamp model in Fig. 2.28. This occurs whenever the amplitudes of one or more input signals are such that the validating inequality in each region is violated over some time intervals. In this case, we probably have to use **all** three regions in Fig. 2.28 and we say that the opamp is "nonlinear." Fortunately, since the

Fig. 2.34 CCCS using opamp

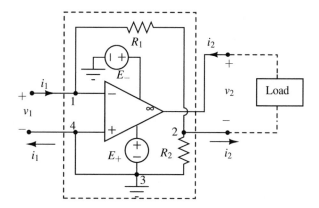

Fig. 2.35 VTC of a voltage follower

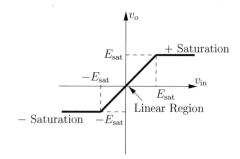

characteristic in Fig. 2.28a is a **PWL** characteristic, the circuit in each region can be easily analyzed as a linear circuit.

Most practical nonlinear opamp circuits involve the use of **positive feedback**: the output is connected to the **noninverting** input. The mindful reader would have noticed that all the circuits in the preceding section involved **negative feedback**: the output was connected to the **inverting** input. Inherently, negative feedback is stable while positive feedback is not. However, stability is a **dynamic** concept and understanding specifically opamp positive feedback requires the use of first-order circuits (to be discussed in Sect. 4.2.1). But, the **fact** that positive feedback **is different** from negative feedback can be easily explained using the PWL model, as discussed below.

Recall the voltage follower from Example 2.5.3. We plot the VTC for the follower in Fig. 2.35.

What happens if we interchange the inverting and noninverting terminals as shown in Fig. 2.36? By inspection, we can find $v_o = v_{in}$ provided $|v_{in}| < E_{sat}$. Hence, in the linear region, the transfer characteristic of this positive feedback circuit is identical to that of the voltage follower VTC in Fig. 2.35. In practice, however, they do not behave in the same way: One functions as a voltage follower, the other **does not**. To uncover the reason, let us derive the transfer characteristics in the remaining regions.

Fig. 2.36 A positive
feedback circuit

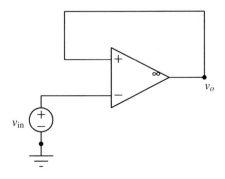

Fig. 2.37 VTC of the
positive feedback circuit from
Fig. 2.36

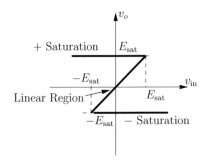

When the opamp is in the $+$ Saturation region, the validating inequality from
Fig. 2.28b for the circuit in Fig. 2.36 becomes: $v_d = E_{sat} - v_{in} > 0$ or $v_{in} < E_{sat}$.
Hence, the transfer characteristic in this region is given by $v_o = E_{sat}$ whenever
$v_{in} < E_{sat}$, as shown in Fig. 2.37.

Conversely, when the opamp is in the $-$ Saturation region, the validating
inequality from Fig. 2.28c becomes: $v_d = -E_{sat} - v_{in} < 0$ or $v_{in} > -E_{sat}$.
Hence, the transfer characteristic in this region is given by $v_o = -E_{sat}$ whenever
$v_{in} > -E_{sat}$, completing the VTC in Fig. 2.37.

Note that complete transfer characteristics in Figs. 2.35 and 2.37 are quite
different. Even if the opamp is operating in the linear region ($|v_{in}| < E_{sat}$), there
are three distinct output voltages for each value of v_{in} for the positive feedback
circuit. Using a more realistic opamp circuit model to be developed in Chap. 4, we
will show that all operating points on the middle segment (linear region) in Fig. 2.37
are **unstable**. The important concepts of **stability** and **instability** will be discussed
in Chap. 4. In the present context, having **unstable operating points** in the middle
region means that even if the voltage $v_{in}(0)$ lies on this segment, it will quickly
move to the $+$ Saturation region if $v_{in}(0) > 0$ or into the negative saturation region
if $v_{in}(0) < 0$.

One may wonder if we can even confirm the VTC in Fig. 2.37 experimentally,
since the linear region is unstable. The lab component for this chapter shows how
to confirm an equivalent VTC for a Schmitt trigger (discussed below), by using an
elegant mathematical trick. The Schmitt trigger is actually a very elegant application

of positive feedback that illustrates the very important concept that stability is a dynamic phenomenon.

2.5.3.1   Schmitt Trigger

Consider the circuit shown in Fig. 2.38.

The Schmitt trigger in Fig. 2.38 is used for signal conditioning in the presence of noise: the output is $\pm E_{\text{sat}}$ depending on the input voltage, but the **physical** circuit also displays **hysteresis or memory**. That is, the output voltage depends on the **derivative** of the input voltage. The advantages offered by the Schmitt trigger when compared to the simple positive feedback circuit in Fig. 2.36 are: we can control the slope of the linear region and the values of the "trip" voltages V^+ and V^- (in the Schmitt trigger VTC in Fig. 2.39), using R_1 and R_2.

Fig. 2.38 The inverting Schmitt trigger

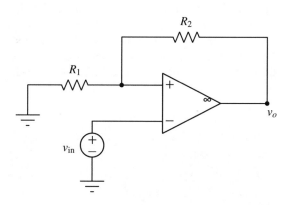

Fig. 2.39 VTC of the inverting Schmitt trigger in Fig. 2.38. Compare to the simple positive feedback VTC in Fig. 2.37

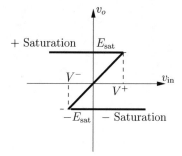

Fig. 2.40 An incorrect VTC
for the inverting Schmitt
trigger in Fig. 2.38. It is
incorrect because the arrows
on the VTC specify dynamic
behavior, whereas the circuit
in Fig. 2.38 does not include
any dynamic elements

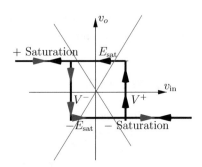

Note the emphasis on the word physical when discussing hysteresis: the justification for the hysteretic behavior is the presence of parasitic components (such as capacitors) in the physical implementation (to be discussed in Sect. 4.2.1). Most people fail to separate the **static VTC characteristic** implied by Fig. 2.38 and **incorrectly** derive the VTC of the Schmitt trigger, shown in Fig. 2.40.

Unfortunately, the VTC in Fig. 2.40 combines both static and dynamic characteristics, whereas Fig. 2.38 does not have any dynamic elements (capacitors, inductors, and memristors). Hence, we will now derive the correct VTC shown in Fig. 2.39 for the Schmitt trigger in Fig. 2.38, by simply using our ideal opamp model from Fig. 2.28.

Example 2.5.9 Derive the v_o-vs-v_{in} expressions for the inverting Schmitt trigger in Fig. 2.38, and hence justify the VTC in Fig. 2.39.

Solution Assuming the opamp is in the linear region of operation and applying KCL at the noninverting input, we get using the inspection method:

$$\frac{0 - v_{in}}{R_1} = \frac{v_{in} - v_o}{R_2} \tag{2.140}$$

Simplifying:

$$v_o = v_{in}\left(1 + \frac{R_2}{R_1}\right) \tag{2.141}$$

Notice that as $R_1 \to \infty$ (an open circuit), we get $v_o = v_{in}$ from Eq. (2.141). This obviously agrees with the slope of the linear region being equal to 1 for the simple positive feedback circuit in Fig. 2.36. Also, notice that as long as $R_2 \not\to \infty$ in Fig. 2.38, the value of R_2 is irrelevant if $R_1 \to \infty$, since the current into the noninverting input is zero and hence there is no voltage drop across R_2. Note that both R_1 and R_2 tending to ∞ means that the circuit

(continued)

Example 2.5.9 (continued)
is physically ill-defined: the noninverting input is floating (not connected to anything).

Applying the validating inequality for the + Saturation region, we get:

$$v_+ - v_- > 0 \tag{2.142}$$

Simplifying:

$$v_{\text{in}} < \left(\frac{R_1}{R_1 + R_2} \right) E_{\text{sat}} \tag{2.143}$$

Thus, $V^+ = \left(\frac{R_1}{R_1 + R_2} \right) E_{\text{sat}}$. Analogously, we can derive an expression for V^-, applying the validating inequality for the − Saturation region, we get:

$$v_{\text{in}} > - \left(\frac{R_1}{R_1 + R_2} \right) E_{\text{sat}} \tag{2.144}$$

Hence, $V^- = - \left(\frac{R_1}{R_1 + R_2} \right) E_{\text{sat}}$.

The example above discussed the inverting Schmitt trigger. Exercise 2.6 explores the noninverting Schmitt trigger.

2.5.3.2   PWL Circuits

Consider the circuit shown in Fig. 2.41a, reproduced from the epigraph to this chapter. Our goal is to derive the DP characteristic. As before, we can use the inspection method.

We note that R_1 and R_2 form a voltage divider so that:

$$e_3 = \frac{R_2}{R_1 + R_2} v_o$$

$$= \beta v_o \tag{2.145}$$

If the opamp is operating in the linear region, $e_3 = v$. Hence, substituting for e_3 in the equation above, we get:

$$v_o = \frac{1}{\beta} v \tag{2.146}$$

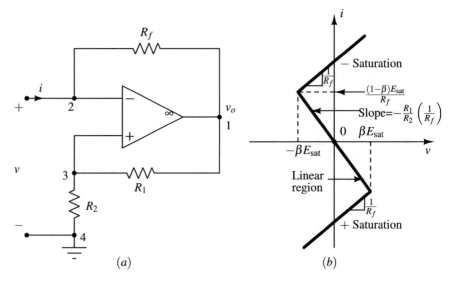

Fig. 2.41 A negative-resistance converter, and its DP characteristic. Here, $\beta \triangleq R_2/(R_1 + R_2)$

Applying KVL around the close node sequence $4 - 1 - 2 - 4$, we get:

$$v = v_o + R_f i \tag{2.147}$$

In Eq. (2.147), we have used that the fact the current into the inverting input of the opamp is zero. Using Eqs. (2.146) and (2.147), we can obtain i-vs-v for the opamp operating in the linear region:

$$i = -\left(\frac{R_1}{R_2}\right)\left(\frac{1}{R_f}\right)v \tag{2.148}$$

Next, we will use the validating inequality to conditions on v for the opamp to be operating in the linear region:

$$-\beta E_{\text{sat}} < v < \beta E_{\text{sat}} \tag{2.149}$$

Notice that obtaining the $i - v$ relationships for the saturation regions is trivial, since Eq. (2.147) is valid for any opamp region of operation, thus:

$$v = \pm E_{\text{sat}} + R_f i \tag{2.150}$$

Specifically, for the $+$ Saturation region:

$$i = \frac{1}{R_f}v - \frac{1}{R_f}E_{\text{sat}} \tag{2.151}$$

For the $-$ Saturation region:

$$i = \frac{1}{R_f} v + \frac{1}{R_f} E_{\text{sat}} \tag{2.152}$$

Equations (2.148), (2.151), and (2.152) complete the DP plot in Fig. 2.41b. The circuit is called a **negative impedance converter (NIC)** because it converts positive resistances R_1, R_2, R_f into a **negative resistance** equal to $-\frac{R_2 R_f}{R_1}$ in the opamp's linear region of operation.

Exercise 2.7 asks the reader to experimentally measure the DP plot for Fig. 2.41a. For more examples of practical negative-resistance opamp circuits, see [5].

We will see in Chapter 4 how this circuit can be used to build a relaxation oscillator. We will also see the enormous advantage of PWL analysis when we cover dynamic nonlinear networks in Chap. 4: PWL techniques allows us to derive closed-form expressions for the period (and frequency), that agree remarkable well with measured values. Moreover, that chapter will show the enormous importance of nonlinear circuit theory in designing a very important class of circuits—oscillators.[10]

2.5.4   A Family of Two-Port Resistors

The functions performed by the opamp circuits discussed so far can be summarized in one statement: They transform input voltage waveforms (functions of time) into some desired output voltage waveforms (functions of time). It is important to observe that the independent variable of the transformation is always time t.

There is another important class of networks which also performs certain transformations, but the independent variable is not time. This class of networks takes the form of a two-port black box, and is in fact a **two-port resistor**. If we connect a nonlinear resistor across port 2 of this two-port resistor, as shown in Fig. 2.42, the resulting two-terminal black box can be interpreted as a new nonlinear resistor because it will have a $v_1 - i_1$ curve different from the original $v - i$ curve. In other words, the function performed by the two-port resistor is that of transforming a given $v - i$ curve into a new $v_1 - i_1$ curve. In this sense, we can generate many new nonlinear resistors from those that are presently available commercially. Of course, an arbitrary transformation is not likely to do us much good. What we need is to discover a few basic transformations from which all others can be obtained. Amazingly, only three types of transformations are necessary, a scaling transformation, a rotation transformation, and a reflection transformation.

---

[10]The linear oscillator, modeled by an LC network, requires zero resistance for sustained oscillations. Since zero resistance is near impossible to obtain physically (save for superconductors), all practical oscillators are nonlinear.

Fig. 2.42 The $v - i$ curve of
a given nonlinear resistor is
transformed into a new
$v_1 - i_1$ curve by connecting
the resistor across one port of
a two-port resistor

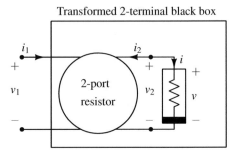

In the scaling transformation, the abscissa or the ordinate of each point on the $v - i$ curve is multiplied by a positive constant k. Such two-port resistors are accordingly called **scalors**.

In the rotation transformation, the original $v - i$ curve is rotated through an angle θ with respect to the origin. Such two-port resistors are accordingly called **rotators**.

In the reflection transformation, the original $v - i$ curve is reflected (the mirror image) with respect to some straight line through the origin. Such two-port resistors are accordingly called **reflectors**.

In the next section, we will mostly give a high-level overview of each of these devices. Implementation details can be found in [3] or in the accompanying online material(s) to this book. However, the reader should notice that these high-level implementations reuse a variety of components (such as controlled sources) from our earlier discussions. Obviously, we realized controlled sources using opamps. Thus, at the implementation level, opamps play a vital role in realizing the family of two-port resistors.

2.5.4.1 Scalors, Rotators, and Reflectors

There are two types of scalors, **voltage scalor** and **current scalor**. As the name implies, a voltage scalor multiplies the voltage (abscissa) of each point on the $v - i$ curve by a prescribed constant k_v, while maintaining the same value of current at the same point. This requirement can be characterized by:

$$v_1 = k_v v_2 \tag{2.153}$$

$$i_1 = -i_2 \tag{2.154}$$

The negative sign in Eq. (2.154) is necessary because we want $i_1 = i$ (i.e., i is unchanged by a **voltage** scalor).

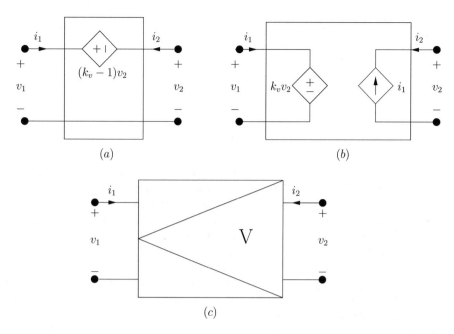

Fig. 2.43 (**a**), (**b**) Two realizations of a voltage scalor using dependent sources. (**c**) Circuit symbol

Example 2.5.10 Show that Fig. 2.43a, b realizes Eqs. (2.153) and (2.154).

Solution For (*a*), KVL gives:

$$v_1 = v_2 + (k_v - 1)v_2$$
$$= k_v v_2 \qquad\qquad (2.155)$$

KCL applied to the dependent source gives:

$$i_1 = -i_2 \qquad\qquad (2.156)$$

For (*b*), KVL applied to port 1 gives:

$$v_2 = k_v v_2 \qquad\qquad (2.157)$$

KCL applied to port 2 gives:

$$i_1 = -i_2 \qquad\qquad (2.158)$$

Fig. 2.44 Symbol of a
rotator

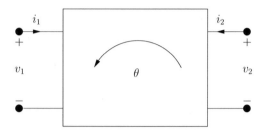

It is important to realize that a scalor is completely different from the opamp scaling circuits that we discussed. The independent variable for a scalor is either a voltage or a current, whereas the independent variable for a scaling circuit is time. Since two relationships must be satisfied by a scalor, in comparison with only one for a scaling circuit, it is more difficult to realize a scalor in practice.

The dual of the voltage scalor is the current scalor, discussed in Exercise 2.8.

Now, we will discuss the rotator, whose circuit symbol is shown in Fig. 2.44. From analytic geometry, we know that the relationship required to rotate a point P with coordinates (v, i) into a point P' with coordinates (v_1, i_1) by $\theta°$ (in the counterclockwise direction) is given by:

$$\begin{pmatrix} v_1 \\ i_1 \end{pmatrix} = \begin{pmatrix} \cos\theta & -\sin\theta \\ \sin\theta & \cos\theta \end{pmatrix} \begin{pmatrix} v \\ i \end{pmatrix} \tag{2.159}$$

Example 2.5.11 Recast Eq. (2.159) in terms of port variables.

Solution From Fig. 2.42, we know that $v = v_2$, $i = -i_2$. Simply substituting for (v, i) in Eq. (2.159), we get the two-port relationships:

$$\begin{pmatrix} v_1 \\ i_1 \end{pmatrix} = \begin{pmatrix} \cos\theta & \sin\theta \\ \sin\theta & -\cos\theta \end{pmatrix} \begin{pmatrix} v_2 \\ i_2 \end{pmatrix} \tag{2.160}$$

In order to allow an arbitrary current scale (since physical currents are usually an order of magnitude less than voltages), we multiply i_1 and i_2 in Eq. (2.160) by a scale factor R, thereby obtaining:

$$\begin{pmatrix} v_1 \\ i_1 \end{pmatrix} = \begin{pmatrix} \cos\theta & (\sin\theta)R \\ \frac{\sin\theta}{R} & -\cos\theta \end{pmatrix} \begin{pmatrix} v_2 \\ i_2 \end{pmatrix} \tag{2.161}$$

A rotator is completely characterized by Eq. (2.161). In the volt-milliampere plane, $R = 10^3$. A physical implementation of a rotator is simply obtained by using a π-network, as Example 2.5.12 shows.

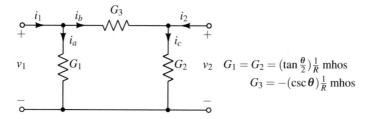

Fig. 2.45 A rotator can be implemented by choosing specific conductances G_n, given the rotation angle θ

Example 2.5.12 Show that the resistive network in Fig. 2.45 implements a rotator, given a specified angle θ.

Solution In order to verify the realization, we need only to drive v_1 and i_1 in terms of v_2 and i_2 for the network and show that they agree with Eq. (2.161).

By inspection, $i_a = G_1 v_1$, $i_b = G_3(v_1 - v_2)$, $i_c = G_2 v_2$. Applying KCL to each port, we obtain:

$$i_1 = G_1 v_1 + G_3(v_1 - v_2) = \left(\tan\frac{\theta}{2}\right)\frac{1}{R}v_1 + (-\csc\theta)\frac{1}{R}(v_1 - v_2)$$

$$(2.162)$$

$$i_2 = G_2 v_2 - G_3(v_1 - v_2) = \left(\tan\frac{\theta}{2}\right)\frac{1}{R}v_2 - (-\csc\theta)\frac{1}{R}(v_1 - v_2)$$

$$(2.163)$$

With the help of the trigonometric identity: $\tan\frac{\theta}{2} = (\csc\theta - \cot\theta)$, we can simplify the above equations to:

$$i_1 = -(\cot\theta)\frac{1}{R}v_1 + (\csc\theta)\frac{1}{R}v_2 \qquad (2.164)$$

$$i_2 = (\csc\theta)\frac{1}{R}v_1 - (\cot\theta)\frac{1}{R}v_2 \qquad (2.165)$$

Solving Eq. (2.165) for v_1, we obtain the v_1 row in Eq. (2.161). Substituting the obtained expression for v_1 in Eq. (2.164), we obtain the i_1 row in Eq. (2.161).

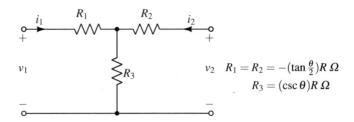

v_2 $R_1 = R_2 = -(\tan \frac{\theta}{2})R \; \Omega$

$R_3 = (\csc \theta)R \; \Omega$

Fig. 2.46 Dual T-network for the π-network from Fig. 2.45

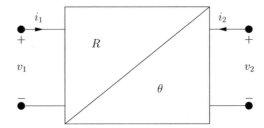

Fig. 2.47 Symbol of a reflector

Note that in the example above, since we have a π-network, we chose to work with conductances instead of resistances. That made our calculations much easier. Analogous to a π-network, we could have also worked with the T-network (recall Example 2.2.1) shown in Fig. 2.46, where working with resistances makes our calculations much simpler. Observe that, depending upon the values of θ, either one or two of the three linear resistors in both realizations may assume negative values. However, only one negative resistor is necessary to realize a rotator with any angle of rotation, provided we choose the π-network whenever $0° < \theta < 180°$, and the T-network whenever $180° < \theta < 360°$. We have already discussed how to synthesize negative resistors in Sect. 2.5.3.2.

A subset of the generic rotator is the reflector, so called because θ is the angle which the line of reflection (through the origin) makes with the horizontal axis. Hence, from analytic geometry, we obtain the characteristic two-port equations for the reflector below:

$$\begin{pmatrix} v_1 \\ i_1 \end{pmatrix} = \begin{pmatrix} \cos 2\theta & -(\sin 2\theta)R \\ \frac{\sin 2\theta}{R} & \cos 2\theta \end{pmatrix} \begin{pmatrix} v_2 \\ i_2 \end{pmatrix} \tag{2.166}$$

We shall denote a reflector by the symbol shown in Fig. 2.47. Exercise 2.9 explores reflector realizations, analogous to rotator realizations discussed previously.

Fig. 2.48 Symbol of a
gyrator

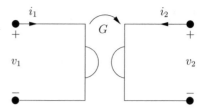

2.5.4.2 Gyrators

There are several angles of reflection which are of special practical importance,
and the corresponding reflectors have been given special names. We will discuss
a particularly important kind, called a gyrator (for other specific reflectors, please
refer to [3]).

When $\theta = 45°$, Eq. (2.166) can be recast into the form:

$$\begin{pmatrix} i_1 \\ i_2 \end{pmatrix} = \begin{pmatrix} 0 & G \\ -G & 0 \end{pmatrix} \begin{pmatrix} v_1 \\ v_2 \end{pmatrix} \tag{2.167}$$

where $G = \frac{1}{R}$ is a constant called the **gyration conductance**. The symbol for a
gyrator is shown in Fig. 2.48. The fundamental property of an ideal gyrator is that it
functions as an "impedance inverter." For instance, if the output port (port 2) of the
gyrator is terminated with an $R_L - \Omega$ linear resistor, the input port's resistance is
given by:

$$\begin{aligned}
\frac{v_1}{i_1} &= \frac{-i_2/G}{Gv_2} \\
&= \frac{1}{G^2} \frac{-i_2}{v_2} \\
&= \frac{G_L}{G^2}
\end{aligned} \tag{2.168}$$

In other words, $R_1 = \frac{G_L}{G^2}$. If $G = 1$, we see that the input port's resistance is the
reciprocal of the output port's resistance. Hence the term "impedance inverter."

The specification definition of "impedance" as a "frequency-dependent resistance" will become clear in Sect. 4.3. Specifically, we will discuss that if the output
port of an ideal gyrator is terminated with a capacitor, the input port behaves like
an inductor. Thus, a gyrator is a useful element in the design of inductorless filters.
This is practically advantageous because physical inductors tend to be bulky and
lossy, when compared to physical capacitors. We will see such an implementation
in the exercises to Chap. 4.

Fig. 2.49 Symbol for a Type
1 $M - R$ mutator

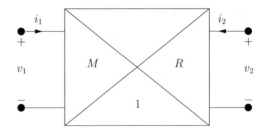

2.5.4.3 Mutators

So far, we have discussed two-port resistors which share the common property that each transforms a nonlinear resistor into another nonlinear resistor. If we liken the four basic network elements, resistors, capacitors, inductors, and memristors, to four distinct species in the generic sense, then the scalor, rotator, and reflector can be said to transform elements belonging to the same species.

In this section, we will show that it is possible to produce a **mutation** from one species into another with the help of a two-port black box called the **mutator**. For example, it is possible to connect a resistor across port 2 of a mutator and produce an inductor across port 1. Conversely, if an inductor is connected across port 1 of the same mutator, a resistor is produced across port 2. For this reason, this class of mutators is called $L - R$ **mutators.**[11]

In the interest of brevity,[12] and the fact that very few physical electronic memristors are commercially available, we will devote this section to the design of $M - R$ mutators for realizing memristors from nonlinear resistors [1]. We chose nonlinear resistors because they can be more easily synthesized and are even commercially available (for example, diodes), as opposed to nonlinear capacitors or inductors.

Figure 2.49 shows the circuit symbol of a Type 1 $M - R$ mutator. In order to transform a resistor into a memristor, it is necessary that the coordinates (v, i) of each point on a $v - i$ curve be transformed into a corresponding point with coordinates (ϕ, q). To accomplish this, first recall the following relationships from our two-port black box in Fig. 2.42: $v_2 = v, i_2 = -i$. Suppose:

$$v_1 \triangleq k_v \frac{dv_2}{dt} \tag{2.169}$$

$$i_1 \triangleq k_i \left(-\frac{di_2}{dt} \right) \tag{2.170}$$

---

[11] We want to clarify mutator terminology. In Chua's seminal book "Introduction to Nonlinear Circuit Theory" [3], Dr. Chua refers to an $L - R$ mutator as an $R - L$ mutator. However, in Dr. Chua's publication defining the mutator [2] and all subsequent works, the terminology is consistent with the one used in this book.

[12] For details on realizing other types of mutators such as $C - R$, $R - L$, etc., please refer to [2].

k_v and k_i are constants that are used for dimensional consistency. We will see in Sect. 2.5.5 when we realize an $M - R$ mutator as to how to compute these constants. Using the definitions in Eqs. (2.169) and (2.170), we have at port 1:

$$\phi_1 \triangleq \int_{-\infty}^{t} v_1(\tau)d\tau$$

$$= k_v \int_{-\infty}^{t} \frac{d}{d\tau} v_2(\tau)d\tau$$

$$= k_v \int_{-\infty}^{t} \frac{d}{d\tau} v(\tau)d\tau$$

$$= k_v v \qquad\qquad (2.171)$$

Similarly:

$$q_1 \triangleq \int_{-\infty}^{t} i_1(\tau)d\tau$$

$$= k_i \int_{-\infty}^{t} \frac{-d}{d\tau} i_2(\tau)d\tau$$

$$= k_i \int_{-\infty}^{t} \frac{d}{d\tau} i(\tau)d\tau$$

$$= k_i i \qquad\qquad (2.172)$$

Thus, our mutator does perform the correct mapping. A high-level realization of a Type 1 $M - R$ mutator using dependent sources is shown in Fig. 2.50.

Fig. 2.50 One realization of an $M - R$ mutator, $k_v = 1, k_i = 1$. For other $M - R$ mutators such as Type 2, please refer to [1]

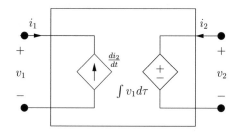

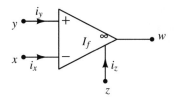

Fig. 2.51 An ideal CFOA. Labels that are prevalent in the literature: output terminal is labeled w, the inverting input is labeled x, noninverting input is labeled y, and the compensation terminal (pin) is labeled z

2.5.5 Current Feedback Opamps

So far, in this section, we have been using opamps that rely on voltage feedback. There is another kind of opamp that uses **current feedback**, appropriately named **current feedback operational amplifier (CFOA)**.

A CFOA is a four-terminal[13] device [10] with the circuit symbol[14] shown in Fig. 2.51.

The terminal behavior of an ideal CFOA is defined below:

$$i_x = i_z \tag{2.173}$$

$$i_y = 0 \tag{2.174}$$

$$v_x - v_y = 0 \tag{2.175}$$

$$v_w - v_z = 0 \tag{2.176}$$

Note that the current through the compensation pin i_z is feedback to the inverting input current i_x, hence the origin of the name "CFOA."

A CFOA is particularly suited for implementing derivative (or integral) operations in controlled sources. Hence, we can easily realize two-ports such as the Type 1 $M - R$ mutator from Fig. 2.50, as the following example shows.

---

[13] Some CFOAs do not have an externally accessible compensation pin z, to maintain pin-compatibility with voltage feedback amplifiers. However, such devices are actually a special class of CFOAs and in this book we will use only the very general CFOAs such as the AD844 that have an externally accessible compensation pin. We will henceforth refer to such CFOAs as an **ideal CFOA**.

[14] A literature search revealed that there is no standard symbol for a CFOA. We are defining this symbol because it closely mimics the symbol of an ideal opamp with the I_f clarifying that we have current feedback.

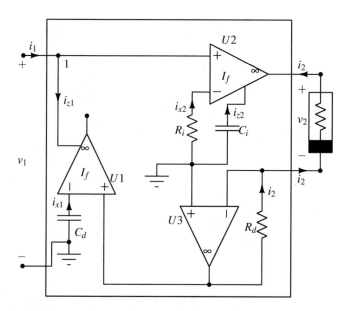

Fig. 2.52 Type 1 $M - R$ mutator realization [1]

Example 2.5.13 Show that the network in Fig. 2.52 implements a Type 1 $M-R$ mutator.

Solution First, we will derive Eq. (2.169): $v_1 = k_v \frac{dv_2}{dt}$. For CFOA $U2$, we have the voltage across C_i equal to v_2. Notice that this is possible because the inverting terminal of opamp $U3$ is at virtual ground. Thus:

$$i_{z2} = -C_i \frac{dv_2}{dt} \tag{2.177}$$

Also, for CFOA $U2$:

$$i_{x2} = i_{z2} \tag{2.178}$$

$$v_1 = v_{x2} \tag{2.179}$$

Using Ohm's law for R_i and simplifying using the above equations:

$$v_1 = -i_{x2} R_i$$

$$= -i_{z2} R_i$$

(continued)

Example 2.5.13 (continued)

$$= R_i C_i \frac{dv_2}{dt} \tag{2.180}$$

Hence, $v_1 = k_v \frac{dv_2}{dt}$, where $k_v = R_i C_i$. For deriving Eq. (2.170): $i_1 = k_i \left(-\frac{di_2}{dt}\right)$, application of KCL at node 1 gives $i_1 = i_{z1}$ (since current into the noninverting input of CFOA $U2$ is zero). From CFOA $U1$:

$$i_{x1} = i_{z1} \tag{2.181}$$

From capacitor C_d:

$$i_{x1} = -C_d \frac{dv_{x1}}{dt} \tag{2.182}$$

Note that output voltage of opamp $U3$ is $R_d i_2$, which is equal to v_{y1}. But, since $v_{x1} = v_{y1}$, we get:

$$v_{x1} = R_d i_2 \tag{2.183}$$

Substituting for v_{x1} in the expression for i_{x1}, and using the fact that $i_{x1} = i_{z1} = i_1$, we get:

$$i_1 = -R_d C_d \frac{di_2}{dt} \tag{2.184}$$

Hence, $i_1 = k_i \frac{di_2}{dt}$, where $k_i = R_d C_d$.

Further detailed discussion of CFOAs is beyond the scope of this book, but the interested reader should consult excellent references such as [10].

2.6 Conclusion

This chapter greatly expanded upon Chap. 1 and we should now have a very good understanding of the laws of elements, for a variety of multi-terminal elements. To summarize:

1. To characterize a multi-terminal black box, we will choose one node as ground (reference). We can then classify the black box as either an n-terminal resistor (involving $n - 1$ terminal voltages and currents), an n-terminal inductor (involving $n - 1$ terminal currents and flux-linkages), an n-terminal capacitor (involving

$n - 1$ terminal voltages and charges), and an n-terminal memristor (involving $n - 1$ terminal flux-linkages and charges).

2. Since most practical devices have single digit values for n (the number of terminals), we can characterize three-terminal ($n = 3$) resistors, inductor, capacitors, and memristors using the six two-port representations: current-controlled, voltage-controlled, hybrid 1, hybrid 2, transmission 1, and transmission 2.

3. We studied the four dependent sources: CCVS, VCCS, CCCS, and VCVS and transformers as linear resistive two-ports.

4. We studied the npn BJT as a nonlinear resistive two-port.

5. A common three-terminal inductor is the physical transformer, usually consisting of two coupled coils wound on a torus of ferromagnetic material.

6. A particularly useful multi-terminal element for redistributing power is the three-port circulator.

7. The opamp is a very versatile multi-terminal nonlinear resistor. We implemented amplifiers and studied nonlinear opamp circuits such as the Schmitt trigger and NICs.

8. The family of two-port resistors: scalors, rotators, reflectors, along with the mutator can be realized using dependent sources. For the mutator, we used a CFOA.

We are now ready to learn about the laws of interconnections, starting with resistive nonlinear networks in Chap. 3.

Exercises

2.1 For the linear two-ports specified by the following equations, find as many representations as you can.

1.
$$-i_1 + 2i_2 + v_2 = 0$$

2.
$$v_1 + v_2 = 0$$
$$v_1 + i_2 + v_2 = 0$$

3.
$$i_1 = 0$$
$$v_1 + v_2 = 0$$
$$i_1 + i_2 = 0$$

2.2 By equating the inductance matrix in Eq. (2.65) (from Example 2.2.5) to the matrix from Eq. (2.55), show that:

$$n = \frac{L_{22}}{M} \qquad L_m = \frac{M^2}{L_{22}} \qquad L_a = L_{11} - \frac{M^2}{L_{22}} \qquad (2.185)$$

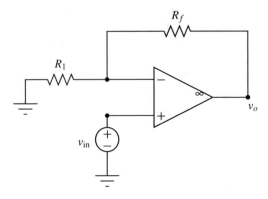

Fig. 2.53 Circuit for problem 2.3

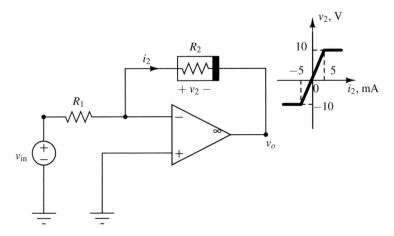

Fig. 2.54 Circuit for problem 2.4

2.3 For the inverting amplifier shown in Fig. 2.53, determine the VTC and also the dynamic range of v_{in} for which the opamp operates in the linear region.

2.4 Consider the circuit in Fig. 2.54, with the nonlinear resistor's DP shown.

1. Compute the nonlinear VTC v_o-vs-v_{in}.
2. Determine the dynamic range for v_{in}, for which the opamp remains in the linear region.
3. Does the opamp operate in the linear region for **all** values of v_{in}, in spite of a nonlinear element in the feedback path? Justify your answer.

2.5 Consider the VTC of the **finite gain opamp model** shown in Fig. 2.55. Using the PWL representation (Eq. (1.52) from Sect. 1.9.1.2), we can describe the finite

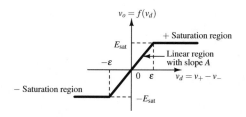

Fig. 2.55 VTC for problem 2.5

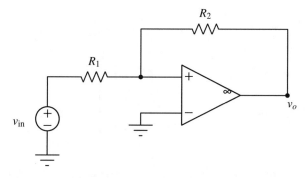

Fig. 2.56 Circuit for problem 2.6

gain model analytically as shown below:

$$i_- = 0 \tag{2.186}$$

$$i_+ = 0 \tag{2.187}$$

$$v_o = \frac{A}{2}|v_d + \epsilon| - \frac{A}{2}|v_d - \epsilon| \tag{2.188}$$

1. Derive circuits similar to Fig. 2.28b, c, and d for the finite gain opamp model. HINT: For the linear region, you will require a VCVS.
2. Now, re-derive the VTC for the inverting amplifier from Example 2.5.4 using the finite gain model.
3. Confirm that as $A \rightarrow \infty$ in your finite gain VTC, we obtain Eq. (2.105).

2.6 Derive the VTC and validating inequalities for the noninverting Schmitt trigger in Fig. 2.56. Also, sketch the VTC.

2.7 Experimentally plot the DP characteristic for Fig. 2.41a. Determine the percent error between the experimental measurements and theoretical calculations for the slopes and breakpoints given by Eqs. (2.148), (2.151), and (2.152).

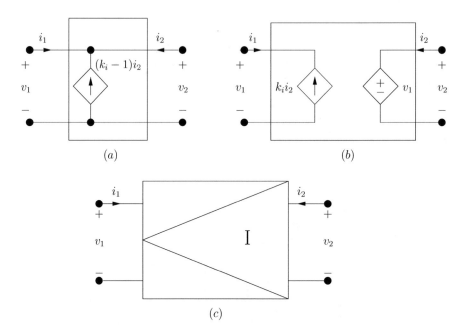

Fig. 2.57 For problem 2.8: (**a**) dependent source implementation 1, (**b**) dependent source implementation 2, (**c**) scalor

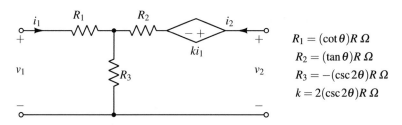

Fig. 2.58 Circuit for problem 2.9

2.8 Derive the following current scalor relationships for the dependent source implementations in Fig. 2.57a, b:

$$v_1 = v_2 \tag{2.189}$$

$$i_1 = -k_i i_2 \tag{2.190}$$

2.9 Show that the circuit in Fig. 2.58 realizes the reflector two-port model in Eq. (2.166).

Lab 2: Noninverting Schmitt Trigger VTC

Objective: To verify the Schmitt trigger characteristics from Exercise 2.6 via QUCS simulation and physical implementation

Theory:

You may be wondering if one can actually physically observe the VTCs for the Schmitt triggers. The answer is yes. In order to do this, consider the circuit shown in Fig. 2.59. We have used the circuit symbol for the finite gain opamp model. Nevertheless, for this lab, we can assume that the opamps are ideal.

The capacitor C_f and resistor R_b provide feedback at higher frequencies (for eliminating instability due to parasitics) and can be ignored for very low frequencies. In fact, at DC ("zero" frequency), the capacitor acts like an open circuit (more on this in Chap. 4) and since the current into the inverting terminal of an opamp is zero, we get Fig. 2.60, that we will use for analysis.

Lab Exercise:

1. First, for the purposes of this lab, we can assume that the upper opamp is operating in the linear region. In fact, you should recognize the upper opamp as a variant of the summing amplifier from Example 2.5.5.

   Using KCL at the inverting input of the upper opamp, write an equation in terms of **conductances** G_u, G_H, G_v and voltages u, v_{out}, v_{in}.

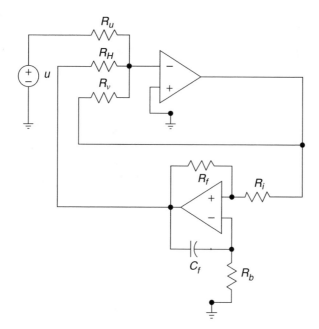

Fig. 2.59 Schematic for experimentally confirming the Schmitt trigger VTC

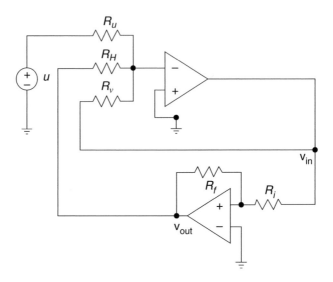

Fig. 2.60 Schematic for analysis

2. Assuming $G_u = G_H = G_v$, find the constraint imposed by the summing amplifier.
3. This step should help you understand the elegant mathematical trick: the constraint that you get from the previous step is essentially shown in Fig. 2.61a, b

    In other words, the derived constraint in step 2 implies that the circuit is **intersecting** the line $v_{out} = -(v_{in} + u)$ with the VTC of the Schmitt trigger. This intersection is the elegant mathematical trick: obviously, only one of the two Fig. 2.61a, b will occur in reality.

    Before proceeding to the simulation and experimental verification, component values that we used are: $R_f = 20\,k\Omega$, $R_i = 10\,k\Omega$, $R_u = R_H = R_v = 20\,k\Omega$, $R_b = 470\,\Omega$, $C_f = 470\,pF$. Opamps are TL074, we chose power supply voltages such that $E_{sat} = \pm 4\,V$. u was varied using a sine function with an amplitude of 8 V. Maximum frequency used was 100 Hz.
4. Use QUCS to confirm the VTC of the noninverting Schmitt trigger by using the schematic in Fig. 2.59.
5. Physically implement your circuit from Fig. 2.59 and experimentally confirm the noninverting Schmitt trigger VTC. Our result is shown in Fig. 2.62. The reference for this lab is [8]. We highly recommend going through Kennedy and Chua's seminal work, if you have access to it. It very clearly dispels common misconceptions about the nature of hysteresis in electronic circuits.

Fig. 2.61 The result of
intersecting the line
$v_{out} = -(v_{in} + u)$ with the
VTC of the Schmitt trigger
will result in either (**a**) or (**b**).
Note that we have used $u = 0$
in this figure

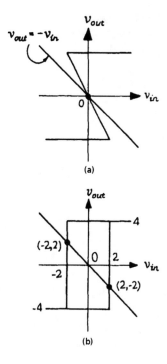

Fig. 2.62 Experimental confirmation of Schmitt trigger VTC

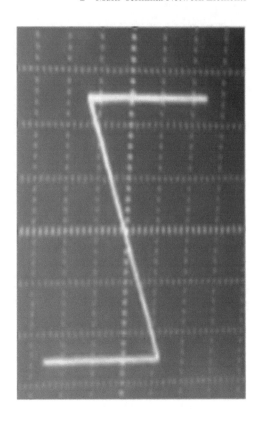

References

1. Biolek, D. et al.: Mutators for transforming nonlinear resistor into memristor. In: 20th European Conference on Circuit Theory and Design (ECCTD), pp. 488–491 (2011)
2. Chua, L.O.: Synthesis of new nonlinear network elements. Proc. IEEE **56**(8), 1325–1342 (1968)
3. Chua, L.O.: Introduction to Nonlinear Network Theory. McGraw-Hill, New York, (1969), (out of print)
4. Chua, L.O.: University of California, Berkeley EE100 Fall 2008 Supplementary Lecture Notes. (2008) Multi-Terminal Resistors: http://inst.eecs.berkeley.edu/~ee100/fa08/lectures/EE100supplementary_notes_4.pdf. Opamps: http://inst.eecs.berkeley.edu/~ee100/fa08/lectures/EE100supplementary_notes_7.pdf. Accessed 12 April 2017
5. Chua, L.O., Ayrom F.: Designing nonlinear single op-amp circuits: a cookbook approach. Int. J. Circuit Theory Appl. **13**, 235–268 (1985)
6. Chua, L.O., Desoer, C.A., Kuh, E.S.: Linear and Nonlinear Circuits. McGraw-Hill, New York (1987), (out of print)
7. Howe, R.T., Sodini, J.T.: Microelectronics: An Integrated Approach. Prentice-Hall, Upper Saddle River (1996)

8. Kennedy, M.P., Chua, L.O.: Hysteresis in electronic circuits: a circuit theorist's perspective. Int. J. Circuit Theory Appl. **19**, 471–515 (1991)

9. Ranganathan, S., Tsividis, Y.: Discrete-time parametric amplification based on three-terminal MOS varactor: analysis and experimental results. IEEE J. Solid-State Circuits **38**(12), 2087–2093 (2003)

10. Senani, R. et al.: Current Feedback Operational Amplifiers and Their Applications. Springer, Berlin (2013)

Chapter 3
Resistive Nonlinear Networks

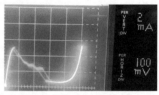

Typical $v - i$ plot $((0, 0)$ is in the lower-left corner) of a negative resistance device, the 1N3716 tunnel-diode. Notice the effects of parasitics are visible in the form of hysteresis in the negative resistance region.

Abstract Having described two-terminal and multi-terminal circuit elements in the "first part" of this book, we have hence discussed the laws of elements. We will now study, in this chapter and the next, KCL/KVL based circuit theoretic techniques that allow us to analyze circuits of varying degrees of "complexity" (a term we make precise in Chap. 4). We will study these techniques by following the classical approach: discussing static (resistive) networks in this chapter and then dynamic (inductive, capacitive, and memristive) networks in Chap. 4. Such a division is not accidental: in terms of circuit variables, dynamic networks usually involve differential equations, unlike static networks. Hence in this chapter we will study simpler resistive networks. We will first discuss the fundamental concept of operating points. Next, we will expand upon graph theoretic concepts and then discuss two of the most important techniques: nodal and tableau analysis. We will conclude the chapter by discussing some general properties of linear resistive networks (superposition, Thévenin-Norton theorems) and nonlinear resistive networks (strict passivity, strict monotonicity).

© Springer International Publishing AG, part of Springer Nature 2019   135
B. Muthuswamy, S. Banerjee, *Introduction to Nonlinear Circuits and Networks*,
https://doi.org/10.1007/978-3-319-67325-7_3

3.1   The Operating Point Concept

Given any circuit, one is interested in determining a solution [3]. For some circuits, there exists a unique solution. This is the case of a circuit containing two-terminal linear passive resistors and an independent current source connected to any two nodes of the circuit serving as input.[1] For other circuits, there may exist a unique solution, multiple solutions, or even no solution at all. This happens with circuits containing nonlinear resistors.

The solutions to a circuit with DC input are called **operating points** or **Quiescent (Q)-point**. The term **DC analysis** refers to the determination of operating points. It will be shown later that DC analysis of general dynamic circuits (with inductors, capacitors, and memristors) is equivalent to finding solutions of a resistive circuit which can be simply derived from the given circuit. The subject is of major importance in circuit theory and electronics. In this section, we will consider DC analysis for simple circuits using a variety of techniques.

The basic concepts of DC analysis can be illustrated with the simple circuit configuration shown in Fig. 3.1, i.e., the back-to-back connection of two one-ports at nodes 1 and 2. What is interesting to note is that this simple configuration, because it includes two unspecified one-ports, covers circuits with great generality. We assume that each one-port is specified by the following DP characteristics in terms of its port voltage and port current, v_a, i_a and v_b, i_b, respectively:

$$f_a(v_a, i_a) = 0 \quad \text{and} \quad f_b(v_b, i_b) = 0 \qquad (3.1)$$

These are the generalizations of the branch characteristics since each one-port is formed by an interconnection of resistors. We are not concerned with what is inside of the one-ports N_a and N_b. Therefore we only need to use KCL and KVL to describe the port interconnection at the two nodes 1 and 2. KCL states:

$$i_a = -i_b \qquad (3.2)$$

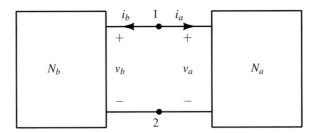

Fig. 3.1 Two resistive one-ports connected in parallel

---

[1]We will discuss existence and uniqueness theorem for general resistive nonlinear networks later in this chapter.

Fig. 3.2 Circuit of Fig. 3.1
with given characteristics of
N_a and N_b

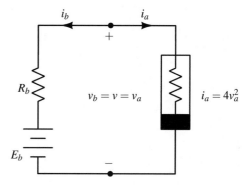

KVL states:

$$v_a = v_b \tag{3.3}$$

Therefore we can eliminate one set of voltage and current by combining Eqs. (3.1)–
(3.3). Let us denote: $i_a = -i_b \triangleq i$ and $v_a = v_b \triangleq v$. The two resulting equations in
terms of v and i are:

$$f_a(v, i) = 0 \quad \text{and} \quad f_b(v, -i) = 0 \tag{3.4}$$

The solutions of the two equations are the operating points that we are looking for.
We will give a number of examples to illustrate the analytic, graphical, and PWL
methods.

Example 3.1.1 (Analytical Method)   Determine the operating points in
Fig. 3.2.

Solution We will consider the nonlinear resistor as N_a, hence the specification $f(v_a, i_a)$ is:

$$i_a - 4v_a^2 = 0 \tag{3.5}$$

Let N_b be the series connection of the DC voltage source and linear resistor,
which can be used to model a real battery connected in series with a resistive
load. The specification $f(v_b, i_b)$ is:

$$v_b - E_b - R_b i_b = 0 \tag{3.6}$$

(continued)

Example 3.1.1 (continued)

As before, we let $i_a = -i_b \triangleq i$ and $v_a = v_b \triangleq v$. Hence $f(v_a, i_a)$ and $f(v_b, i_b)$ become:

$$i = 4v^2 \tag{3.7}$$

$$v = E_b - R_b i \tag{3.8}$$

These two equations lead to a quadratic equation in terms of v:

$$4R_b v^2 + v - E_b = 0 \tag{3.9}$$

The equation above can be solved for specific values of E_b and R_b. For instance, with $E_b = 2$ V and $R_b = 0.25$ Ω, the two solutions (and hence operating points) are: $v = 1$ V, -2 V and $i = 4$ A, 16 A, respectively.

In practice one rarely encounters problems in nonlinear circuits which can be solved analytically. Hence we will next see probably one of the most powerful graphical analysis techniques, the load line method.

Example 3.1.2 (Graphical (Load Line) Method)   Determine the operating points in Fig. 3.2 graphically.

Solution The circuit in Fig. 3.2 represents a typical **biasing circuit** in DC design, i.e., a simple nonlinear circuit which consists of a battery, a resistor and an electronic device modeled by a nonlinear resistor with a specified $v - i$ characteristic.

The way to find the solution is to transcribe the characteristic of the battery and the resistor in the $v_b - i_b$ plane to the $v_a - i_a$ plane, where the characteristic of the device is plotted. We could of course transcribe the characteristic of the device from the $v_a - i_a$ plane onto the battery-resistor characteristic in the $v_b - i_b$ plane. But it is always easier to transcribe a linear equation.

Since $v_b = v_a$ and $i_b = -i_a$, the transcribed curve is the mirror image with respect to the v axis of the curve in the $v_b - i_b$ plane. This is superimposed with the characteristic of the nonlinear one-port N_a, as shown in Fig. 3.3, for $E_b = 2$ V, $R_b = 0.25$ Ω from Example 3.1.1. There are two intersections of the two curves, and these give the operating points, equal to the values we obtained in Example 3.1.1.

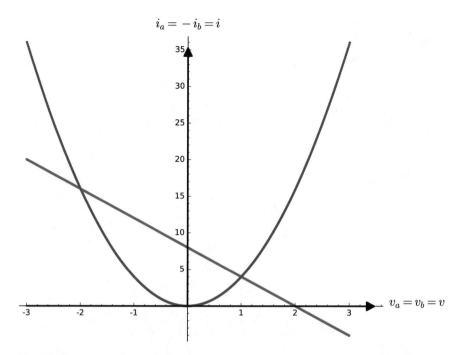

Fig. 3.3 The two one-port characteristics are superimposed on the $v - i$ plane

The transcribed battery-resistor characteristic in Fig. 3.3 is called the **load line**. It is a straight line which has E_b as its v-axis intercept and has E_b/R_b as its i-axis intercept. The load line method for determining the operating point(s) is used in practice because the $v - i$ characteristic of the one-port N_a is often given as a measured curve.

The third example shows how to use a PWL numerical method.

Example 3.1.3 (PWL Method) Consider the tunnel-diode circuit shown in Fig. 3.4 where the nonlinear characteristic N_a has been changed to that of a tunnel-diode with PWL characteristics. This device exhibits negative resistance characteristics. Determine the operating points of the circuit.

Solution Using the ideas from Sect. 1.9.1.2, we assume that the PWL characteristics for the tunnel-diode can be written as:

$$i = a_0 + a_1v + b_1|v - E_1| + b_2|v - E_2| \qquad (3.10)$$

(continued)

Example 3.1.3 (continued)

The parameters are: $a_0 = -\frac{1}{2}, a_1 = 2, b_1 = -\frac{5}{2}, b_2 = \frac{3}{2}, E_1 = 1, E_2 = 2$. Let the battery-resistor characteristic be given by $E_b = 6$ V, $R_b = 2\ \Omega$. The superimposed curves in the $v - i$ plane via the load line method are shown in Fig. 3.4b. Thus we know that the three operating points are at the three intersections Q_1, Q_2, Q_3. However, for the present, we wish to determine them analytically by using Eq. (3.10).

As we have shown in the Sect. 1.9.1.2 on PWL characteristics, the v axis can be divided into three regions:

$$\text{Region 1:} \quad v \le E_1 = 1 \tag{3.11}$$

$$\text{Region 2:} \quad 1 < v \le E_2 = 2 \tag{3.12}$$

$$\text{Region 3:} \quad v > 2 \tag{3.13}$$

In the three regions, Eq. (3.10) can be replaced by equations without absolute value signs as follows:

$$\text{Region 1:} \quad i = a_0 + a_1 v - b_1(v - E_1) - b_2(v - E_2) \tag{3.14}$$

$$\text{Region 2:} \quad i = a_0 + a_1 v + b_1(v - E_1) - b_2(v - E_2) \tag{3.15}$$

$$\text{Region 3:} \quad i = a_0 + a_1 v + b_1(v - E_1) + b_2(v - E_2) \tag{3.16}$$

For the battery-resistor combination, the equation is:

$$v = E_b - R_b i$$

$$= 6 - 2i \tag{3.17}$$

First, solving Eqs. (3.14) and (3.17) for the solution in region 1, we obtain $V_{Q1} = \frac{6}{7}$. Similarly, solving Eqs. (3.15) and (3.17) for the solution in region 2, we obtain $V_{Q2} = \frac{4}{3}$. Finally, solving Eqs. (3.16) and (3.17), we get $V_{Q3} = \frac{8}{3}$ for region 3.

It is crucial to remember that we must check these calculated solutions to see whether they fall in the assumed regions. If they indeed do, they are valid solutions, otherwise they are called virtual solutions. They do not corresponding to reality, they are artifacts of the method. In the present case, we see that all three voltages are valid solutions because they do indeed fall in the respective regions: $V_{Q1} = \frac{6}{7} \le 1$, $V_{Q2} = \frac{4}{3}$ falls in region 2 ($1 < v \le 2$) and $V_{Q3} = \frac{8}{3}$ falls in region 3 ($v > 2$). Since all voltage solutions are confirmed to be valid, we can find the corresponding currents from Eq. (3.17). Thus, the operating points are : $\left(\frac{6}{7}, \frac{18}{7}\right), \left(\frac{4}{3}, \frac{7}{3}\right), \left(\frac{8}{3}, \frac{5}{3}\right)$.

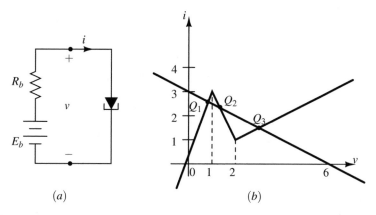

Fig. 3.4 Operating points of a tunnel-diode circuit determined by the PWL method (**a**) Circuit (**b**) load line

3.1.1 Small Signal Analysis

There is a good reason to call the solutions to DC analysis "operating points." When a circuit is used, some input signal (example, a sinusoidal waveform) is applied to it so that we get a useful output. An operating point specifies a region in the $v - i$ plane in the neighborhood of which the actual voltage and current in the circuit vary as a function of time. If the applied signal has a sufficiently small voltage or current (in magnitude), the circuit can be analyzed to a good approximation by using **small-signal analysis**.

Consider the tunnel-diode circuit shown in Fig. 3.5 where, in addition to the circuit elements treated earlier, there is a sinusoidal voltage source:

$$v_s(t) = V_m \cos \omega t \tag{3.18}$$

First we assume that the biasing circuit, i.e., the circuit without the signal source $v_s(t)$ has been designed properly so that there is only one operating point Q as shown. To be specific, assume that it lies where the slope is negative. As $v_s(t)$ varies with time, we may imagine that the load line is being moved parallel to the biasing load line as shown in the figure. Thus the solution of the circuit driven by the input signal $v_s(t)$ can be determined graphically point by point as the intersection point of the characteristic of the tunnel diode and the moving load line. This gives us a mental picture of the influence of the signal source $v_s(t)$ as t changes.

Let the $v - i$ characteristic of the tunnel diode be specified by:

$$i = \hat{i}(v) \tag{3.19}$$

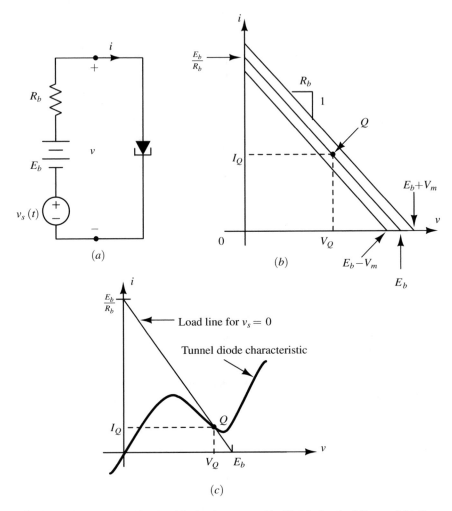

Fig. 3.5 (**a**) Tunnel-diode circuit with signal source $v_s(t)$. (**b**) Moving load line, and (**c**) linear approximation to the diode characteristic at the operating point Q

KCL states that all branch currents in the circuit are the same. KVL for the single loop in the circuit yields the following equation:

$$v(t) = v_s(t) + E_b - R_b i(t) \tag{3.20}$$

Combining Eqs. (3.19) and (3.20) we obtain a single equation with $v(\cdot)$ as the unknown to be solved for:

$$v(t) = v_s(t) + E_b - R_b \hat{i}[v(t)] \tag{3.21}$$

This cannot be solved readily since we only know the curve given by the tunnel-diode data sheet. Of course for each value of t, we can find $v(t)$, thus $v(\cdot)$ can be determined point by point.

As seen in Fig. 3.5b, the actual signal voltage $v(t)$ and signal current $i(t)$ lie on the characteristic in the neighborhood[2] of Q. Therefore, let us denote:

$$v(t) \triangleq V_Q + \tilde{v}(t) \tag{3.22}$$

$$i(t) \triangleq I_Q + \tilde{i}(t) \tag{3.23}$$

where (V_Q, I_Q) is the operating point. This, in essence, shifts the coordinates from the origin to the operating point. The two equations (3.19) and (3.20) are satisfied with the signal $v_s(t) = 0$, i.e.,

$$I_Q = \hat{i}(V_Q) \tag{3.24}$$

$$V_Q = E_b - R_b I_Q \tag{3.25}$$

Note that $\tilde{v}(t)$ and $\tilde{i}(t)$ book keep the displacement of the instantaneous operating point away from (V_Q, I_Q) when the signal is applied. The pertinent concept above can be illustrated with the two circuits shown in Fig. 3.6. Figure 3.6a gives the DC equivalent circuit which is specified by Eqs. (3.24) and (3.25). We can eliminate E_b in Eq. (3.20) by using Eq. (3.25):

$$v(t) = v_s(t) + V_Q + R_b(I_Q - i(t)) \tag{3.26}$$

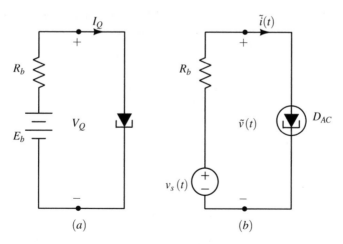

(a) (b)

Fig. 3.6 The circuit shown in Fig. 3.5a can be viewed in terms of (**a**) its DC equivalent circuit and (**b**) its AC equivalent circuit, where the diode characteristic has its origin at (V_Q, I_Q). D_{AC} denotes the diode with the origin shifted

---

[2]Figure 3.5b has been exaggerated for clarity.

Using the definitions of V_Q and I_Q from Eqs. (3.22) and (3.23), respectively, we can eliminate $v(t)$ and $i(t)$ in the equation above to obtain:

$$\tilde{v}(t) = v_s(t) - R_b\tilde{i}(t) \tag{3.27}$$

This equation can be represented by the circuit shown in Fig. 3.6b, where D_{AC} represents the AC behavior of the diode measured with respect to the operating point Q. To determine $(\tilde{v}(t), \tilde{i}(t))$, we substitute Eqs. (3.22) and (3.23) into (3.19) to obtain:

$$I_Q + \tilde{i}(t) = \hat{i}[V_Q + \tilde{v}(t)] \tag{3.28}$$

Up to now our analysis is general. At this juncture, let us assume that the amplitude of the sinusoidal voltage $v_s(t)$ is small, i.e., $V_M \ll E$. Thus the voltage $\tilde{v}(t)$ is "small" in comparison with V_Q. What follows below is **small-signal analysis**.

Taking the first two terms of the Taylor series expansion of $\hat{i}[V_Q + \tilde{v}(t)]$ about the points (V_Q, I_Q), we get:

$$
\begin{aligned}
i(t) &= I_Q + \tilde{i}(t) \\
&= \hat{i}[V_Q + \tilde{v}(t)] \\
&\approx \hat{i}[V_Q] + \left(\left.\frac{d\hat{i}}{dv}\right|_{V_Q}\right)\tilde{v}(t) \quad \forall t
\end{aligned} \tag{3.29}
$$

Geometrically (see Fig. 3.5c), the approximation carried out in Eq. (3.29) amounts to replacing the nonlinear diode characteristic by its linear approximation about the operating point Q. In other words:

$$\tilde{i}(t) \approx \left(\left.\frac{d\hat{i}}{dv}\right|_{V_Q}\right)\tilde{v}(t) \tag{3.30}$$

The term $\left.\frac{d\hat{i}}{dv}\right|_{V_Q}$ is the slope of the diode characteristic at the operating point Q; note that in the present case it is negative. Let us denote:

$$G \triangleq \left.\frac{d\hat{i}}{dv}\right|_{V_Q} \tag{3.31}$$

where G is negative. The quantity $\left.\frac{d\hat{i}}{dv}\right|_{V_Q}$ is called the **small-signal conductance** of the diode at the operating point Q. In other words, we simply have:

$$\tilde{i}(t) = G\tilde{v}(t) \tag{3.32}$$

Fig. 3.7 Small-signal equivalent circuit for the tunnel-diode

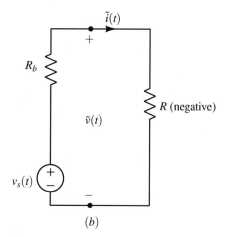

(b)

Hence D_{AC} from Fig. 3.6 is interpreted as a **negative** small-signal resistance $R = \frac{1}{G}$ as shown in Fig. 3.7. The small-signal equivalent circuit is a linear circuit because the two resistors are linear. Note that the resistance R is negative, thus we have a linear **active** resistor in the circuit. The solution can be obtained immediately from the small-signal equivalent circuit:

$$\tilde{i}(t) = \frac{V_m}{R_b + R} \cos \omega t \tag{3.33}$$

$$\tilde{v}(t) = \frac{R V_m}{R_b + R} \cos \omega t \tag{3.34}$$

Since R is negative, the factor $|R/(R_b + R)|$ can be made very large. From the equations above, we can define the **small-signal power gain** as:

$$\mathscr{P} \triangleq \left| \frac{\tilde{v}\tilde{i}}{v_s \tilde{i}} \right|$$

$$= \left| \frac{R}{R_b + R} \right| \tag{3.35}$$

We will now derive the small-signal (linearized) hybrid two-port representation of the npn bipolar transistor (recall Sect. 2.2.2). In other words, we are extending small-signal analysis above from a two-terminal element (diode) to a three-terminal element (transistor). The procedure is the same, the only difference being we will obtain a matrix for our small-signal hybrid parameter(s).

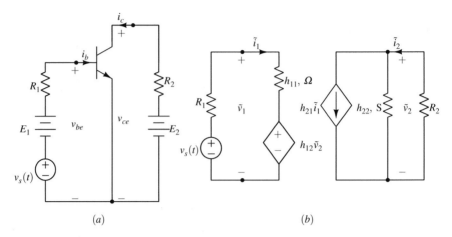

Fig. 3.8 (**a**) BJT Common-Emitter (CE) amplifier. (**b**) Small-signal model

Example 3.1.4 Derive the small-signal model for the CE amplifier shown in Fig. 3.8a.

Solution We will assume the input is a small-signal source, $v_s(t) = V_m \cos \omega t$. We will then see that the output voltage v_{ce} contains an amplified waveform at the same angular frequency ω.

The hybrid representation of the CE amplifier is repeated below:

$$v_{be} = \hat{v}_{be}(i_b, v_{ce}) \qquad (3.36)$$

$$i_c = \hat{i}_c(i_b, v_{ce}) \qquad (3.37)$$

Obviously, without the small-signal source v_s, the operating point (V_{beQ}, I_{bQ}), (V_{ceQ}, I_{cQ}) satisfies not only the above two equations but also Eqs. (3.40) and (3.41). They are written as follows:

$$V_{beQ} = \hat{v}_{be}\left(I_{bQ}, V_{ceQ}\right) \qquad (3.38)$$

$$I_{cQ} = \hat{i}_c\left(I_{bQ}, V_{ceQ}\right) \qquad (3.39)$$

$$V_{beQ} = E_1 - R_1 I_{bQ} \qquad (3.40)$$

$$V_{ceQ} = E_2 - R_2 I_{cQ} \qquad (3.41)$$

(continued)

Example 3.1.4 (continued)

When the signal source $v_s(t)$ is present in the circuit, we may express the four signal variables $v_{be}(t), i_b(t).v_{ce}(t), i_c(t)$ for all t as:

$$v_{be}(t) = V_{beQ} + \tilde{v}_1(t) \tag{3.42}$$

$$i_b(t) = I_{bQ} + \tilde{i}_1(t) \tag{3.43}$$

$$v_{ce}(t) = v_{ceQ} + \tilde{v}_2(t) \tag{3.44}$$

$$i_c(t) = I_{cQ} + \tilde{i}_2(t) \tag{3.45}$$

where $\tilde{v}_{1,2}(t), \tilde{i}_{1,2}(t)$ represent the "small" displacements of voltages and currents from the fixed operating point Q. At this juncture it remains only to determine these small-signal voltages and currents.

First substituting Eqs. (3.42) through (3.45) into Eqs. (3.36) and (3.37), we obtain:

$$v_{be}(t) = V_{beQ} + \tilde{v}_1(t)$$

$$= \hat{v}_{be}[I_{bQ} + \tilde{i}_1(t), V_{ceQ} + \tilde{v}_2(t)] \tag{3.46}$$

$$i_c(t) = I_{cQ} + \tilde{i}_2(t)$$

$$= \hat{i}_c[I_{bQ} + \tilde{i}_1(t), V_{ceQ} + \tilde{v}_2(t)] \tag{3.47}$$

No approximation has been introduced up to this step. Next we assume that the signal $v_s(t)$ is "small" and take the first two terms of the Taylor series expansions of $\hat{v}_{be}(\cdot, \cdot)$ and $\hat{i}_c(\cdot, \cdot)$ about the operating point Q. We obtain the following approximation:

$$v_{be}(t) \approx \hat{v}_{be}(I_{bQ}, V_{ceQ}) + \left.\frac{\partial \hat{v}_{be}}{\partial i_b}\right|_Q \tilde{i}_1(t) + \left.\frac{\partial \hat{v}_{be}}{\partial v_{ce}}\right|_Q \tilde{v}_2(t) \tag{3.48}$$

$$i_c(t) \approx \hat{i}_c(I_{bQ}, V_{ceQ}) + \left.\frac{\partial \hat{i}_c}{\partial i_b}\right|_Q \tilde{i}_1(t) + \left.\frac{\partial \hat{i}_c}{\partial v_{ce}}\right|_Q \tilde{v}_2(t) \tag{3.49}$$

(continued)

Example 3.1.4 (continued)
Comparing Eqs. (3.46) and (3.47) with Eqs. (3.48) and (3.49), and using
Eqs. (3.36) and (3.37), we obtain the following approximations:

$$\tilde{v}_1(t) \approx \left.\frac{\partial \hat{v}_{be}}{\partial i_b}\right|_Q \tilde{i}_1(t) + \left.\frac{\partial \hat{v}_{be}}{\partial v_{ce}}\right|_Q \tilde{v}_2(t) \tag{3.50}$$

$$\tilde{i}_2(t) \approx \left.\frac{\partial \hat{i}_c}{\partial i_b}\right|_Q \tilde{i}_1(t) + \left.\frac{\partial \hat{i}_c}{\partial v_{ce}}\right|_Q \tilde{v}_2(t) \tag{3.51}$$

The two equations can be viewed as hybrid equations relating the small
signals $\tilde{i}_1$ and $\tilde{v}_2$ to $\tilde{v}_1$ and $\tilde{i}_2$. Hence we have:

$$\begin{bmatrix} \tilde{v}_1 \\ \tilde{i}_2 \end{bmatrix} = \mathbf{H} \begin{bmatrix} \tilde{i}_1 \\ \tilde{v}_2 \end{bmatrix} \tag{3.52}$$

where:

$$\mathbf{H} = \begin{bmatrix} h_{11} & h_{12} \\ h_{21} & h_{22} \end{bmatrix} \triangleq \begin{bmatrix} \dfrac{\partial \hat{v}_{be}}{\partial i_b} & \dfrac{\partial \hat{v}_{be}}{\partial v_{ce}} \\[2ex] \dfrac{\partial \hat{i}_c}{\partial i_b} & \dfrac{\partial \hat{i}_c}{\partial v_{ce}} \end{bmatrix}_Q \tag{3.53}$$

Figure 3.8b shows the small-signal model for the CE amplifier. Exercise 3.1 asks
the reader to derive the small-signal voltage gain.

If we can build an amplifier with a single transistor, as Exercise 3.1 shows, why
then do we have more than one transistor in the schematic for μA741 in Fig. 2.25?
One answer to this question is the concept of **gain-bandwidth product**. A detailed
discussion is beyond the scope of this book, but conceptually, we need to ensure
that the gain of the amplifier is **ideally** maintained across a range of frequencies
(the **bandwidth** of the amplifier). In other words, the gain-bandwidth product is
a constant. In a nutshell, we need the transistors shown in Fig. 2.25 to ensure a
constant gain-bandwidth product. Nevertheless, the fact is: we are able to achieve a
constant gain across a large bandwidth with so **few** transistors.

In Sect. 4.2.2.3, we will extend small signal analysis to nonlinear dynamic
networks.

3.2   Matrix Formulation of Kirchhoff's Laws

So far, we have been using KCL and KVL to describe simple circuits. For more complicated circuits, other formal circuit techniques of circuit analysis exist. These methods will help us systematically derive the circuit equations. Before discussing these methods, we need to expand upon the graph theoretic concepts and matrix formulation of Kirchhoff's laws, that were introduced in Chap. 1. This will help us in the discussion of nodal and tableau analysis techniques later in this chapter.

3.2.1   Cut Sets, Hinged Graphs, and Linear Independence

Definition 3.1   Given a network graph $\mathscr{G}$, a **cut set** is a set of branches $\mathscr{C}$ of $\mathscr{G}$ having the property that if we "cut" (as if with scissors) each branch in the set once, $\mathscr{G}$ gets separated into two disconnected **subgraphs** $\mathscr{G}_1$ and $\mathscr{G}_2$, and if we leave any one branch of the set uncut, $\mathscr{G}$ remains connected in one piece by that branch.

For instance, consider the digraph shown in Fig. 3.9. The set of branches $\{3, 4, 8, 6\}$ is a cut set, since cutting these branches once separates the graph into two subgraphs. Similarly, the set $\{3, 4, 8, 5, 7\}$ is also a cut set.

Some remarks about cut sets:

1. Any cut set creates a partition of the set of nodes in the graph into two subsets
2. To any cut set corresponds a gaussian surface (recall Definition 1.9) which cuts precisely the same branches
3. Similarly, to any gaussian surface corresponds either one cut set or a union of cut sets. For example, $\mathscr{S}_1$ in Fig. 3.10.
4. To each cut set we can define arbitrarily a **reference direction**, as shown by the arrows attached to the cut sets in Fig. 3.10.

Fig. 3.9 A digraph associated with a bridge circuit

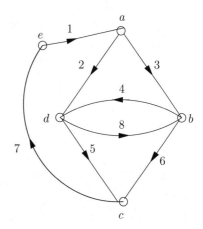

Fig. 3.10 Digraph
illustrating cut sets

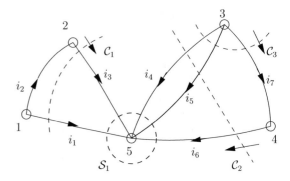

Fig. 3.11 Digraph for
Example 3.2.1

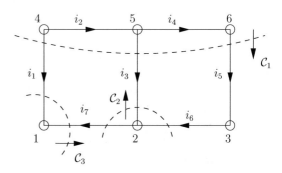

Definition 3.2 KCL (Cut Set Law): For all lumped circuits, for all time t, the
algebraic sum of the currents associated with any cut set is equal to zero.

Example 3.2.1 The digraph in Fig. 3.11 shows some example cut sets. Write
the KCL equation associated with those cut sets.

Solution Cut set $\mathscr{C}_1$ consists of the set of branches $\{\beta_1, \beta_3, \beta_5\}$. Since all
currents are in the direction of the cut set, the KCL associated with $\mathscr{C}_1$ is:

$$i_1 + i_3 + i_5 = 0 \tag{3.54}$$

For $\mathscr{C}_2$, we have:

$$i_7 - i_3 - i_6 = 0 \tag{3.55}$$

$-i_3, -i_6$ are because both those currents are going in, while the cut set
direction is pointing out. Similarly, for $\mathscr{C}_3$, we have:

$$-i_1 - i_7 = 0 \tag{3.56}$$

Combined with the KCL definitions from Chap. 1, we have learned three forms of KCL, namely, in terms of (1) gaussian surfaces, (2) nodes, and (3) cut sets.

Theorem 3.1 (KCL Equivalence Theorem) *The three forms of KCL are equivalent: (1) KCL gaussian surface $\Leftrightarrow$ (2) KCL node law $\Leftrightarrow$ (3) KCL cut sets*

Proof We will only prove the implication as the other direction can be proved in a similar manner.

- (1) $\Rightarrow$ (2) Simply use the gaussian surface that surrounds only the node in question. For example, consider node 5 in Fig. 3.10. For the gaussian surface $\mathscr{S}_1$, KCL applied to $\mathscr{S}_1$ is identical to KCL applied to node 5:

$$i_1 - i_3 - i_4 - i_5 - i_6 = 0 \tag{3.57}$$

- (2) $\Rightarrow$ (3) Any cut set partitions the set of nodes into two subsets. Writing the KCL equation for each node in such a subset and adding the results, we obtain the cut set equation, except for maybe a -1 factor. For example, consider the cut set $\mathscr{C}_2$ in Fig. 3.10. If we add the KCL equations applied to nodes 3 and 4, we obtain:

$$i_4 + i_5 + i_6 = 0 \tag{3.58}$$

Note that i_7 cancels out in the addition, resulting in the KCL cut set equation for $\mathscr{C}_2$.

- (3) $\Rightarrow$ (1) It is easy to demonstrate that the set of branches cut by a gaussian surface is either a cut set or a disjoint union of cut sets. So given any gaussian surface, let us write the KCL equation for each of these cut sets; then adding or subtracting these equations, we obtain the KCL equation for the gaussian surface. For example, consider gaussian surface $\mathscr{S}_1$ in Fig. 3.10. It is the union of the cut set $\{\beta_1, \beta_3\}$ and cut set $\{\beta_4, \beta_5, \beta_6\}$ whose equations are, respectively,

$$-i_1 - i_3 = 0 \tag{3.59}$$

$$-i_4 - i_5 - i_6 = 0 \tag{3.60}$$

Adding the two equations above, we get:

$$-i_1 - i_3 - i_4 - i_5 - i_6 = 0 \tag{3.61}$$

which is the KCL equation for the gaussian surface $\mathscr{S}_1$. $\qquad\square$

Up to now, we have assumed the circuit is connected. But, recall from our discussion of transformers in Sects. 2.2.1.3 and 2.2.3.1 that a circuit with a physical transformer is not connected. It turns out that we can easily take care of this situation. We first generalize the element graph representation from one-port to a two-port, by using two branches and four nodes for its element graph as shown in Fig. 3.12.

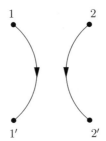

Fig. 3.12 The element graph of a two-port

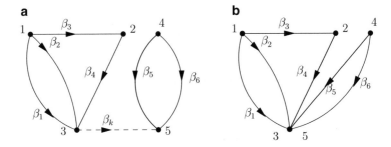

Fig. 3.13 (**a**) Connecting nodes 3 and 5 by a branch k. (**b**) Soldering together nodes 3 and 5 to obtain a hinged graph

Next, we need to understand the element graph of a two-port consists of two branches which are not connected, because it signifies that port voltages or port currents at different ports are not related because of connections but rather are **coupled** because of physical phenomena within the element. For example, recall from Sect. 2.2.3.1 that physical transformer port voltages are coupled magnetically via the flux linkages among the various windings.

To avoid an unconnected digraph (circuit graph), we can tie together the two separate ports of a digraph at two arbitrary nodes by a branch. This is illustrated in Fig. 3.13a where nodes 3 and 5 are tied together by a branch k. This connection does not change any branch voltage or current in the original circuit. This is easily seen because, by using KCL with a gaussian surface which encloses one of the separate parts of the graph and which cuts branch k, the current i_k is zero. If $i_k = 0$, it amounts to an open circuit or no connection; thus we have not changed the behavior of the circuit. Next, since voltages are measured between nodes, we choose a ground node for the separate parts. If we choose nodes 3 and 5 as the ground nodes for the separate parts, we may "solder" together node 3 and node 5 as shown in Fig. 3.13b to make them the common ground node. The graph so obtained is called a **hinged graph**. With the introduction of the concept of a hinged graph, we have generalized our treatment so far to include two-ports and multi-ports, that is, we can always assume without loss of generality that any lumped circuit and its associated digraph are connected.

We now have all the graph theoretic concepts that we need for this chapter. But, before discussing independent KCL and KVL equations, we need to discuss the concept of **linear independence**.

Consider a set of m linear algebraic equations in n unknowns. For $j = 1, 2, \cdots, m$

$$f_j(x_1, x_2, \ldots, x_n) \triangleq \alpha_{j1}x_1 + \alpha_{j2}x_2 + \cdots + \alpha_{jn}x_n = 0 \tag{3.62}$$

where the α_{jk}'s are real or complex numbers. It is important to decide whether or not each equation brings new information not contained in the others; equivalently, it is important to decide whether the equations are linearly independent. These m equations are said to be **linearly dependent** iff there are constants $k_1, k_2, \cdots, k_m$ and **not all zero** such that:

$$\sum_{j=1}^{m} k_j f_j(x_1, x_2, \ldots, x_n) = 0 \quad \forall\, x_1, x_2, \ldots, x_n \tag{3.63}$$

Clearly if these m equations are linearly dependent, then at least one equation may be written as a linear combination of the others; in other words, that equation repeats the information contained in the others.

It is crucial to note that the LHS of Eq. (3.63) must be zero **for all** values of $x_1, x_2, \ldots, x_n$.

Example 3.2.2 Are the equations below ($m = 3$ and $n = 4$) linearly dependent?

$$x_1 - x_2 + x_3 + 3x_4 = 0$$
$$2x_1 + 3x_2 - x_3 - 4x_4 = 0$$
$$-4x_1 - 11x_2 + 5x_3 + 18x_4 = 0$$

Solution Direct calculation shows that with $k_1 = 2$, $k_2 = -3$ and $k_3 = -1$ the condition for Eq. (3.63) holds; in other words, these three equations are linearly dependent.

A set of m linear algebraic equations is said to be **linearly independent** iff it is not linearly dependent. In practice, we use gaussian elimination to decide whether or not a given set of linear equations is linearly dependent.

3.2.2 Independent KCL Equations

For a given circuit, we can write KCL equations by the node law or the cut set law, or using gaussian surfaces. How many of the KCL equations are linearly independent and how to write a complete set that contains all the necessary information as far as KCL is concerned are the subjects of this subsection. We will give a systematic treatment by means of the digraph of the circuit under consideration: in particular, a list of nodes, a list of branches, and for each branch the specification of the node that the branch leaves and enters. This is done by the **incidence matrix** $\mathbf{A}_a$ of the digraph.

Let the digraph $\mathscr{G}$ have n nodes and b branches, then $\mathbf{A}_a$ has n rows—one row for each node—and b columns—one column for each branch. To see how the matrix is built up consider the four-node six-branch digraph shown in Fig. 3.14. Let us write the KCL equations for each node:

$$
\begin{aligned}
i_1 + i_2 \qquad\qquad\quad - i_6 &= 0 \\
-i_1 \quad\ - i_3 + i_4 \qquad\qquad &= 0 \\
- i_2 + i_3 \qquad + i_5 \quad &= 0 \\
- i_4 - i_5 + i_6 &= 0
\end{aligned}
\tag{3.64}
$$

Fig. 3.14 A digraph with four nodes and six branches

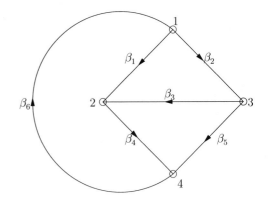

In matrix form, it reads:

$$
\begin{array}{c}
\\
\text{Node 1}\rightarrow \\
\text{Node 2}\rightarrow \\
\text{Node 3}\rightarrow \\
\text{Node 4}\rightarrow
\end{array}
\overset{\text{Branch 1}\qquad\qquad\qquad\text{Branch 6}}{
\begin{bmatrix}
1 & 1 & 0 & 0 & 0 & -1 \\
-1 & 0 & -1 & 1 & 0 & 0 \\
0 & -1 & 1 & 0 & 1 & 0 \\
0 & 0 & 0 & -1 & -1 & 1
\end{bmatrix}}
\begin{pmatrix}
i_1 \\ i_2 \\ i_3 \\ i_4 \\ i_5 \\ i_6
\end{pmatrix}
=
\begin{pmatrix}
0 \\ 0 \\ 0 \\ 0
\end{pmatrix}
\qquad (3.65)
$$

Since each row corresponds to a node and each column corresponds to a branch, we have the 4×6 incidence matrix $\mathbf{A}_a$. For example, for node 4 we have i_4, i_5 coming in and i_6 going out and hence the 4th row in the matrix has two -1 s and one $+1$. Similarly, branch β_1 that connects node 1 to 2 has one $+1$ and one -1 in column 1.

In general, for any n-node b-branch **connected** digraph $\mathcal{G}$ which does **not** contain **self-loops**,[3] the matrix $\mathbf{A}_a$ is specified as follows: For $i = 1, 2, \cdots, n$ and $k = 1, 2, \cdots b$:

$$
a_{ik} =
\begin{cases}
+1 & \text{if branch } k \text{ leaves node } i \\
-1 & \text{if branch } k \text{ enters node } i \\
0 & \text{if branch } k \text{ does not touch node } i
\end{cases}
\qquad (3.66)
$$

and the node n node equations of $\mathcal{G}$ read:

$$
\mathbf{A}_a \mathbf{i} = \mathbf{0} \qquad (3.67)
$$

where $\mathbf{i} = (i_1, i_2, \cdots, i_b)^T$ is called the **branch current vector**.

Example 3.2.3 Is the incidence matrix in Eq. (3.65) full row rank?

Solution Equivalently, the question posed is asking whether the KCL equations corresponding to the incidence matrix are linearly dependent or independent. We could transform the incidence matrix to row-echelon form. Instead, simple observation shows that with $k_1 = k_2 = k_3 = k_4 = 1$, the condition for Eq. (3.63) holds; in other words, the incidence matrix is **not** full row rank.

Example 3.2.3 shows that each column of $\mathbf{A}_a$ has precisely a single $+1$ and a single -1; consequently, if we add together n equations in Eq. (3.67), all the

---

[3]A self-loop contains precisely one node and one branch, they are not loops according to Definition 1.6 (of a loop).

variables $i_1, i_2, \cdots, i_b$ cancel out; equivalently, the n KCL equations are linearly dependent.

But, suppose that for a **connected**[4] digraph $\mathscr{G}$ we choose a ground node and we throw away the corresponding KCL equation, then the remaining $n - 1$ equations are linearly independent. This is the defining property of this subsection, hence we state it formally as a theorem and prove it:

Theorem 3.2 (Independence Property of KCL Equations) *For any connected digraph $\mathscr{G}$ with n nodes, the KCL equations for **any** $n - 1$ of these nodes form a set of $n - 1$ **linearly independent equations**.*

Proof We prove it by contradiction. Suppose that the first k of these $n - 1$ equations are linearly dependent. More precisely, there are k real constants $\gamma_1, \gamma_2, \cdots, \gamma_k$ not all zero such that:

$$\sum_{j=1}^{k} \gamma_j f_j(i_1, i_2, \ldots, i_n) = 0 \quad \forall i_1, i_2, \ldots, i_n \qquad (3.68)$$

Consider the two sets of nodes in $\mathscr{G}$, namely, the set which corresponds to the k equations and that of the remaining nodes. Since the digraph is connected, there is at least one branch which connects a node in the first set to a node in the second set. Clearly the current in that branch appears only once in the first k node equations, hence that current cannot cancel out in the sum of Eq. (3.68). This contradiction shows that for any $k \leq n-1$, it is not the case that a subset of k of the KCL equations is linearly independent. That is, these $n - 1$ equations are linearly independent. □

If in $\mathbf{A}_a$, the incidence matrix of the connected digraph $\mathscr{G}$, we delete the row corresponding to the ground node, we obtain the **reduced incidence matrix A** which is of dimension $(n - 1) \times b$. The corresponding **linearly independent** KCL equations read:

$$\mathbf{Ai} = \mathbf{0} \qquad (3.69)$$

As a consequence of the independence property proved in Theorem 3.2, we may equivalently state that the matrix $\mathbf{A}$ is full rank.

---

[4]If a digraph is not connected, there are two simple solutions to the problem: one approach would be to treat each graph separately. In this case, each part would have its own incidence matrix and ground node. The other approach would be to use a hinged graph, as described in Sect. 3.2.1. We will use both approaches in this book.

3.2.3 Independent KVL Equations

Similarly, to write a set of linearly independent KVL equations in a systematic way is of crucial importance. Let us write the KVL equations for the four-node six-branch digraph of Fig. 3.14. Using associated reference directions and choosing node 4 as the ground node, we obtain:

$$v_1 = e_1 - e_2$$
$$v_2 = e_1 - e_3$$
$$v_3 = -e_2 + e_3$$
$$v_4 = e_2$$
$$v_5 = e_3$$
$$v_6 = -e_1 \tag{3.70}$$

or in matrix form:

$$\mathbf{v} = \mathbf{M}\mathbf{e} \tag{3.71}$$

where $\mathbf{v} = (v_1, v_2, \ldots, v_b)^T$ is the **branch voltage vector**, $\mathbf{e} = (e_1, e_2, \ldots, e_{n-1})^T$ is the **node-to-ground voltage vector**, and $\mathbf{M}$ is a $b \times (n-1)$ matrix. Thinking in terms of KVL, we see that for $k = 1, 2, \ldots, b$ and $i = 1, 2, \ldots, n-1$:

$$m_{ki} = \begin{cases} +1 & \text{if branch } k \text{ \textbf{leaves} node } i \\ -1 & \text{if branch } k \text{ \textbf{enters} node } i \\ 0 & \text{if branch } k \text{ does not touch node } i \end{cases} \tag{3.72}$$

Comparing Eq. (3.72) with (3.66) we conclude that:

$$\mathbf{M} = \mathbf{A}^T \tag{3.73}$$

and more usefully, KVL is expressed by the equation:

$$\mathbf{v} = \mathbf{A}^T \mathbf{e} \tag{3.74}$$

With a connected digraph $\mathscr{G}$, $\mathbf{A}$ has $n-1$ linearly independent rows (full row rank) and consequently $\mathbf{A}^T$ has $n-1$ linearly independent columns (full column rank).

Thus, to summarize, in order to obtain linearly independent KCL and KVL equations from a digraph representation of a circuit:

1. We choose current reference directions

2. We choose a ground node and define the reduced incidence matrix $\mathbf{A}$
3. We write KCL as $\mathbf{Ai} = \mathbf{0}$
4. We then use associated reference directions (or passive sign convention, recall Definition 1.2) to find that KVL reads $\mathbf{v} = \mathbf{A}^T \mathbf{e}$.

Note that we are assuming we use the same ground node for writing KCL and KVL.

3.2.4 A Proof of Tellegen's Theorem

We can now state and prove Tellegen's theorem.

Theorem 3.3 (Tellegen's Theorem) *Consider an arbitrary circuit. Let the associated digraph $\mathscr{G}$ have b branches. Using passive sign convention, let $\mathbf{v} = (v_1, v_2, \ldots, v_b)^T$ be **any** set of branch voltages satisfying KVL for $\mathscr{G}$ and let $\mathbf{i} = (i_1, i_2, \ldots, i_b)^T$ be **any** set of branch currents satisfying KCL for $\mathscr{G}$. Then:*

$$\sum_{k=1}^{b} v_k i_k = 0 \tag{3.75}$$

Equivalently:

$$\mathbf{v}^T \mathbf{i} = 0 \tag{3.76}$$

Proof For a connected digraph $\mathscr{G}$, choose a ground node; hence, a reduced matrix $\mathbf{A}$ is defined unambiguously. Since $\mathbf{i}$ satisfies KCL, we have:

$$\mathbf{Ai} = \mathbf{0} \tag{3.77}$$

Since $\mathbf{v}$ satisfies KVL and since we use associated reference directions, for some node-to-ground voltages $\mathbf{e}$, we have:

$$\mathbf{v} = \mathbf{A}^T \mathbf{e} \tag{3.78}$$

Using the two equations above, we successively obtain:

$$\mathbf{v}^T \mathbf{i} = (\mathbf{A}^T \mathbf{e})^T \mathbf{i}$$
$$= \mathbf{e}^T (\mathbf{A}^T)^T \mathbf{i}$$
$$= \mathbf{e}^T (\mathbf{Ai})$$
$$= 0 \tag{3.79}$$

$\square$

Note how the proof essentially uses our discussion from Sect. 1.6.1. Now the idea from Sect. 1.6.1 should be very clear: $\mathbf{v}$ and $\mathbf{i}$ in the theorem need not bear any relation to each other: $\mathbf{v}$ must only satisfy KVL and $\mathbf{i}$ must only satisfy KCL (using associated reference directions). We will use Tellegen's theorem to prove some very general results for nonlinear resistive networks in Sect. 3.7.

3.2.5 The Relation Between Kirchhoff's Laws and Tellegen's Theorem

In circuit theory, there are two fundamental postulates: KCL and KVL. We have proved that KCL and KVL imply Tellegen's theorem. It is interesting to note that any one of Kirchhoff's laws together with Tellegen's theorem implies the other. More precisely, we have the following theorem:

Theorem 3.4 (Tellegen's Theorem and Kirchhoff's Laws)

1. If, for all $\mathbf{v}$ satisfying KVL, $\mathbf{v}^T\mathbf{i} = 0$, then $\mathbf{i}$ satisfies KCL.
2. If, for all $\mathbf{i}$ satisfying KCL, $\mathbf{v}^T\mathbf{i} = 0$, then $\mathbf{v}$ satisfies KVL.

Proof For 1: Since $\mathbf{v}$ satisfies KVL, we know that $\mathbf{v} = \mathbf{A}^T\mathbf{e}$ for all $\mathbf{e}$. But, given that Tellegen's theorem is also satisfied, we have:

$$\mathbf{v}^T\mathbf{i} = \mathbf{e}^T(\mathbf{Ai}) = 0 \qquad (3.80)$$

Since $\mathbf{e}$ is an arbitrary node-to-ground voltage, the last equality implies $\mathbf{Ai} = \mathbf{0}$, that is, $\mathbf{i}$ satisfies KCL.

For 2: Let $\mathscr{L}$ be an arbitrary loop in the graph $\mathscr{G}$. Consider the $\mathbf{i}$ obtained by assigning zero current to all branches of $\mathscr{G}$ except for those of loop $\mathscr{L}$; depending on whether the reference direction of branch j in loop $\mathscr{L}$ agrees with that of loop $\mathscr{L}$, we assign i_j to be 1 A or -1 A. The resulting $\mathbf{i}$ satisfies KCL at all nodes of $\mathscr{G}$. Tellegen's theorem applied only to the branches in loop $\mathscr{L}$ gives:

$$\sum \pm v_j = 0 \qquad (3.81)$$

Thus the algebraic sum of branch voltages around loop $\mathscr{L}$ is zero, i.e., KVL holds for loop $\mathscr{L}$. Since $\mathscr{L}$ is arbitrary, we have shown that KVL holds for all loops of $\mathscr{G}$.
□

3.3 An Introduction to General Resistive Circuit Analysis

We can now embark on a more general and definitive study of resistive circuits. Our aim for the rest of this chapter is to develop general methods of analysis, for both linear and nonlinear resistive circuits, and to derive general properties of such circuits. A common theme would be to start with (an example of) the linear case because equations for linear circuits can almost be derived by "inspection."

The term **resistive circuit** applies to all circuits containing **two-terminal resistors, multi-terminal resistors, multi-port resistors** and **independent voltage** and **current sources**. Common circuit elements such as ideal transformers, rotators, gyrators, controlled sources, transistors and opamps modeled by resistive circuits etc. **are all included**. To avoid clutter, all of these garden variety circuit elements will be lumped under the umbrella "multi-terminal and multi-port resistors". However, independent sources will always be singled out separately, because, as will be clear shortly, they play a fundamentally different role.

The importance of resistive circuits cannot be understated. The analysis of many general nonresistive circuits reduces to the analysis of the associated resistive circuit. Secondly, many computer algorithms for simulating dynamic circuits require at each step the analysis of a resistive circuit.

Recall that a **physical** circuit is an interconnection of **real** electric devices. For purposes of analysis and design, each electric device is replaced by a **device model** made of ideal circuit elements[5] (e.g., ideal diodes, batteries, linear and nonlinear resistors, controlled sources, etc.). The interconnection of these models gives the electric circuit. Since, the detailed but important study of **device modeling** is beyond the scope of this book, our point of departure for analysis would be a circuit. Whether the circuit arises from models of physical devices, or from the figment of one's imagination is irrelevant. In fact, it is often through the **introduction of hypothetical**, and **sometimes pathological circuits**, that one gains an **in-depth understanding of this subject**.

A few words concerning some general technical terms to be used throughout this book. A resistive circuit is said to be linear iff, after settings all independent sources to zero, it contains only **linear** (recall Exercise 1.9 for the superposition definition of linearity) two-terminal, multi-terminal, and/or multi-port resistors. A resistive circuit is said to be nonlinear iff it contains at least one nonlinear resistor besides independent sources.

Finally, we need to caution the reader to "not lose sight of the forest for its trees." In other words, one should not be so consumed by the systematic techniques that we lose total insight into the circuit at hand. After all, only a computer circuit simulation program "blindly" applies the techniques, without any insight.

---

[5]Of course, **all circuit elements are ideal**. We will, nevertheless, occasionally throw in the word "ideal" to remind the reader that "nonphysical" answers (e.g., the Schmitt trigger VTC) are quite possible and even expected. When they do occur, the culprit is not the theory, but the model. Such situations can only be rectified by returning to the drawing board to come up with a more detailed circuit model. Again, in the case of the Schmitt trigger, we will account for physical parasitics to explain the observed behavior.

3.4 Nodal Analysis for Resistive Circuits

The simplest method for analyzing a resistive circuit is to solve for its **node-to-ground** voltages. Once these node voltages have been calculated, we can solve for the branch voltages trivially via KVL: $\mathbf{v} = \mathbf{A}^T \mathbf{e}$. They in turn can be used to calculate the branch currents, provided all the elements in the circuit other than current sources are voltage-controlled. In this section, we will consider only the subclass of resistive circuits which are amenable to this common analysis method, henceforth called **node analysis**. The goal would be to determine the corresponding **node equation** for the circuit in question.

For simple resistive circuits, the node equation can be formulated almost by inspection, as illustrated in the following example.

Example 3.4.1 Determine the node equation for the circuit in Fig. 3.15.

Solution The circuit shown in Fig. 3.15 contains only linear (two-terminal) resistors and independent current sources. Choosing (arbitrarily) node 4 as the ground node, each branch current can be expressed in terms of at most two node voltages, simply by using Ohm's law since we have linear resistors. Thus:

$$i_1 = G_1 v_1 = G_1 e_1 \qquad i_4 = G_4 v_4 = G_4(e_1 - e_2)$$

$$i_2 = G_2 v_2 = G_2(e_2 - e_1) \quad i_5 = G_5 v_5 = G_5(e_3 - e_2)$$

$$i_3 = G_3 v_3 = G_3(-e_2) \qquad i_6 = G_6 v_6 = G_6 e_3 \qquad (3.82)$$

It follows from Sect. 3.2.2 that we can write three linearly independent KCL equations in terms of e_1, e_2, and e_3, namely:

Node 1 : $G_1 e_1 - G_2(e_2 - e_1) + G_4(e_1 - e_2) = i_{s1}(t)$

Node 2 : $G_2(e_2 - e_1) - G_3(-e_2) - G_4(e_1 - e_2) - G_5(e_3 - e_2) = -i_{s3}(t)$

Node 3 : $G_5(e_3 - e_2) + G_6 e_3 = i_{s3}(t) - i_{s2}(t) \qquad (3.83)$

Recasting in matrix form:

$$\begin{bmatrix} (G_1 + G_2 + G_4) & -(G_2 + G_4) & 0 \\ -(G_2 + G_4) & (G_2 + G_3 + G_4 + G_5) & -G_5 \\ 0 & -G_5 & (G_5 + G_6) \end{bmatrix} \begin{bmatrix} e_1 \\ e_2 \\ e_3 \end{bmatrix} = \begin{bmatrix} i_{s1}(t) \\ -i_{s2}(t) \\ i_{s3}(t) - i_{s2}(t) \end{bmatrix}$$

$$(3.84)$$

(continued)

Example 3.4.1 (continued)
In other words, we have:

$$\mathbf{Y}_n \mathbf{e} = \mathbf{i}_s(t) \qquad (3.85)$$

henceforth called the **node equation**. $\mathbf{Y}_n$ is a square matrix called the **node-admittance matrix** and $\mathbf{i}_s(t)$ is called the equivalent source vector.

We will shortly show that a large class of linear resistive circuits is described by a form like Eq. (3.85). But first, an inspection of Fig. 3.15 and Eq. (3.84) reveals the following properties.

We will prove the properties in Table 3.1, once we obtain the node-admittance matrix in terms of the reduced incidence matrix $\mathbf{A}$, which we will do so below.

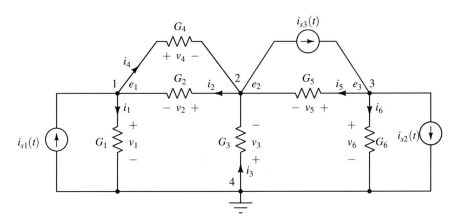

Fig. 3.15 Circuit for example 3.4.1. Here, G_j denotes the conductance in S for the jth resistor

Table 3.1 Properties of Eq. (3.85)

For any circuit made of linear two-terminal resistors and independent sources
1. The kth diagonal element of $\mathbf{Y}_n$ is equal to the sum of all conductances attached to node k
2. The jkth off-diagonal element of $\mathbf{Y}_n$ is equal to the **negative** of the sum of all conductances between node j and node k
3. The matrix $\mathbf{Y}_n$ is **symmetric**: $\mathbf{Y}_n = \mathbf{Y}_n^T$
4. The kth element of $\mathbf{i}_s(t)$ is equal to the algebraic sum of currents of all independent current sources entering node k

3.4.1 Formulation in Terms of Reduced Incidence Matrix

Let $\mathcal{N}$ denote any connected linear resistive circuit containing only two-terminal, multi-terminal and/or multi-port linear voltage-controlled resistors, and independent current sources. For example, $\mathcal{N}$ may contain gyrators because they are defined by a voltage-controlled linear equation, Eq. (2.167). On the other hand, $\mathcal{N}$ may not contain ideal transformers because it is not voltage-controlled, that is, it is impossible to solve for i_1, i_2 from the defining Eq. (2.37) in terms of only v_1, v_2. Controlled sources other than VCCS are also disallowed for the same reason.[6] Note that although independent voltage sources are not allowed in our present formulation, they can be included later through equivalent circuit transformations.

If the terminals and/or ports of all circuit elements which are not independent current sources are labeled consecutively, and if $\mathbf{v} = (v_1, v_2, \ldots, v_b)^T$ and $\mathbf{i} = (i_1, i_2, \ldots, i_n)^T$ denote the respective branch voltage and branch current vectors, then $\mathcal{N}$ is precisely the class where $\mathbf{i}$ can be described as a linear function of $\mathbf{v}$; namely,

$$
\begin{bmatrix} i_1 \\ i_2 \\ . \\ . \\ . \\ i_b \end{bmatrix} = \begin{bmatrix} y_{11} & y_{12} & \cdots & y_{1b} \\ y_{21} & y_{22} & \cdots & y_{2b} \\ . & . & \cdots & . \\ . & . & \cdots & . \\ . & . & \cdots & . \\ y_{b1} & y_{b2} & \cdots & y_{bb} \end{bmatrix} \begin{bmatrix} v_1 \\ v_2 \\ . \\ . \\ . \\ v_b \end{bmatrix} \tag{3.86}
$$

or simply,

$$
\mathbf{i} = \mathbf{Y}_b \mathbf{v} \tag{3.87}
$$

Equation (3.87) is the **branch equation**, where $\mathbf{Y}_b$ is called the **branch admittance matrix**. In general, $\mathbf{Y}_b$ is a $b \times b$ nonsymmetric and nondiagonal real matrix, where b is the number of branches, excluding the independent current sources, in the associated digraph.

We have deliberately left out the independent current sources because they can be easily accounted for separately. In particular, the contribution of current sources can be represented by a single vector:

$$
\mathbf{i}_s(t) = \begin{bmatrix} i_{s1}(t) & i_{s2}(t) & \cdots & i_{s(n-1)}(t) \end{bmatrix}^T \tag{3.88}
$$

where $1_{sk}(t)$ denotes the algebraic sum of currents of all independent current sources entering node k, $k = 1, 2, \ldots, n - 1$ and n denotes the number of nodes in the

---

[6]We will however be able to use tableau analysis from Sect. 3.5 to analyze circuits with any resistive element.

connected circuit $\mathcal{N}$. To avoid violating KCL, it is necessary to assume that no cut sets are made exclusively of independent current sources.

Equivalently, the digraph associated with the reduced circuit obtained by open-circuiting all independent current sources is connected. Let $\mathbf{A}$ denote the reduced incidence matrix of this connected digraph. It follows that:

$$\mathbf{Ai} = \mathbf{i}_s(t) \tag{3.89}$$

constitutes a system of $n - 1$ linearly independent KCL equations. It is important to note that the KCL Eq. (3.89) differs from the usual form $(\mathbf{Ai} = \mathbf{0})$ because here, the reduced incidence matrix $\mathbf{A}$ pertains to the reduced digraph obtained by open-circuiting all independent current sources from the digraph associated with the circuit.

Substituting the branch Eq. (3.87) in place of $\mathbf{i}$ in Eq. (3.89), we obtain:

$$\mathbf{AY}_b\mathbf{v} = \mathbf{i}_s(t) \tag{3.90}$$

Rewriting the branch voltage $\mathbf{v}$ in terms of the node voltage $\mathbf{e}$ via KVL (Eq. (3.74)), we get:

$$(\mathbf{AY}_b\mathbf{A}^T)\mathbf{e} = \mathbf{i}_s(t) \tag{3.91}$$

Comparing Eqs. (3.85) and (3.91), we have derived the **node-admittance matrix**:

$$\mathbf{Y}_n = \mathbf{AY}_b\mathbf{A}^T \tag{3.92}$$

We have hence derived the following general result, our first systematic circuit analysis technique:

Nodal Analysis for Linear Resistive Circuits

For any connected circuit containing two-terminal, multi-terminal, and/or multi-port linear voltage-controlled resistors and independent current sources which do not form cut sets, the node equation is given explicitly by:

$$\mathbf{Y}_n(t)\mathbf{e}(t) = \mathbf{i}_s(t) \tag{3.93}$$

$\mathbf{Y}_n(t) \triangleq \mathbf{AY}_b(t)\mathbf{A}^T$, $\mathbf{Y}_n(t)$ is the node-admittance matrix, $\mathbf{i}_s(t)$ denotes the equivalent source vector whose kth element $i_{sk}(t)$ is equal to the algebraic sum of the current of all independent current sources **entering** node k, and $\mathbf{A}$ denotes the reduced incidence matrix of the digraph associated with the **reduced** circuit obtained by deleting all **independent** current sources.

The astute reader would have noticed that we have defined Eq. (3.93) for time-varying elements as well. This is fine because if branch k is a time-varying resistor described by $i_k(t) = G_k(t)v_k(t)$, then $y_{kk} = G_k(t)$.

Note that the dimension of A and $Y_b(t)$ are $(n-1) \times b$ and $b \times b$, respectively, where n is the number of nodes in the circuit, and b is the number of branches in the digraph associated with the reduced circuit. Consequently, the dimensions of the node-admittance matrix $\mathbf{Y}_n(t)$ is $(n-1) \times (n-1)$.

In other words, the node Eq. (3.93) always contains $n-1$ linear equations in terms of $n-1$ node voltages $e_1, e_2, \ldots, e_{n-1}$.

To find the solution of the circuit, we simply solve Eq. (3.93) at each instant of time t by any convenient method, say gaussian elimination. If n is very large, say $n > 100$, and if the matrix $\mathbf{Y}_n(t)$ contains only a small percentage of nonzero entries as is typical in practice ($\mathbf{Y}_n(t)$ is said to be sparse), there exist specially efficient computer algorithms for solving the equation. If the circuit is time-invariant and contains only DC current sources, then $\mathbf{Y}_n(t)$ is a constant matrix and $i_s(t)$ is a constant vector. In this case, Eq. (3.93) need to be solved only once.

Unlike several other methods of analysis (e.g., tableau analysis, modified nodal analysis) to be studied later, the number of equations to be solved in node analysis does not depend on the number of circuit elements. Hence for a 100-element circuit containing only 10 nodes, we only need to solve 9 equations.

Once $\mathbf{e}(t)$ has been found, the branch voltages can be calculated by substitution into the time-varying case for Eq. (3.74): $\mathbf{v}(t) = \mathbf{A}^T \mathbf{e}(t)$—and the branch currents can be calculated by substitution into the time-varying case for branch Eq. (3.87): $\mathbf{i}(t) = \mathbf{Y}_b(t)\mathbf{v}(t)$.

Example 3.4.2 Prove the properties in Table 3.1.

Solution Let us begin by expanding Eq. (3.92), for a circuit with three nodes and three resistors:

$$\mathbf{Y_n} = \mathbf{AY}_b\mathbf{A}^T$$

$$= \begin{bmatrix} a_{11} & a_{12} & a_{13} \\ a_{21} & a_{22} & a_{23} \end{bmatrix} \begin{bmatrix} G_1 & 0 & 0 \\ 0 & G_2 & 0 \\ 0 & 0 & G_3 \end{bmatrix} \begin{bmatrix} a_{11} & a_{21} \\ a_{12} & a_{22} \\ a_{13} & a_{23} \end{bmatrix}$$

$$= \begin{bmatrix} (a_{11}a_{11}G_1 + a_{12}a_{12}G_2 + a_{13}a_{13}G_3) & (a_{11}a_{21}G_1 + a_{12}a_{22}G_2 + a_{13}a_{23}G_3) \\ (a_{21}a_{11}G_1 + a_{22}a_{12}G_2 + a_{23}a_{13}G_3) & (a_{21}a_{21}G_1 + a_{22}a_{22}G_2 + a_{23}a_{23}G_3) \end{bmatrix}$$

$$(3.94)$$

(continued)

Example 3.4.2 (continued)

If we denote the jth element of $\mathbf{Y}_n$ by $(\mathbf{Y}_n)_{jk}$ and the kth diagonal of $\mathbf{Y}_b$ by G_k, then in general, we have:

$$(\mathbf{Y}_n)_{jk} = \sum_{l=1}^{b} a_{jk} a_{kl} G_l \qquad (3.95)$$

provided that $\mathbf{Y}_b$ is a diagonal matrix, i.e., provided the circuit contains only two-terminal resistors and independent current sources.

If we let $j = k$ in Eq. (3.95), we find the kth diagonal element is given by:

$$(\mathbf{Y}_n)_{kk} = \sum_{l=1}^{b} a_{kl}^2 G_l$$

$$= \sum_{\beta_k} G_l \qquad (3.96)$$

where $\sum_{\beta_k}$ is defined as the sum over all branches connected to node k. This is true because of the observation that $a_{kl} = 1, -1$ or 0, and $a_{kl} \neq 0$ if and only if branch G_l is connected to node k. Hence we have proved property 1 of Table 3.1 holds for any circuit described by a diagonal branch admittance matrix $\mathbf{Y}_b$.

Observe next that if $a_{jl} \neq 0$, i.e., G_l is connected to node j, then

$$a_{kl} = -a_{jl} \qquad (3.97)$$

if G_l is connected between nodes j and k, and

$$a_{kl} = 0 \qquad (3.98)$$

if G_l is connected between node j and the ground node. It follows from Eqs. (3.95) and (3.98) that each off-diagonal element ($j \neq k$) of $\mathbf{Y_n}$ can be simplified as follows:

$$(\mathbf{Y}_n) = -\sum_{\beta_{jk}} G_l \qquad (3.99)$$

(continued)

Example 3.4.2 (continued)

where $\sum\limits_{\beta_{jk}}$ is defined to be the sum over all branches connected between nodes j and k. Hence we have proved property 2 of Table 3.1. Moreover, Eq. (3.99) implies that:

$$(\mathbf{Y}_n)_{jk} = (\mathbf{Y}_n)_{kj} \quad \text{or} \quad \mathbf{Y}_n = \mathbf{Y}_n^T \tag{3.100}$$

This proves property 3 of Table 3.1. Note that this symmetry property has nothing to do with whether the circuit is symmetrical or not. It is actually a consequence of an important circuit-theoretic property called **reciprocity** that will be discussed in Sect. 4.6.1.

The last property of Table 3.1 follows by definition and is therefore true regardless of whether $\mathbf{Y}_n$ is diagonal or not.

3.4.2   Existence and Uniqueness of Solutions

When we talked about various methods for solving the linear node Eq. (3.93) in the previous section, we implicitly assumed that Eq. (3.93) had a unique solution for any time t. To show that this assumption is not always satisfied even by simple circuits, consider the circuit shown in Fig. 3.16a.

Using the properties from Table 3.1, we obtain the following node equation by inspection (note the resistances are given in ohms):

$$\begin{bmatrix} 1 & -2 \\ -2 & 4 \end{bmatrix} \begin{bmatrix} e_1 \\ e_2 \end{bmatrix} = \begin{bmatrix} i_{s1}(t) \\ i_{s2}(t) \end{bmatrix} \tag{3.101}$$

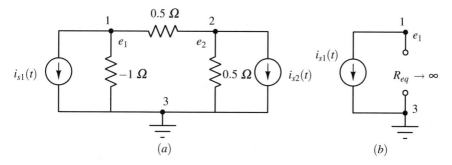

Fig. 3.16 A circuit containing a negative resistance

Since the determinant of $\mathbf{Y}_n$ is zero, Eq. (3.101) either has no solution or has infinitely many solutions. The latter occurs if and only if, $\forall\, t$, $i_{s1}(t) = -\frac{1}{2}i_{s1}(t)$.

To give a circuit interpretation of the above conclusion, let us assume for simplicity that $i_{s2}(t) = 0$ for all t so that the current source on the right-hand side can be deleted without affecting the circuit's solution. The resulting circuit can be further simplified to that shown in Fig. 3.16b, where the three resistors in Fig. 3.16a are replaced by an equivalent resistor R_{eq}. Since the current source $i_s(t)$ flows into an open circuit, it follows that the circuit does not have a solution if $i_{s1}(t) \neq 0$. On the other hand, if $i_{s1}(t) = 0$ for all t, then the circuit is satisfied by any node voltage e_1, and hence it admits an infinite number of solutions.

The following result gives a sufficient (but not necessary) condition for a circuit to have a unique solution.

Existence and Uniqueness Condition

Any resistive circuit containing only two-terminal linear positive conductances and independent current sources which do not form cut sets has a unique solution.

Proof Note that linear positive conductances or strictly passive resistors will be defined in Sect. 3.7. The above hypotheses guarantee that the node equation given by Eq. (3.93) is well-defined. Moreover, $\mathbf{Y}_b$ is a positive-definite diagonal matrix (since for all j, $G_j > 0$); i.e., $\mathbf{v}^T \mathbf{Y}_b \mathbf{v} > 0$, $\forall\, \mathbf{v} \neq \mathbf{0}$.

Now, for any node voltage vector $\mathbf{e} \neq \mathbf{0}$:

$$\mathbf{e}^T \mathbf{Y_n} \mathbf{e} = \mathbf{e}^T (\mathbf{A}\mathbf{Y}_b\mathbf{A}^T)\mathbf{e}$$
$$= (\mathbf{A}^T \mathbf{e})^T \mathbf{Y}_b (\mathbf{A}^T \mathbf{e})$$
$$= \mathbf{v}^T \mathbf{Y}_b \mathbf{v}$$
$$> 0 \qquad\qquad (3.102)$$

Hence $\mathbf{Y}_n$ is a positive-definite matrix and thus is full rank. Therefore Eq. (3.93) has a unique solution given by $\mathbf{e} = \mathbf{Y}_n^{-1}\mathbf{i}_s(t)$. $\square$

3.4.3   Node Equation Formulation: Nonlinear Resistive Circuits

When the circuit contains one more nonlinear resistors, the procedure for writing the node equation discussed in Sect. 3.4 in terms of the node voltage vector $\mathbf{e}$ still holds provided all nonlinear resistors are voltage-controlled. For example, consider the two linear resistors G_2 and G_5 in Fig. 3.15 replaced by a *pn*-junction diode

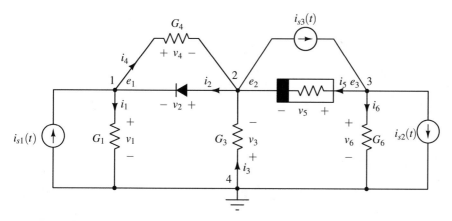

Fig. 3.17 A nonlinear circuit

described by $i_2 = I_s \left(e^{(v_2/V_T)} - 1 \right)$ and an $\mathcal{N}_R$ described by $i_5 = v_5^3$ as shown in Fig. 3.17.

Our first step as usual is to express the branch currents of the resistors in terms of the node voltages e_1, e_2, and e_3:

$$i_1 = G_1 v_1 = G_1 e_1 \qquad\qquad i_4 = G_4 v_4 = G_4(e_1 - e_2)$$

$$i_2 = I_s \left(e^{(v_2/V_T)} - 1 \right) = I_s \left(e^{\frac{e_2 - e_1}{V_T}} - 1 \right) \quad i_5 = v_5^3 = (e_3 - e_2)^3$$

$$i_3 = G_3 v_3 = G_3(-e_2) \qquad\qquad i_6 = G_6 v_6 = G_6 e_3 \qquad (3.103)$$

Note that this step is possible as long as the nonlinear resistors are voltage-controlled, i.e., the branch currents are functions of branch voltages.

Our next step is to apply KCL at each node (excluding the ground node):

$$\text{Node 1}: \quad G_1 e_1 - I_s \left(e^{\frac{e_2 - e_1}{V_T}} - 1 \right) + G_4(e_1 - e_2) = i_{s1}(t)$$

$$\text{Node 2}: \quad I_s \left(e^{\frac{e_2 - e_1}{V_T}} - 1 \right) - G_3(-e_2) - G_4(e_1 - e_2) - (e_3 - e_2)^3 = -i_{s3}(t)$$

$$\text{Node 3}: \quad (e_3 - e_2)^3 + G_6 e_3 = i_{s3}(t) - i_{s2}(t) \qquad (3.104)$$

The equations above constitute the node equation of the circuit in Fig. 3.17. But since these equations are nonlinear, they cannot be described by a node-admittance matrix.

Consider now the general case where the circuit may contain two-terminal, multi-terminal, and/or multi-port nonlinear voltage-controlled resistors, in addition to independent current sources. In this case, the branch equations now assume the following form:

$$i_1 = g_1(v_1, v_2, \ldots, v_b)$$
$$i_2 = g_2(v_1, v_2, \ldots, v_b)$$
$$\vdots$$
$$i_b = g_b(v_1, v_2, \ldots, v_b) \tag{3.105}$$

In vector notation we have:

$$\mathbf{i} = \mathbf{g}(\mathbf{v}) \tag{3.106}$$

called the **nonlinear branch equation**. Since independent current sources do not form cut sets (by assumption), Eq. (3.89) remains valid. Substituting Eq. (3.106) for **i** in Eq. (3.89), we get:

$$\mathbf{A}\mathbf{g}(\mathbf{v}) = \mathbf{i_s}(t) \tag{3.107}$$

Substituting next Eq. (3.74) for **v**, we get the **nonlinear node equation**:

$$\mathbf{A}\mathbf{g}(\mathbf{A}^T\mathbf{e}) = \mathbf{i}_s(t) \tag{3.108}$$

For each solution of **e** in Eq. (3.108), we can calculate the corresponding branch voltage vector **v** by direct substitution into Eq. (3.74), namely, $\mathbf{v} = \mathbf{A}^T\mathbf{e}$. This in turn can be used to calculate the branch current vector **i** by direct substitution into Eq. (3.106). Hence the basic problem is to solve the nonlinear node Eq. (3.108). The rest is trivial. In general, nonlinear equations do not have closed form solutions. Consequently, they must be solved by numerical techniques, that are beyond the scope of this book. The most widely used method is the Newton-Raphson algorithm, the reader is referred to excellent sources such as [3] for details.

3.5   Tableau Analysis for Resistive Circuits

The only, albeit major, shortcoming of node analysis is that it disallows many standard circuit elements from the class of allowable circuits, e.g., the voltage source, ideal transformer, ideal op amp, CCCS, CCVS, VCVS, current-controlled nonlinear resistor, etc. In this section, we overcome this issue by presenting a completely general analysis method—one that works for all resistive circuits. Conceptually, this method is simpler than node analysis. It consists of writing out the complete list of linearly independent KCL equations, linearly independent KVL equations, and the branch equations. For obvious reasons, this list of equations is called **tableau equations** [2].

Since no variables are eliminated[7] in listing the tableau equations, all three vectors $\mathbf{e}$, $\mathbf{v}$, and $\mathbf{i}$ are present as variables. Since we must have as many tableau equations as there are variables, it is clear that the price we pay for the increased generality is that tableau analysis involves many more equations than node analysis does. In our era of computer-aided circuit analysis, however, this objection turns out to be a blessing in disguise because the matrix associated with tableau analysis is often extremely sparse, thereby allowing highly efficient numerical algorithms to be used.

The significance of tableau analysis actually transcends the above more mundane numerical considerations. As the reader will gather while reading this and other advanced textbooks on nonlinear circuits, tableau analysis is a powerful analytic tool which allows us to derive many profound results with almost no pain at all—at least compared to other approaches.

To write the tableau equation for any linear resistive circuit, we simply use the following algorithm[8]:

1. Draw the digraph of the circuit and hinge it if necessary so that the resulting digraph is connected. Pick an arbitrary ground node and formulate the reduced incidence matrix $\mathbf{A}$.

2. Write a complete set of linearly independent KCL equations:

$$\mathbf{A}\mathbf{i}(t) = \mathbf{0} \tag{3.109}$$

   Note that unlike Eq. (3.89), tableau analysis deals with the original digraph where each independent current source is represented by a branch.

3. Write a complete set of linearly independent KVL equations:

$$\mathbf{v}(t) - \mathbf{A}^T \mathbf{e}(t) = \mathbf{0} \tag{3.110}$$

4. Write the branch equations. Since the circuit is linear, these equations can always be recast into the form:

$$\mathbf{M}(t)\mathbf{v}(t) + \mathbf{N}(t)\mathbf{i}(t) = \mathbf{u}_s(t) \tag{3.111}$$

Together Eqs. (3.109)–(3.111) constitute the tableau equations. If the digraph has n nodes and b branches, Eqs. (3.109)–(3.111) will contain $n - 1$, b and b equations, respectively. Since the vectors e, v, and i also contain $n - 1$, b and b variables, respectively, the tableau equation for a linear resistive circuit always consists of $(n - 1) + 2b$ linear equations in $(n - 1) + 2b$ variables.

---

[7] Recall both $\mathbf{v}$ and $\mathbf{i}$ must be eliminated in node analysis, leaving $\mathbf{e}$ as the only variable.

[8] The reader may wish to scan Example 3.5.1 after each step in order to get familiarized first with the notations used in writing the tableau equation.

Example 3.5.1 Write the tableau equations for the linear circuit in Fig. 3.18.

Solution The circuit only contains three elements: a voltage source, an ideal transformer, and a time-varying resistor. The first two elements are not allowed in nodal analysis because they are not voltage-controlled. The third element, which would normally be acceptable, is also disallowed here because its conductance $G(t) = 1/(R_0 \sin t) \to \infty$ at $t = 0, 2\pi, 4, \pi, \cdots$ and is therefore not defined for all t.

Applying the preceding recipe, we hinge nodes 3 and 4 and draw the connected digraph shown in Fig. 3.18b. Choosing the hinged node as ground, the tableau equations are formulated below.

$$\text{KCL}: \quad \mathbf{AI} = \mathbf{0} \Leftrightarrow \begin{bmatrix} 1 & 0 & 0 & 1 \\ 0 & 1 & 1 & 0 \end{bmatrix} \begin{bmatrix} i_1 \\ i_2 \\ i_3 \\ i_4 \end{bmatrix} = \begin{bmatrix} 0 \\ 0 \end{bmatrix} \tag{3.112}$$

$$\text{KVL}: \quad \mathbf{v} - \mathbf{A}^T \mathbf{e} = \mathbf{0} \Leftrightarrow \begin{bmatrix} v_1 \\ v_2 \\ v_3 \\ v_4 \end{bmatrix} - \begin{bmatrix} 1 & 0 \\ 0 & 1 \\ 0 & 1 \\ 1 & 0 \end{bmatrix} \begin{bmatrix} e_1 \\ e_2 \end{bmatrix} = \begin{bmatrix} 0 \\ 0 \\ 0 \\ 0 \end{bmatrix} \tag{3.113}$$

$$\text{Branch Equations}: \quad \left. \begin{matrix} n_2 v_1 - n_1 v_2 = 0 \\ n_1 i_1 + n_2 i_2 = 0 \\ v_3 - R(t) i_3 = 0 \\ v_4 = E \cos \omega t \end{matrix} \right\} \Leftrightarrow \begin{bmatrix} n_2 & -n_1 & 0 & 0 \\ 0 & 0 & 0 & 0 \\ 0 & 0 & 1 & 0 \\ 0 & 0 & 0 & 1 \end{bmatrix} \begin{bmatrix} v_1 \\ v_2 \\ v_3 \\ v_4 \end{bmatrix}$$

$$+ \begin{bmatrix} 0 & 0 & 0 & 0 \\ n_1 & n_2 & 0 & 0 \\ 0 & 0 & -R(t) & 0 \\ 0 & 0 & 0 & 0 \end{bmatrix} \begin{bmatrix} i_1 \\ i_2 \\ i_3 \\ i_4 \end{bmatrix} = \begin{bmatrix} 0 \\ 0 \\ 0 \\ E \cos \omega t \end{bmatrix} \tag{3.114}$$

$n = 3, b = 4$ for the digraph in Fig. 3.18b. Consequently, we expect the tableau equation to contain $(n - 1) + 2b = 10$ equations involving 10 variables, namely $e_1, e_2, v_1, v_2, v_3, v_4, i_1, i_2, i_3, i_4$. An inspection of Eqs. (3.112), (3.113), and (3.114) shows that indeed we have 10 equations involving these 10 variables.

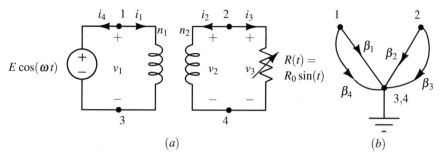

Fig. 3.18 All three elements in this circuit are disallowed in node analysis

Example 3.5.1 simply illustrated how to apply the tableau analysis algorithm. Had one encountered the circuit in Fig. 3.18a in practice, assuming enough experience in circuit analysis, one can quickly write the necessary equations "on the back of an envelope." The point we wish to emphasize again is that one should use insight, along with technique.

The vector $\mathbf{u}_s(t)$ on the right-hand side of Eq. (3.111) does not depend on any variable e_j, v_j or i_j and is therefore due to only independent voltage and current sources in the circuit. Consequently, element k of $\mathbf{u}_s(t)$ will be zero whenever branch k is not an independent source. Note that controlled source coefficients always appear in the matrices $\mathbf{M}(t)$ and/or $\mathbf{N}(t)$, never in $\mathbf{u}_s(t)$.

An inspection of Eq. (3.114) reveals that each row k of $\mathbf{M}(t)$ and $\mathbf{N}(t)$ contains coefficients or time functions which define uniquely the linear relation between v_k and i_k of branch k in the digraph, assuming branch k corresponds to a resistor. If branch k happens to be an independent source, then the kth diagonal element is equal to one in $\mathbf{M}(t)$ (for a voltage source) or $\mathbf{N}(t)$ (for a current source), while all other elements in row k are zeros. In this case, the kth element of $\mathbf{u}_s(t)$ will contain either a constant (for a DC source) or a time function which specifies uniquely this independent source. On the other hand, if branch k is not an independent source, then the k element of $\mathbf{u}_s(t)$ is always zero. It follows from the above interpretation that both $\mathbf{M}(t)$ and $\mathbf{N}(t)$ are $b \times b$ matrices and $\mathbf{u}_s(t)$ is a $b \times 1$ vector, where b is the number of branches in the digraph.

Finally note that we can state that a resistive circuit is linear iff its branch equations can be written in the form stipulated in Eq. (3.111), and it is time-invariant iff both $\mathbf{M}(t)$ and $\mathbf{N}(t)$ are constant real matrices.

In the general case, it is more illuminating to write Eqs. (3.109)–(3.111) as a single matrix equation, the **linear tableau equation**, shown in Eq. (3.115).

$$
\underbrace{\begin{bmatrix} \mathbf{0} & \mathbf{0} & \mathbf{A} \\ -\mathbf{A}^T & \mathbf{I} & \mathbf{0} \\ \mathbf{0} & \mathbf{M}(t) & \mathbf{N}(t) \end{bmatrix}}_{\mathbf{T}(t)} \underbrace{\begin{bmatrix} \mathbf{e}(t) \\ \mathbf{v}(t) \\ \mathbf{i}(t) \end{bmatrix}}_{\mathbf{w}(t)} = \underbrace{\begin{bmatrix} \mathbf{0} \\ \mathbf{0} \\ \mathbf{u}_s(t) \end{bmatrix}}_{\mathbf{u}(t)}
\tag{3.115}
$$

It is natural to call $\mathbf{T}(t)$ the **tableau matrix** associated with the linear resistive circuit. If the circuit is time-invariant, $\mathbf{T}(t) = \mathbf{T}$, a constant real matrix.

Every linear resistive circuit is associated with a **unique** $[(n-1)+2b] \times [(n-1)+2b]$ square tableau matrix $\mathbf{T}(t)$, and a **unique** $[(n-1)+2b] \times 1$ vector $\mathbf{u}(t)$.[9] Note the significance of the tableau matrix is the fact that, if and only if, $\det[\mathbf{T}(t_0)] \neq 0$ at any time t_0, a **unique** solution to the linear circuit exists in the form of $\mathbf{w}(t_0) = \mathbf{T}^{-1}(t_0)\mathbf{u}(t_0)$.

3.5.1 Tableau Equation Formulation: Nonlinear Resistive Circuits

Exactly the same principle is used to formulate the tableau equation for **nonlinear** resistive circuits: Simply list the linearly independent KCL and KVL equations, and the branch equations, which are now nonlinear. Hence, the first three steps of the algorithm at the beginning of Sect. 3.5 remain unchanged. Only step 4 needs to be modified because Eq. (3.111) is valid only for linear resistive circuits. Example 3.19 illustrates and suggests the modified form of Eq. (3.111).

Example 3.5.2 Write the tableau equations for the nonlinear circuit in Fig. 3.19. The *npn* transistor is modeled by the following nonlinear Ebers-Moll equation (see Eqs. (2.40) and (2.41) from Chap. 2):

$$
i_1 = -I_{ES}\left(e^{\frac{-v_1}{V_T}} - 1\right) + \alpha_R I_{CS}\left(e^{\frac{-v_2}{V_T}} - 1\right)
\tag{3.116}
$$

$$
i_2 = \alpha_F I_{ES}\left(e^{\frac{-v_1}{V_T}} - 1\right) - I_{CS}\left(e^{\frac{-v_2}{V_T}} - 1\right)
\tag{3.117}
$$

(continued)

---

[9]The "uniqueness" is of course relative to a particular choice of element and node numbers.

Example 3.5.2 (continued)

Solution Since the digraph for this circuit is identical to that shown in Fig. 3.18b, the same KCL Eq. (3.112) and KVL Eq. (3.113) also apply for this circuit. However, instead of Eq. (3.114), we have the following branch equations:

$$h_1(v_1, v_2, i_1) \triangleq i_1 + I_{ES}\left(e^{\frac{-v_1}{V_T}} - 1\right) - \alpha_R I_{CS}\left(e^{\frac{-v_2}{V_T}} - 1\right) = 0$$

$$h_2(v_1, v_2, i_1) \triangleq i_2 - \alpha_F I_{ES}\left(e^{\frac{-v_1}{V_T}} - 1\right) + I_{CS}\left(e^{\frac{-v_2}{V_T}} - 1\right) = 0$$

$$h_3(v_3, i_3, t) \triangleq v_3 - R(t)i_3 = 0$$

$$h_4(v_4, t) \triangleq v_4 - E\cos\omega t = 0 \tag{3.118}$$

Note that $h_1(\cdot, \cdot, \cdot)$ and $h_2(\cdot, \cdot, \cdot)$ are nonlinear functions of (v_1, v_2, i_1) and (v_1, v_2, i_2), respectively; $h_3(\cdot, \cdot, \cdot)$ is a linear function of v_3 and i_3 but a nonlinear function of t; and $h_4(\cdot, \cdot)$ is a function of only v_4 and t. Even for this simple circuit, we see that there is really no simple form analogous to Eq. (3.114). To avoid keeping track of which variables are present in each function, we will simply denote Eq. (3.118) as follows:

$$h_1(v_1, v_2, i_1, i_2, t) = 0$$

$$h_2(v_1, v_2, i_1, i_2, t) = 0$$

$$h_3(v_1, v_2, i_1, i_2, t) = 0$$

$$h_4(v_1, v_2, i_1, i_2, t) = 0 \tag{3.119}$$

or in vector form, we simply write:

$$\mathbf{h}(\mathbf{v}, \mathbf{i}, t) = 0 \tag{3.120}$$

It is understood that some variables may not be present in each component equation.

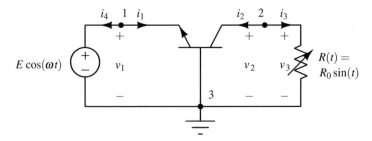

Fig. 3.19 A nonlinear resistive circuit

It follows from Example 3.5.2 that **every** nonlinear resistive circuit is described by a unique system of $(n-1)+2b$ nonlinear algebraic equations in $(n-1)+2b$ variables, called the **nonlinear tableau equation**:

$$\mathbf{A}\mathbf{i}(t) = \mathbf{0}$$

$$\mathbf{v}(t) - \mathbf{A}^T\mathbf{e}(t) = \mathbf{0}$$

$$\mathbf{h}(\mathbf{v}(t), \mathbf{i}(t), t) = \mathbf{0} \tag{3.121}$$

We usually resort to numerical methods to solve Eq. (3.121), which is beyond the scope of this book.

3.6   General Properties of Linear Resistive Circuits

In this section, we state and prove two general theorems for linear time-invariant resistive circuits,[10] namely the **superposition theorem** and the **Thévenin-Norton theorem**. Intelligent use of these theorems often results in a dramatic simplification of an otherwise much more difficult problem.

Both these theorems are valid if and only if the associated circuit is **uniquely solvable**, equivalently, if and only if the associated tableau matrix **T** is nonsingular. Although these theorems are stated only for time-invariant circuits for simplicity, both theorems are valid also for time-varying circuits by simply allowing all parameters and coefficients to vary with time.

---

[10]Recall that a **linear** resistive circuit may contain, in addition to two-terminal resistors and independent sources, any multi-terminal or multi-port linear resistors (for example, ideal transformers, gyrators, and all four types of linear-dependent sources).

3.6.1   Superposition Theorem

Theorem 3.5 (Superposition Theorem for Linear Time-Invariant Circuits)
*Let $\mathcal{N}$ be any linear time-invariant uniquely solvable resistive circuit driven
by α independent voltage sources $v_{s1}(t), v_{s2}(t), \ldots, v_{s\alpha}(t)$ and β independent
current sources $i_{s1}(t), i_{s2}(t), \ldots, i_{s\beta}(t)$.*

*Then any node voltage $e_j(t)$, any branch voltage $v_j(t)$, or any branch
current $i_j(t)$ is given by an expression of the form*

$$H_1 v_{s1}(t) + \cdots + H_\alpha v_{s\alpha}(t) + K_1 i_{s1}(t) + \cdots + K_\beta i_{s\beta}(t) \qquad (3.122)$$

*where the coefficients H_k, $k = 1, 2, \ldots, \alpha$ and K_k, $k = 1, 2, \ldots, \beta$ are
constants which depend only on the circuit parameters of $\mathcal{N}$ and the choice of
the output variable (i.e., e_j, v_j or i_j) but **not** on the independent sources.*

Before we prove Theorem 3.5, it is instructive to give some circuit interpretations
and an example. The circuit interpretations are:

1. Each term $y(v_{sk}) \triangleq H_k v_{sk}$ in Eq. (3.122) is equal to the response of y when all
   independent sources in $\mathcal{N}$ except $v_{sk}(t)$ are set to zero.
2. Each term $y(i_{sk}) \triangleq K_k i_{sk}$ in Eq. (3.122) is equal to the response of y when all
   independent sources in $\mathcal{N}$ except $i_{sk}(t)$ are set to zero.
3. Equation (3.122) shows that the response due to several independent voltage and
   current sources is equal to the **sum** of the responses due to **each** independent
   source **acting alone**, i.e., with all other **independent** voltage sources replaced
   by short circuits, and all other **independent** current sources replaced by open
   circuits.[11]
4. Equation (3.122) also shows that in applying the superposition theorem, **controlled sources are left intact**.
5. The response at **any time** $t = t_0$ depends **only** on the value of the independent
   sources at the **same time** $t = t_0$. In other words, linear **resistive** circuits have **no
   memory**.

---

[11] Compare this description to the definition of superposition from Exercise 1.9.

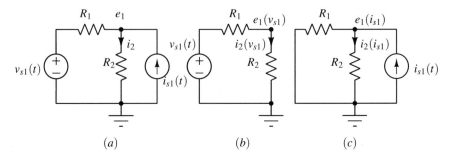

Fig. 3.20 (**a**) Circuit for superposition. (**b**) Voltage divider. (**c**) Current divider

Example 3.6.1 Use the superposition theorem to calculate the node voltage e_1 and resistor current i_2 in Fig. 3.20a.

Solution The contributions to e_1 and i_2 due to $v_{s1}(t)$ acting alone (with $i_{s1}(t) = 0$) can be found by inspection of the **voltage-divider** circuit in Fig. 3.20b, obtained by replacing the current source in Fig. 3.20a with an open circuit:

$$e_1(v_{s1}) = \frac{R_1}{R_1 + R_2} v_{s1}(t) \tag{3.123}$$

$$i_2(v_{s1}) = \frac{1}{R_1 + R_2} v_{s1}(t) \tag{3.124}$$

Here, the "input" v_{s1} is shown as the "argument" of $e_1(\bullet)$ and $i_2(\bullet)$ to remind the reader that the node voltage e_1 given by Eq. (3.123) and the branch current i_2 given by Eq. (3.124) are due to v_{s1} acting alone, and are therefore functions of v_{s1} only.

The contributions to e_1 and i_2 due to $i_{s2}(t)$ acting alone (with $v_{s1}(t) = 0$) can be found by inspection of the **current-divider** circuit shown in Fig. 3.20c, obtained by replacing the voltage source in Fig. 3.20a with a short circuit:

$$e_1(i_{s1}) = \frac{R_1 R_2}{R_1 + R_2} i_{s1}(t) \tag{3.125}$$

$$i_2(i_{s1}) = \frac{R_1}{R_1 + R_2} i_{s1}(t) \tag{3.126}$$

Adding the respective contributions, we obtain:

$$e_1 = e_1(v_{s1}) + e_2(i_{s1}) = H_1 v_{s1}(t) + K_1 i_{s1}(t) \tag{3.127}$$

(continued)

Example 3.6.1 (continued)

where $H_1 \triangleq \frac{R_2}{R_1+R_2}$, $K_1 \triangleq \frac{R_1 R_2}{R_1+R_2}$. and:

$$i_2 = i_2(v_{s1}) + i_2(i_{s1}) = H_1 v_{s1}(t) + K_1 i_{s1}(t) \tag{3.128}$$

where $H_1 \triangleq \frac{1}{R_1+R_2}$, $K_1 \triangleq \frac{R_1}{R_1+R_2}$.

As expected, both e_1 and i_2 are of the form specified by Eq. (3.122) where H_1 and K_1 are constants depending **only** on the circuit parameters R_1, R_2 and the chosen output variable. They do not depend on $v_{s1}(t)$ or $i_{s1}(t)$. Of course, for different choices of output variables, we get different H_1's and K_1's, as seen in the expressions for e_1 and i_2.

Proof of the Superposition Theorem Since $\mathcal{N}$ is linear and time-invariant, it is described by the linear tableau equation:

$$\mathbf{T}\mathbf{w}(t) = \mathbf{u}(t) \tag{3.129}$$

where $\mathbf{T}$ is an $[(n-1)+2b] \times [(n-1)+2b]$ constant real tableau matrix. However, since $\mathcal{N}$ is uniquely solvable (by assumption), $\mathbf{T}^{-1}$ exists and the unique solution is given by:

$$\mathbf{w}(t) = \mathbf{T}^{-1}\mathbf{u}(t) \tag{3.130}$$

where:

$$\mathbf{u}(t) \triangleq \left[\underbrace{\mathbf{0}^T}_{n-1} \ \underbrace{\mathbf{0}^T}_{b} \ \underbrace{0 \cdots 0}_{\text{resistors}} \ \underbrace{v_{s1}(t) \cdots v_{s\alpha}(t)}_{\text{voltage sources}} \ \underbrace{i_{s1}(t) \cdots i_{s\beta}(t)}_{\text{current sources}} \right]^T \tag{3.131}$$

Here we have assumed without loss of generality that all independent sources are labeled last in the order depicted above.

Since each component of $\mathbf{w}(t)$ (i.e., e_j, v_j or i_j) is obtained by multiplying the corresponding row of $\mathbf{T}^{-1}$ with $\mathbf{u}(t)$, it follows that each response e_j, v_j or i_j is given by an expression in the form of Eq. (3.122). Moreover, since $\mathbf{T}^{-1}$ is a constant matrix which does not involve any independent source terms, so are the constant coefficients H_k and K_k. $\square$

3.6.2 Thévenin-Norton Theorem

Definition 3.3 A one-port N is said to be **well-defined** iff it does not contain any circuit element which is **coupled**, electrically or nonelectrically, to some physical variable **outside** of N.

An example of an ill-defined N would be if it contains a photoresistor coupled to an external light source.

Theorem 3.6 *Any well-defined linear time-invariant resistive one-port N which satisfies the following **unique solvability** condition can be replaced by the following **equivalent one-ports** N_{eq} without affecting the solution of any **external** circuit (not necessarily linear or resistive) connected across N.*

1. *Thévenin equivalent one-port N_{eq}*
   > *unique solvability condition: The circuit $\mathcal{N}$ obtained by connecting a current source i across N has a unique solution for all i.*

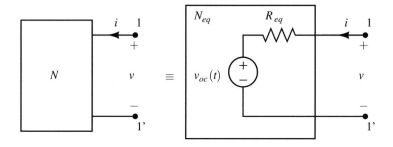

$$
\begin{aligned}
R_{eq} &\triangleq \; \textit{Thévenin-equivalent resistance in ohms} \\
&\triangleq \; \textit{DP or input resistance across N} \\
& \; \textit{after all independent sources inside N are set to zero} \\
v_{oc}(t) &\triangleq \; \textit{open-circuit voltage} \\
&\triangleq \; \textit{voltage v across terminals 1 and 1' when the port} \\
& \; \textit{1, 1' is left open-circuited}
\end{aligned}
$$

2. *Norton equivalent one-port N_{eq}*
   > *unique solvability condition: The circuit $\mathcal{N}$ obtained by connecting a voltage source v across N has a unique solution for all v.*

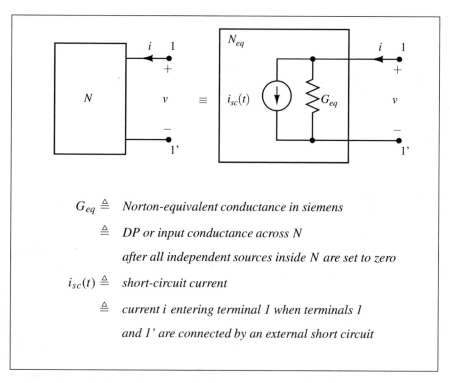

$$G_{eq} \triangleq \text{Norton-equivalent conductance in siemens}$$

$$\triangleq \text{DP or input conductance across } N$$

$$\text{after all independent sources inside } N \text{ are set to zero}$$

$$i_{sc}(t) \triangleq \text{short-circuit current}$$

$$\triangleq \text{current } i \text{ entering terminal } 1 \text{ when terminals } 1$$

$$\text{and } 1' \text{ are connected by an external short circuit}$$

As before, we will consider first some circuit interpretations and an example, before proving the theorem. The circuit interpretations are:

1. The main value of Thévenin's and Norton's theorem is that it allows us to replace any part of a circuit which forms a linear resistive one-port, by only two circuit elements, without affecting the solution of the remainder of the circuit. Conceptually this works because a linear circuit is described by a linear equation. Graphically, in the $i-v$ $(v-i)$ plane, we need only two points to fully characterize the linear equation. Thévenin and Norton theorems say we choose the intercepts (i_{sc}, v_{oc}) $((v_{oc}, i_{sc}))$ as the two points (see 3. below).

2. Let $R_{eq} \neq 0$. If we short-circuit the Thévenin equivalent circuit N_{eq} and solve for the current i, we would obtain

$$i_{sc} = -\frac{v_{oc}}{R_{eq}} \tag{3.132}$$

If $i_{sc} \neq 0$, we can calculate the Thévenin equivalent resistance by

$$R_{eq} = -\frac{v_{oc}}{i_{sc}} \tag{3.133}$$

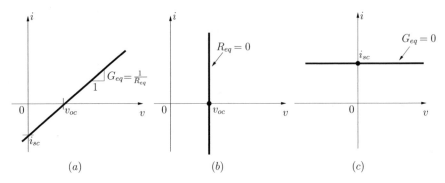

Fig. 3.21 (a) DP characteristic of N with $v_{oc} > 0$ and $G_{eq} > 0$. (b) DP characteristic with $v_{oc} > 0$ and $R_{eq} = 0$. (c) DP characteristic with $i_{sc} > 0$ and $G_{eq} = 0$

3. When $R_{eq} \neq 0$ and $G_{eq} \neq 0$, the one-port N is equivalent to both its Thévenin and its Norton equivalent one-ports: Its DP characteristic at any time t is defined by:

$$v = R_{eq}i + v_{oc}(t) \qquad (3.134)$$

$$i = G_{eq}v + i_{sc}(t) \qquad (3.135)$$

This DP characteristic consists of a straight line with a slope R_{eq} and voltage intercept $v_{oc}(t)$ in the $i - v$ plane, or with a slope G_{eq} and current intercept $i_{sc}(t)$ in the $v - i$ plane (shown in Fig. 3.21a).

4. The limiting case of $R_{eq} = 0$ is shown in Fig. 3.21b. The Thévenin equivalent one-port in this case consists of just a battery of v_{oc} volts. The corresponding Norton equivalent one-port does not exist because $G_{eq} \to \infty$. Indeed, the unique solvability condition fails in this case—KVL is violated when a voltage source $v \neq v_{oc}$ is applied.

   The "dual" limiting case $G_{eq} = 0$ is shown in Fig. 3.21c.

5. A one-port which has neither a Thévenin nor Norton equivalent is shown in Fig. 3.22a.

   Its DP characteristic is defined by:

$$v = 0 \quad i = 0 \qquad (3.136)$$

and consists therefore of only one point, namely, the origin. Note that the "virtual short circuit" characterizing the input port of an **ideal** opamp operating in the linear region has precisely this property. Such a one-port is called a **nullator**.

6. It follows from the above observations that if N is not current-controlled, then it does **not** possess a Thévenin equivalent. Dually, if N is not voltage-controlled, then it does **not** possess a Norton equivalent. Hence, in applying Thévenin's or Norton's theorem, we can ignore checking for the "unique solvability condition"

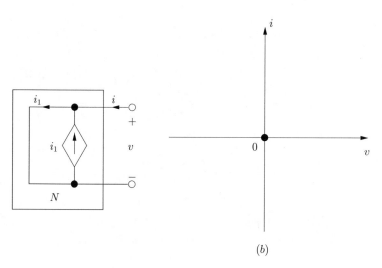

Fig. 3.22   A one-port characterized by only one point (**a**) Circuit (**b**) DP plot

since this generally entails the difficult task of checking if the associated tableau matrix **T** is invertible. Instead, we simply proceed to calculate R_{eq} or G_{eq}. Failure to obtain a unique finite value for R_{eq} (respectively G_{eq}) would then imply that N does not have a Thévenin (respectively Norton) equivalent.

Example 3.6.2 Find the Thévenin and Norton equivalent one-ports for the circuit shown in Fig. 3.23a.

Solution   Let us calculate R_{eq} and G_{eq} first using the simplified circuit shown in Fig. 3.23b. For any applied voltage v, we find $i_1 = v/R$ so that $i = -4i_1 = -(4/R)v$. Hence,

$$R_{eq} = \frac{1}{G_{eq}} = -\frac{R}{4} \qquad (3.137)$$

Since both R_{eq} and G_{eq} are finite numbers, we know that N has a Thévenin and a Norton equivalent one-port.

We proceed therefore to calculate v_{oc} using the circuit shown in Fig. 3.23c. Applying KCL we obtain $i_1 - 5i_1 + I_s = 0$ or $i_1 = I_s/4$. Hence,

$$v_{oc} = E + \frac{R}{4}I_s \qquad (3.138)$$

(continued)

Example 3.6.2 (continued)
To calculate i_{sc}, we use Eqs. (3.132) and (3.138) to get:

$$i_{sc} = -\frac{v_{oc}}{R_{eq}} = \frac{4E}{R} + I_s \qquad (3.139)$$

As an independent check, let us derive i_{sc} using the circuit in Fig. 3.23d.
Since $i_1 = -\frac{E}{R}$ in this case, KCL implies:

$$i_{sc} = i_1 - 5i_1 + I_s = \frac{4E}{R} + I_s \qquad (3.140)$$

which agrees with our first i_{sc} equation (as it should).

Proof of Norton's Theorem We will prove only Norton's theorem, as the dual proof
then applies to Thévenin's theorem. Let N denote the one-port in question, and let
the remaining part of the circuit $\mathscr{N}$ be denoted by N_L, as shown in Fig. 3.24a. By
hypotheses, N contains only linear time-invariant resistors and independent sources,
whereas N_L need not be linear or resistive.

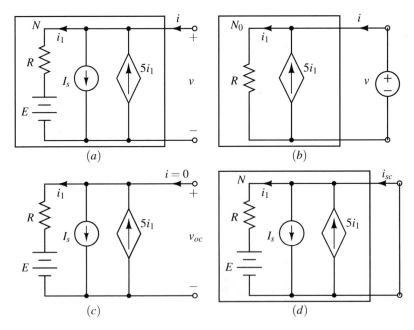

Fig. 3.23 (a) One-port N. (b) Simplified one-port N_0 obtained by setting all independent sources
inside N to zero. (c) Circuit used for calculating v_{oc}. (d) Circuit used for calculating i_{sc}

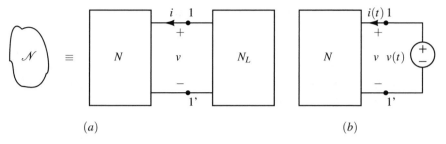

Fig. 3.24 (**a**) Partitioning arbitrary circuit $\mathscr{N}$ into a linear resistive one-port N and a not necessarily linear or resistive one-port N_L. (**b**) Driving N with a voltage source $v(t)$

Since N is purely resistive, it is completely specified by its DP characteristic at each instant of time. Hence, as far as N_L is concerned, its solution depends only on this DP characteristic: The elements inside N which give rise to this DP characteristic are completely irrelevant. For example, we don't care if N consists of a 2 Ω resistor or two 1 Ω resistors in series, as long as we have a 2 Ω **equivalent** DP resistance. It suffices therefore to prove that both N and its Norton equivalent one-port have identical DP characteristics.

Let us drive N with an independent voltage source $v(t)$ as shown in Fig. 3.24b. Let us label this voltage source, together with the independent voltage sources inside N by $v_{s0}(t), v_{s1}(t), \ldots, v_{s\alpha}(t)$, where $v_{s0}(t) \triangleq v(t)$. Similarly, let us label the independent current sources inside N by $i_{s1}(t), \ldots, i_{s\beta}(t)$.

It follows from the unique solvability condition that the linear time-invariant resistive circuit in Fig. 3.24b has a unique solution for all values of the independent sources, at all times. Hence we can apply the superposition theorem and conclude that the port current $i(t)$ in Fig. 3.24b must assume the form:

$$i(t) = H_0 v(t) + \overset{\alpha}{\underset{k=1}{\sum}} H_k v_{sk}(t) + \overset{\beta}{\underset{k=1}{\sum}} K_k i_{sk}(t) \tag{3.141}$$

Now if $v(t) = 0 \; \forall t$, $i(t)$ is by definition $i_{sc}(t)$. Hence the last two sums in Eq. (3.141) add up to $i_{sc}(t)$.

If we set to zero all independent sources inside N, we are left with $i(t) = H_0 v(t)$, i.e., $H_0 = G_{eq}$. Hence Eq. (3.141) can be written in the form:

$$i(t) = G_{eq} v(t) + i_{sc}(t) \tag{3.142}$$

where G_{eq} and $i_{sc}(t)$ are as defined in the theorem. Equation (3.142) gives the DP characteristic of the given one-port N. Since this is the same equation which defines the Norton equivalent one-port N_{eq}, it follows that N can indeed be replaced by a Norton equivalent N_{eq} without affecting the solution inside N_L. □

3.7 Some General Properties of Nonlinear Resistive Circuits

The behavior of linear resistive circuits is intimately related to linear algebraic equations. As a consequence of linearity, we were able to derive several rather general properties in the preceding section. Precisely because their proofs depend on linearity in a crucial way, none of these properties holds even if the circuit contains only one nonlinear resistor.

The behavior of nonlinear resistive circuits is far more complicated. For example, multiple solutions are frequent. Even describing a two-terminal nonlinear resistor alone can be complicated. To specify it analytically we need to use a function which may require many parameters (for example, the *pn*-junction diode).

In spite of its greatly increased complexity, many useful properties can be proved for various subclasses of nonlinear resistive circuits. Our objective in this section is to state only those properties which we are in a position to prove, in a remarkably elegant manner. These general properties, derived from the fundamental concepts of passivity and monotonicity, form only a small albeit important subset of our "nonlinear tool kits." We hope this final section will whet the reader's appetite into a more advanced study of this subject.

3.7.1 Strict Passivity

Definition 3.4 A two-terminal resistor is said to be **strictly passive** iff $vi > 0$ for all points (v, i) on its characteristic, except the origin $(0, 0)$.

Geometrically, this means that the $v - i$ curve of a strictly passive resistor must lie only in the first and third quadrants and stay clear of the v and i axis, except the origin.

Most of the nonlinear resistors we have encountered so far as strictly passive. However, the ideal diode concave resistor, and convex resistor are passive but not strictly passive.

In this section we will state and prove three general theorems for circuits containing only strictly passive resistors and independent sources.

Theorem 3.7 (Strict Passivity Property) *A one-port made of strictly passive two-terminal resistors is itself strictly passive.*

Proof Consider the one-port N shown in Fig. 3.25, which is driven by a voltage source. Let N contain m strictly passive resistors. Applying Tellegen's theorem and noting that the current **entering** the positive terminal of the voltage source is equal

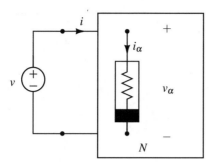

Fig. 3.25 One-port N

to $-i$ (passive sign convention), we obtain:

$$vi = \sum_{\alpha=1}^{m} v_\alpha i_\alpha \tag{3.143}$$

Since the m resistors are strictly passive for all α, $v_\alpha i_\alpha \geq 0$, hence $vi \geq 0$.

Suppose $v > 0$, then by KVL some of the v_α's must be nonzero. Thus by strict passivity, the corresponding i_α's are also nonzero and of the same sign. Hence whenever $v > 0$ at least one term, say $v_k i_k$, is positive. So we have $v > 0$ implies $i > 0$.

A similar argument shows that $v < 0$ implies $i < 0$. Hence $vi > 0$ for all points on the driving point characteristic except the origin (where $vi = 0$). Therefore N is strictly passive. □

Theorem 3.8 (Maximum Node-Voltage Property) *Let $\mathcal{N}$ be a connected circuit made of strictly passive two-terminal resistors and driven by a single DC voltage source of E volts, $E > 0$. Then, with the negative voltage-source terminal chosen as ground, no node-to-ground voltage can exceed E volts.*

Proof Since $\mathcal{N}$ is connected, all node-to-ground voltages $e_1, e_2, \cdots, e_{n-1}$ are well-defined.

Suppose there exists a node m with the highest potential $e_m > E$. Since $e_a, e_b, \ldots, e_k \leq e_m$, we have $v_a, v_b, \ldots, v_k \geq 0$. Since all resistors are strictly passive, this implies that $i_a, i_b, \ldots, i_k \geq 0$. But for KCL to be satisfied at node m, we must have $i_a = i_b = \cdots = i_k = 0$. By strict passivity, this implies that $v_a = v_b = \cdots = v_k = 0$. Thus, we have $e_a = e_b = \cdots = e_k = e_m > E$ (Fig. 3.26).

Hence we can move on to nodes $a, b, \ldots, k$ and repeat the above reasoning. We must eventually reach node 1 of the voltage source, where our reasoning would still

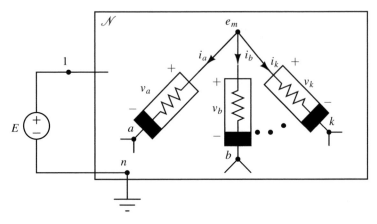

Fig. 3.26 KCL at node m implies $i_a + i_b + \cdots + i_k = 0$

imply that $e_1 = e_m > E$, which is false. Hence our assumption that $e_m > E$ is wrong and thus $e_m \le E$. $\square$

Theorem 3.9 (Transfer Characteristic Bounding Region) *The v_o vs. v_{in} transfer characteristic of any connected circuit made of strictly passive two-terminal resistors must lie within the wedge-shaped region*

$$|v_o| \le |v_{in}| \tag{3.144}$$

as shown in Fig. 3.27b.

Proof Consider in Fig. 3.27b the right-half plane with $v_{in} > 0$. Suppose the output voltage v_o is measured between node k and node l so that:

$$v_o = e_k - e_l \tag{3.145}$$

where e_k and e_l are measured with respect to the ground node shown in Fig. 3.27a.

Since $\mathcal{N}$ contains only strictly passive two-terminal resistors, it follows from the maximum node-voltage property in Theorem 3.8 that:

$$0 \le e_k \le v_{in} \tag{3.146}$$

$$0 \le e_l \le v_{in} \tag{3.147}$$

Inequality (3.147) can be rewritten as:

$$-v_{in} \le -e_l \le 0 \tag{3.148}$$

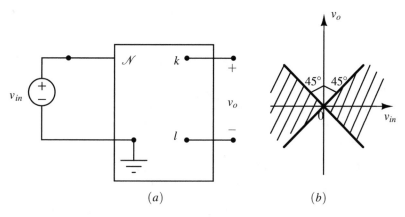

Fig. 3.27 Output voltage bounding region

Adding both inequalities (3.146) and (3.148), we get:

$$-v_{in} \leq e_k - e_l \leq v_{in} \tag{3.149}$$

Using our earlier definition: $v_o = e_k - e_l$ and simplifying we get:

$$|v_o| \leq v_{in} \tag{3.150}$$

Note that we had assumed $v_{in} > 0$ and used the right-half plane. A similar proof for the left-half plane ($v_{in} < 0$) would give: $|v_o| \leq -v_{in}$. We thus have: $|v_o| \leq |v_{in}|$.

□

3.7.2   Strict Monotonicity

Strict passivity does not impose any constraint on the slope of the resistor characteristic. It only requires that the product vi be positive except at the origin. For example, the tunnel diode described earlier is strictly passive. Yet the slope of its characteristics can assume both positive and negative values, depending on the operating point. Such characteristics are said to be **nonmonotonic**.

It is clear that resistive circuits made of nonmonotonic resistors would in general also give rise to a nonmonotonic DP and transfer characteristics. Hence in order to derive properties involving constraints on the slope of the DP and transfer characteristics, it is necessary to impose stronger conditions on the resistor characteristics. The strictly monotone-increasing, or **strictly increasing** for brevity, is one such condition which we investigate in this final section.

Strictly increasing means roughly that the slope of the characteristic is positive everywhere. More precisely, a two-terminal resistor is said to be **strictly increasing** iff, for all pairs of (distinct) points on its characteristic, say (v', i') and (v'', i'') $(v' > v'', i' > i'')$ we have:

$$(v' - v'')(i' - i'') > 0 \tag{3.151}$$

Note that a strictly increasing characteristic is not restricted to lie in the first and third quadrants only. Hence, a strictly increasing resistor need not be strictly passive, and a strictly passive resistor need not be strictly increasing.

Theorem 3.10 *Any circuit made of strictly increasing two-terminal resistors and independent sources has at most one solution.*

Proof Suppose there are two distinct operating points Q and Q', at some time t, corresponding to $(v_1, v_2, \ldots, v_b; i_1, i_2, \ldots, i_b)$ and $(v'_1, v'_2, \ldots, v'_b; i'_1, i'_2, \ldots, i'_b)$, respectively. Here we assume passive sign convention for all elements.

Since each of these two solutions satisfies Tellegen's theorem, so does their difference:

$$\sum_{k=1}^{b}(v_k - v'_k)(i_k - i'_k) = 0 \tag{3.152}$$

Observe that each term in Eq. (3.152) which corresponds to either a voltage source $(v_k = v'_k)$ or a current source $(i_k = i'_k)$ vanishes. However, since these are two distinct solutions and since all resistors are strictly increasing, there must exist at least one branch such that $(v_k - v'_k)(i_k - i'_k) > 0$ for this branch. This contradicts Eq. (3.152). Hence there cannot be two distinct operating points Q and Q'.   □

Theorem 3.11 *A one-port made of strictly increasing two-terminal resistors is itself strictly increasing.*

Proof Suppose the one-port N in Fig. 3.25 contains only strictly increasing resistors. Then for any two distinct DP voltages v and v', let (v_k, i_k) and (v'_k, i'_k), $k = 1, 2, \ldots, b$ denote the corresponding unique branch voltage and current solutions, for all b resistors inside N. It follows from Tellegen's theorem that:

$$(v - v')(i - i') = \sum_{k=1}^{b}(v_k - v'_k)(i_k - i'_k) \tag{3.153}$$

where the input term appears on the left of the equation because the input current i in Fig. 3.25 is defined as leaving the positive terminal of the voltage source.

Since $v \neq v'$, KVL requires that at least one of the $(v_k - v'_k)$ differs from 0; hence, at least one term on the right-hand side of Eq. (3.153) is positive, while all the others are ≥ 0 (since all resistors are strictly increasing) where the equality sign holds whenever $v_k = v'_k$ or $i_k = i'_k$. Consequently,

$$(v - v')(i - i') > 0 \tag{3.154}$$

whenever $v \neq v'$, i.e., the DP characteristic of N is strictly increasing. □

3.8 Conclusion

This chapter has given an **overview** of techniques for analysis of nonlinear networks. But, unlike dynamic nonlinear networks (the subject of Chap. 4), the realm of resistive nonlinear networks does have a general theory. Once the reader has mastered the concepts summarized below from this chapter, they can pick up this general theory from excellent references such as [1].

1. For resistive circuits, nodal analysis is applicable if the circuit contains only voltage-controlled resistors and independent current sources (which do not form cut sets among themselves).
2. The node equation for a linear resistive circuit is given by:

$$\mathbf{Y}_n \mathbf{e}(t) = \mathbf{i}_s(t) \tag{3.155}$$

where $\mathbf{Y}_n \triangleq \mathbf{A} \mathbf{Y}_b \mathbf{A}^T$ is called the node-admittance matrix; $\mathbf{A}$ is the reduced incidence matrix of the reduced digraph obtained by open-circuiting all branches corresponding to independent current sources from the original digraph; $\mathbf{Y}_b$ is the branch-admittance matrix; $\mathbf{i}_s(t)$ is the source vector whose kth entry is equal to the algebraic sum of all independent current sources entering node k.

   For a reduced digraph with n nodes and b branches, $\mathbf{Y}_b$ is a $b \times b$ matrix, $\mathbf{Y}_n$ is an $(n-1) \times (n-1)$ matrix, $\mathbf{A}$ is an $(n-1) \times b$ matrix; both $\mathbf{e}$ and $\mathbf{i}_s(t)$ are $n-1$ vectors.
3. A nonlinear resistive circuit driven only by independent current sources has a node equation given by:

$$\mathbf{A} \mathbf{g}(\mathbf{A}^T \mathbf{e}) = \mathbf{i}_s(t) \tag{3.156}$$

where $\mathbf{i} = \mathbf{g}(\mathbf{v})$ denotes the characteristics of all (voltage-controlled) resistors.
4. Both the linear and nonlinear node equations consist of $n-1$ equations in terms of the node voltage vector $\mathbf{e}$, where n is the number of nodes in the circuit. Hence the number of equations in nodal analysis does not depend on the number of branches in the circuit.

5. Every linear time-invariant resistive circuit has a tableau equation of the form:

$$
\underbrace{\begin{bmatrix} \mathbf{0} & \mathbf{0} & \mathbf{A} \\ -\mathbf{A}^T & \mathbf{I} & \mathbf{0} \\ \mathbf{0} & \mathbf{M}(t) & \mathbf{N}(t) \end{bmatrix}}_{\mathbf{T}(t)} \underbrace{\begin{bmatrix} \mathbf{e}(t) \\ \mathbf{v}(t) \\ \mathbf{i}(t) \end{bmatrix}}_{\mathbf{w}(t)} = \underbrace{\begin{bmatrix} \mathbf{0} \\ \mathbf{0} \\ \mathbf{u}_s(t) \end{bmatrix}}_{\mathbf{u}(t)}
\tag{3.157}
$$

The entries of $\mathbf{M}$ and $\mathbf{N}$ contain constant coefficients defining the resistors; the entries of $\mathbf{u}_s(t)$ contain constant or time functions defining the independent sources.

6. A linear time-invariant resistive circuit has a unique solution iff the tableau matrix $\mathbf{T}$ is nonsingular.

7. Every nonlinear resistive circuit has a tableau equation of the form:

$$
\mathbf{A}\mathbf{i}(t) = \mathbf{0}
$$

$$
\mathbf{v}(t) - \mathbf{A}^T\mathbf{e}(t) = \mathbf{0}
$$

$$
\mathbf{h}(\mathbf{v}(t), \mathbf{i}(t), t) = \mathbf{0}
\tag{3.158}
$$

8. A resistive circuit is said to be uniquely solvable iff Kirchhoff's laws and the branch equations are simultaneously satisfied by a unique set of branch voltages and a unique set of branch currents for all t.

9. The superposition theorem is applicable to any linear uniquely solvable resistive circuit. It allows us to find the solution by calculating first the solutions due to each independent source acting alone, and then adding them.

10. A one-port N is said to be well-defined iff it does not contain any circuit element which is coupled, electrically or nonelectrically, to some physical variable outside of N.

11. The Thévenin (Norton) theorem allows us to replace any well-defined linear current-controlled (voltage-controlled) resistive one-port by an equivalent one-port consisting of an equivalent Thévenin resistance R_{eq} (equivalent Norton conductance G_{eq}) in series (parallel) with an open-circuit (short-circuit) voltage source $v_{oc}(t)$ (current source $i_{sc}(t)$).

12. In applying the superposition, Thévenin and Norton theorems, all dependent sources must be left intact.

13. A two-terminal resistor is strictly passive iff $vi > 0$, for all points in its characteristic except the origin.

14. We studied the strict passivity, maximum node-voltage and transfer characteristic bounding regions for strictly passive networks.

15. A two-terminal resistor is strictly increasing iff $(v'-v'')(i'-i'') > 0$ for all pairs of distinct points (v', i') and (v'', i'') on its characteristic $(v' > v'', i' > i'')$.

16. We studied the uniqueness and strictly increasing closure properties for networks with strictly increasing resistors.

Lab 3: DC Simulation in QUCS

Objective: To understand DC simulation in QUCS
Theory:
Unlike the previous chapters, we first encourage you to do the lab component to this chapter. In other words, now that you have an understanding of the techniques for nonlinear resistive circuit analysis, be sure to simulate the circuits from this chapter (and the exercises) below in QUCS. In this lab, you will perform DC analysis (DC simulation in QUCS) for the nonlinear circuit shown in Fig. 3.28.

1. Suppose $E = 6$ V, $R = 2\ \Omega$. Solve analytically for the DC solution, specifically i_{QR} and v_{QR}. Although the circuit equations are trivial to set up, we recommend that you use tableau analysis so that you become familiar with the method.
2. Now, let $E = 2$ V, $R = 2\ \Omega$. Again analytically find the DC solution: i_{QR}, v_{QR}.

We will now simulate the circuit in QUCS.
Lab Exercise:

1. The circuit.[12] to be entered in QUCS is shown in Fig. 3.29. Use the Equation Defined Device (EDD) for specifying $\mathscr{N}_R$. This device can be found under nonlinear components.
2. Simulate the circuit for both $E = 6$ V and $E = 2$ V. Discuss the results. Specifically, what do you notice about the solution when $E = 2$ V. Explain the solution.

Fig. 3.28 Circuit for lab 3

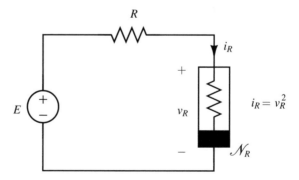

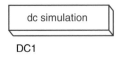

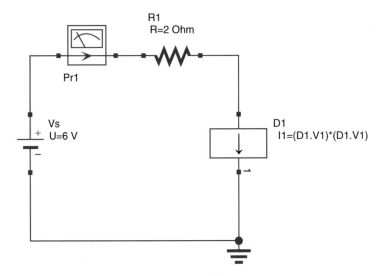

Fig. 3.29 QUCS schematic for circuit in Fig. 3.28

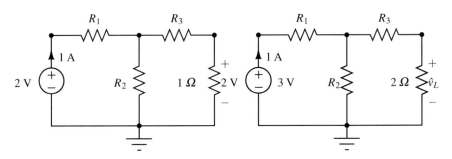

Fig. 3.30 Circuit for problem 3.2

Exercises

3.1 Show that the voltage gain of the CE amplifier in Fig. 3.8b is given by:

$$\frac{\tilde{v}_2}{v_s} = \frac{-h_{21}}{(h_{11} + R_1)(h_{22} + 1/R_2) - h_{12}h_{21}} \qquad (3.159)$$

3.2 Figure 3.30 shows two distinct networks $\mathcal{N}$ and $\hat{\mathcal{N}}$. Determine the value of $\hat{v}_L$.

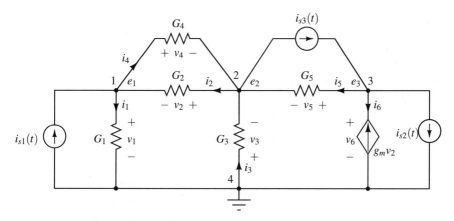

Fig. 3.31 Circuit for problem 3.3

3.3 Write the node equations for the circuit in Fig. 3.31, in terms of the reduced incidence matrix **A**.

3.4 Show that branch admittance matrix Y_b in Eq. (3.87) is a diagonal matrix, if $\mathcal{N}$ contains only two-terminal linear resistors and independent current sources.

3.5 To appreciate the benefits of superposition and its domain of applicability, consider a nonlinear resistor $v = \hat{v}(i) = i^3$ driven by two current sources $i_{s1}(t) = I_1 \cos \omega_1 t$ and $i_{s2}(t) = I_2 \cos \omega_2 t$ connected in parallel, where $I_1, \omega_1, I_2, \omega_2$ are constants. Calculate the voltage v when each source acts alone, and when they act together. In each case, reduce your answer to a sum of pure sine waves.

1. Does superposition hold for this circuit?
2. What are the frequency components of the output waveform for each case?

Exercise 4.13 explores further the frequency behavior of linear vs. nonlinear systems.

3.6 Find the Thévenin and Norton equivalent circuits for the one-ports shown in Fig. 3.32. If a particular circuit fails to have a Thévenin and/or Norton equivalent, explain.

3.7 In this exercise, we will derive the **maximum power transfer theorem for linear resistive circuits**.

Consider the circuit shown in Fig. 3.33. R_L models a loudspeaker in a concert hall. In order to maximize the output power delivered by the power amplifier (modeled by $v_s(t)$ in series with internal resistance R_1), a transformer with an appropriate turns ratio n is sandwiched between the amplifier and the loudspeaker.

1. Simplify the circuit by first finding the Thévenin equivalent at terminals $2, 2'$. Your v_{oc} and R_{eq} expressions should include a function of the transformer turns ratio n.

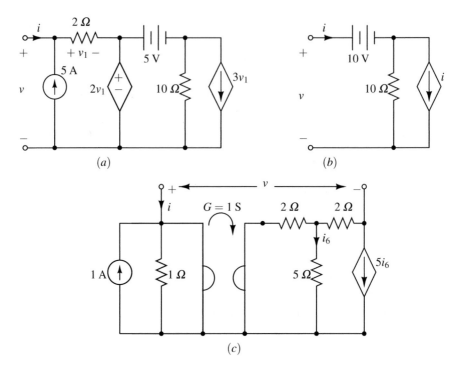

(a)   (b)

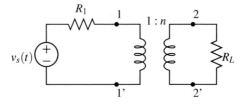

(c)

Fig. 3.32 (a-c) Circuits for problem 3.6

Fig. 3.33 Circuit for problem 3.7

2. From the answer in 1. above, determine the value of R_L (in terms of R_{eq}) that would maximize the power dissipated **in** R_L. To do this, you would have to find an expression for the power associated with R_L and use calculus.

The answer to 2. above is the maximum power transfer theorem for linear resistive circuits.

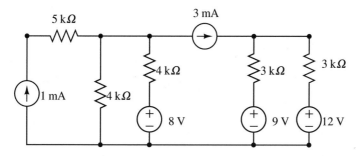

Fig. 3.34 Circuit for problem 3.8

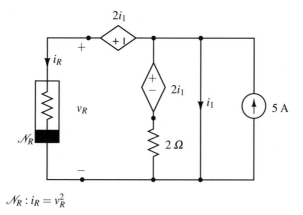

$$\mathcal{N}_R : i_R = v_R^2$$

Fig. 3.35 Circuit for problem 3.9

3.8 Using repeated application of Thévenin and Norton theorems, simplify the circuit in Fig. 3.34 to a single loop and then determine the voltage across current source 3 mA current source. This repeated simplification of a circuit by switching between Thévenin and Norton equivalents is called **source transforms**.

3.9 Find all possible values for i_R and v_R for the circuit shown in Fig. 3.35.

References

1. Chua, L.O.: Introduction to Nonlinear Network Theory. McGraw-Hill, New York (1969) (out of print)
2. Chua, L.O.: University of California, Berkeley EE100 Fall 2008 Supplementary Lecture Notes on Tableau Analysis (2008). Available, online: http://inst.eecs.berkeley.edu/~ee100/fa08/lectures/EE100supplementary_notes_12.pdf. Last accessed November 25th 2017
3. Chua, L.O., Desoer, C.A., Kuh, E.S.: Linear and Nonlinear Circuits. McGraw-Hill, New York (1987) (out of print)

Chapter 4
Dynamic Nonlinear Networks

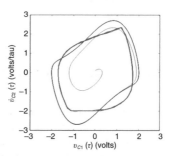

Simulated (blue) and experimental (red) limit cycle of a Van der Pol oscillator derived from Chua's circuit [1]

Abstract We will now learn about techniques for analyzing dynamic circuits, that are governed by differential equations. We will emphasize fundamental concepts behind dynamic nonlinear networks, time domain analysis of nth-order nonlinear networks, frequency response concepts, circuit analysis techniques for memristive networks and energy approaches (Lagrangian, Hamiltonian). We cannot hope to cover all the analysis techniques for dynamic nonlinear networks in detail in one chapter. Nevertheless, this chapter should prepare the reader for picking up advanced techniques for analyzing dynamic nonlinear networks from any specialized references.

4.1 Basic Concepts of Dynamic Nonlinear Networks

Definition 4.1 A network $\mathscr{D}$ consisting of an arbitrary interconnection of a finite number of four fundamental circuit elements, is called a **dynamic nonlinear network**.

Before we begin, we ask the reader recall from Sect. 3.3, to not "lose sight of the forest for its trees". That is, one should not be so consumed by the systematic

© Springer International Publishing AG, part of Springer Nature 2019
B. Muthuswamy, S. Banerjee, *Introduction to Nonlinear Circuits and Networks*,
https://doi.org/10.1007/978-3-319-67325-7_4

techniques that we lose total insight into circuit behavior. Also recall that it is often through the introduction of hypothetical, and sometimes pathological circuits, that one gains an in-depth understanding of this subject.

By Definition 4.1, $\mathscr{D}$ represents the class of all nonlinear networks other than resistive networks [3]. Since this class of dynamic networks is so much larger than the class of resistive networks, it is virtually impossible for us to formulate a general theory that is applicable to the solution of all dynamic networks. After all, it took us an entire chapter just to give an **overview** of the analysis techniques for resistive nonlinear circuits.

Hence in this chapter, we will primarily use **two-terminal dynamic elements** and also restrict our discussion to **fundamental concepts**, starting with the order of complexity.

4.1.1 Order of Complexity

Since the basic problem in dynamic nonlinear networks is to find the solution to a system of nonlinear ordinary differential equations, it is more appropriate to classify dynamic networks according to the "complexity" of their system of differential equations. It is well known that the solution to any system of differential equations can be found only to within a number of arbitrary constants $k_1, k_2, \cdots, k_n$. In order to determine the n arbitrary constants, we must specify n **independent initial conditions**.

Definition 4.2 A set of initial conditions is said to be **independent** if its values can be arbitrarily chosen.

Two systems of differential equations requiring different numbers of initial conditions are usually solved by quite different methods. Hence one meaningful basis for classification of $\mathscr{D}$ can be stated in terms of the number of independent initial conditions that must be specified in order to uniquely solve for the solution of the network.

Definition 4.3 The **order of complexity** of a dynamic network is the **minimum** number n of independent initial conditions that **must** be specified in terms of the circuit variables in $\mathscr{D}$, for **completely** describing the behavior of the network.

For convenience we shall refer to $\mathscr{D}$ as a first-order network if $n = 1$ and a second-order network if $n = 2$. Since $n \geq 1$ for any dynamic network, we might, for the sake of completeness, refer to any resistive network as a zero-order network.

It is important to observe that Definition 4.3 requires that the number of initial conditions be independent of one another. Definition 4.2 implies that none of the specified initial conditions can be derived from the rest, as Example 4.1.1 illustrates.

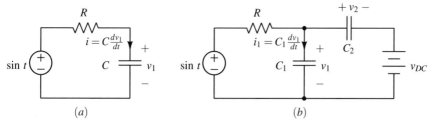

Fig. 4.1 Circuits for Example 4.1.1

Example 4.1.1 Determine the order of the two networks in Fig. 4.1.

Solution Since $\mathcal{D}_a$ contains only one storage element, we can easily infer that $\mathcal{D}_a$ is a first-order network. Now, since $\mathcal{D}_b$ contains two storage elements, it appears at first sight that we can specify two initial conditions, namely, the voltage $v_1(t_0)$ across capacitor C_1 and the voltage $v_2(t_0)$ across capacitor C_2 at some time t_0. However, since by KVL $v_2 = v_1 - v_{DC}$, the two initial conditions are dependent because once $v_1(t_0)$ is specified, $v_2(t_0)$ is constrained by $v_1(t_0) - v_{DC}$, and hence $v_2(t_0)$ cannot be specified arbitrarily. Therefore $\mathcal{N}_b$ is a first-order network.

From the theory of differential equations in the **normal form**, it is known that a system of n differential equations requires exactly n initial conditions for its solution. Therefore, it is important that we understand Definition 4.4 for the normal form.

Definition 4.4 The system of n first-order differential equations:

$$\frac{dx_1}{dt} = f_1(x_1, x_2, \cdots, x_n)$$

$$\frac{dx_2}{dt} = f_2(x_1, x_2, \cdots, x_n)$$

$$\cdots\cdots\cdots$$

$$\frac{dx_n}{dt} = f_n(x_2, x_2, \cdots, x_n) \tag{4.1}$$

is said to be in **normal form** because:

1. Only first-order time derivatives appear on the left-hand side of the equations.
2. No time derivatives appear on the right-hand side of the equations.
3. The dependent variables coincide with the **state variables** that appear on the left-hand side.

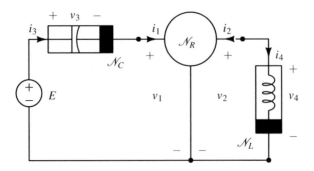

Fig. 4.2 Circuit for Example 4.1.2

Since the order of complexity is equal to the number of state variables when the system equations are written in normal form, one approach to determining the order of complexity would be to always write normal form equations for $\mathscr{D}$, as shown in Example 4.1.2.

Example 4.1.2 Determine the dynamic equations for the network in Fig. 4.2. The characteristics of the various circuit elements are:

$$\mathscr{N}_C: \quad q(v) = 2 - 3v^3 + 5v^5$$

$$\mathscr{N}_L: \quad \phi(i) = 1 + 2i - 3i^2 + i^3$$

$$\mathscr{N}_R: \quad i_1 = 1 + v_1 + 3v_1 i_2^3 - 4i_2^5$$

$$v_2 = 4 - i_2 v_1 - 2i_2^2 v_1^5 + v_1^3 \tag{4.2}$$

Solution For this circuit, we can determine the dynamic equations by inspection, without resorting to advanced techniques like MNA that will be discussed later in this chapter. Recall the memory property for inductors and capacitors from Chap. 1: Eq. (1.63) implies that a current $i_L(t_0)$ through an inductor is an initial condition. By duality, Eq. (1.71) implies that a voltage $v_C(t_0)$ across a capacitor is another suitable initial condition. Hence let us choose v_3 and i_4 to be the state variables. Hence the order of complexity is 2. Thus we need to obtain the following normal form:

$$\frac{dv_3}{dt} = f_1(v_3, i_4)$$

$$\frac{di_4}{dt} = f_2(v_3, i_4) \tag{4.3}$$

(continued)

Example 4.1.2 (continued)

For $\mathcal{N}_C$, in terms of circuit variables the $q_3 - v_3$ characteristic is $q_3(v_3) = 2 - 3v_3^3 + 5v_3^5$. Differentiating with respect to time and applying the chain rule we get:

$$i_3 = -9v_3^2 \frac{dv_3}{dt} + 25v_3^4 \frac{dv_3}{dt} \tag{4.4}$$

Since by KCL $i_3 = i_1$ and by KVL $v_1 = E - v_3$, we can simplify the equation above as:

$$\frac{dv_3}{dt} = \frac{i_1}{v_3^2 \left(25v_3^2 - 9\right)}$$

$$= \frac{1 + v_1 + 3v_1 i_2^3 - 4i_2^5}{v_3^2 \left(25v_3^2 - 9\right)}$$

$$= \frac{1 + (E - v_3) + 3(E - v_3)i_2^3 - 4i_2^5}{v_3^2 \left(25v_3^2 - 9\right)} \tag{4.5}$$

From KCL at the output port of $\mathcal{N}_R$: $i_2 = -i_4$. Thus we have the $\frac{dv_3}{dt}$ equation as:

$$\frac{dv_3}{dt} = \frac{1 + (E - v_3) - 3(E - v_3)i_4^3 + 4i_4^5}{v_3^2 \left(25v_3^2 - 9\right)} \tag{4.6}$$

With respect to the second state equation, for $\mathcal{N}_L$ in terms of circuit variables the $\phi_4 - i_4$ characteristic is: $\phi_4(i_4) = 1 + 2i_4 - 3i_4^2 + i_4^3$. Taking the derivative of this characteristic with respect to time and applying the chain rule:

$$v_4 = 2\frac{di_4}{dt} - 6i_4 \frac{di_4}{dt} + 3i_4^2 \frac{di_4}{dt} \tag{4.7}$$

Rewriting in terms of $\frac{di_4}{dt}$ and using the fact that by KVL $v_4 = v_2$, along with the v_2 definition from $\mathcal{N}_R$, we get:

$$\frac{di_4}{dt} = \frac{4 - i_2 v_1 - 2i_2^2 v_1^5 + v_1^3}{2 - 6i_4 + 3i_4^2} \tag{4.8}$$

Applying KCL: $i_2 = -i_4$, KVL: $v_1 = E - v_3$ and simplifying we get the second state equation:

$$\frac{di_4}{dt} = \frac{4 + i_4(E - v_3) - \left(2i_4^2(E - v_3)^2 + 1\right)(E - v_3)^3}{2 - 6i_4 + 3i_4^2} \tag{4.9}$$

We need to be aware that it may not be possible to write normal form equations, given a **specific choice** of state variables. To demonstrate the difficulty involved, let us examine Eq. (4.4) more closely. Observe that we were able to express i_3 in terms of v_3 and $\dot{v}_3$ because $\mathcal{N}_C$ was voltage controlled. Suppose instead $\mathcal{N}_C$ was charge-controlled: $v_3(q_3) = q_3^3 - q_3$. In this case, it is necessary that we express q_3 in terms of v_3 before applying the chain rule (to evaluate $\frac{dq_3}{dv_3}$). Unfortunately, this is not possible because q_3 is a multivalued function of v_3. This is equivalent to saying that the inverse function does not exist. In this case, the normal form equations cannot be obtained, if we insist on v_3 as the state variable.

There is, of course, no reason why we should insist on choosing only voltages and currents as state variables. Any other set of variables $x_1, x_2, \cdots, x_n$ is just as valid, provided Definition 4.4 is satisfied.

Although we could always determine the order of complexity by writing the state equations for $\mathcal{D}$, we shall now develop a simple technique for determining the order of complexity for a particular class[1] of $\mathcal{D}$ by inspection, i.e., without writing down any equation. In order to understand how this method works, it is important for us to obtain a deeper understanding of why initial conditions are necessary from the network's point of view, and to understand which electrical variables qualify as an appropriate set of initial conditions.

From the mathematical point of view, initial conditions are introduced as a "gimmick" for determining the values of the arbitrary constants associated with the solution to a system of differential equations. From the network's point of view, initial conditions are introduced because of our ignorance or incomplete knowledge of the past history of excitations that have been applied to the network. In order to understand the above reason, let us consider an arbitrary capacitor C_j of an arbitrary network $\mathcal{D}$. Suppose we want to find the charge $q_j(t)$ of this capacitor at time t, namely,

$$q_j(t) = \int_{-\infty}^{t} i_j(\tau)d\tau \qquad (4.10)$$

From Eq. (4.10) it is clear that $q_j(t)$ can be found only if we know the exact waveform of the capacitor current $i_j(t)$ from $t \to -\infty$ up to the present time t, that is, from the time the capacitor was manufactured. However, practically speaking, in any physical network excitations are applied at some finite time in the past, say $t = t_0$. Hence we would usually have information on the excitation of waveforms only for $t \geq t_0$. This ignorance of the past history of $i_j(t)$ prevents us from determining $q_j(t)$. However, let us rewrite Eq. (4.10) in the form:

$$q_j(t) = \int_{-\infty}^{t_0} i_j(\tau)d\tau + \int_{t_0}^{t} i_j(\tau)d\tau \qquad (4.11)$$

---

[1] We mean a network containing only two-terminal fundamental circuit elements and independent sources. No dependent sources, ideal transformers, gyrators, etc. are allowed.

The second integral can be found because we know $i_j(t)$ for $t \geq t_0$. It is the first integral that is giving us trouble. Observe, however, that at $t = t_0$ Eq. (4.10) becomes:

$$q_j(t_0) = \int_{-\infty}^{t_0} i_j(\tau)d\tau \tag{4.12}$$

Hence Eq. (4.11) becomes:

$$q_j(t) = q_j(t_0) + \int_{t_0}^{t} i_j(\tau)d\tau \tag{4.13}$$

where $t \geq t_0$. Equation (4.13) tells us that, provided we are interested only in knowing $q_j(t)$ for $t \geq t_0$, it is not necessary to know the entire past history of $i_j(t)$ for $t < t_0$. Instead, we need to know only the value of the charge q_j in the capacitor at the initial time t_0. This value $q_j(t_0)$ is called the **initial condition**.

Let us now recall that a capacitor is characterized by a curve in the $v - q$ plane, and if we know $v(t)$, we can find $q(t)$ and vice versa. Since it is necessary to know the initial condition $q(t_0)$ in order to find $q(t)$ for $t \geq t_0$, it follows that it is necessary to know $v(t_0)$ in order to find $v(t)$ for $t \geq t_0$. However, since given $v(t_0)$ we can find $q(t_0)$ and vice versa, it is sufficient to specify an initial condition either in terms of capacitor charge or voltage at time t_0. But, notice an examination of Eq. (4.13) shows that specifying capacitor current $i_j(t_0)$ would not do any good because one cannot determine $q_j(t_0)$ from this information alone. We conclude therefore that the current in a capacitor is not an appropriate initial condition.

By exact dual arguments, we find that for an inductor:

$$\phi_j(t) = \phi_j(t_0) + \int_{t_0}^{t} v_j(\tau)d\tau \tag{4.14}$$

Thus we can specify either the flux linkage $\phi(t_0)$ or inductor current $i(t_0)$ as appropriate initial conditions. The voltage across an inductor at t_0 is not an appropriate initial condition.

Let us now explore the concept of independent initial conditions in more detail. We have already seen in Example 4.1.1 that the order of complexity of a dynamic network may not be equal to the number of energy storage elements, because some initial conditions may not be independently specified. In order to diagnose the source of "dependency," let us consider the more complicated network $\mathscr{D}$ in Fig. 4.3.

Since $\mathscr{D}$ contains ten energy-storage elements (six capacitors and four inductors), it appears that we can specify 10 initial conditions, $v_{C1}, v_{C2}, v_{C3}, v_{C4}, v_{C5}, v_{C6}$, i_{L1}, i_{L2}, i_{L3}, and i_{L4}. However a more careful inspection of the network shows that not all these initial conditions are independent. For example, the loop consisting of

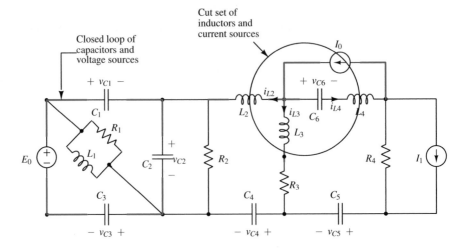

Fig. 4.3 An example of the two possible sources of dependent initial conditions, namely, a loop of capacitors and voltage sources, and a cut set of inductors and current sources

capacitors C_1, C_2, C_3 and voltage source E_0 imposes a constraint due to KVL:

$$v_{C1} + v_{C2} + v_{C3} = E_0 \tag{4.15}$$

This equation implies that only two of the three initial conditions v_{C1}, v_{C2}, and v_{C3} can be specified arbitrarily. We conclude that although there are six capacitors, only five capacitor voltages are independent. Similarly, the cut set consisting of inductors L_2, L_3, L_4 and current source I_0 imposes a constraint due to KCL:

$$i_{L2} + i_{L3} + i_{L4} = I_0 \tag{4.16}$$

Thus only two of three initial conditions i_{L2}, i_{L3}, i_{L4} can be specified arbitrarily. Hence we conclude that although there are four inductors, only three inductor currents are independent. The maximum number of initial conditions that can be specified is therefore equal to $5 + 3 = 8$.

Based on our discussion above, it is clear that a dependency exists whenever it is possible to write a constraint involving only capacitor voltages and voltage sources; therefore, we must subtract one initial condition from the total number of energy-storage elements. Similarly, it is clear that a dependency exists whenever it is possible to write a constraint involving only inductor currents and current sources; therefore, we must likewise subtract one initial condition from the total number of energy-storage elements. The first constraint involving only capacitor voltages and voltage sources occurs if and only if there exists a loop in the network containing **only** capacitors and independent voltage sources. A dual argument applies to inductors and current sources: a constraint occurs if and only if there exists a cut set in the network containing **only** inductors and current sources.

Hence, we have the following theorem [4] for the order of complexity:

Theorem 4.1 (Order of Complexity)   *Let $\mathscr{D}$ be a network containing only two-terminal fundamental circuit elements and independent sources. Then the order of complexity m of $\mathscr{D}$ is given by:*

$$m = (b_L + b_C + b_M) - (n_M + n_{CE} + n_{LM}) - (\hat{n}_M + \hat{n}_{LJ} + \hat{n}_{CM}) \qquad (4.17)$$

where:

1. b_L *is the total number of inductors*
2. b_C *is the total number of capacitors*
3. b_M *is the total number of memristors*
4. n_M *is the number of independent loops containing only memristors*
5. n_{CE} *is the number of independent loops containing only capacitors and voltage sources*
6. n_{LM} *is the number of independent loops containing only inductors and memristors*
7. $\hat{n}_M$ *is the number of independent cut sets containing only memristors*
8. $\hat{n}_{LJ}$ *is the number of independent cut sets containing only inductors and current sources*
9. $\hat{n}_{CM}$ *is the number of independent cut sets containing only capacitors and memristors*

Proof[2]
We have just discussed the order of complexity for $\mathscr{D}$ without memristors: $m = (b_L + b_C) - n_{CE} - \hat{n}_{LJ}$.

From the definition of a memristor, for a $\mathscr{D}$ with $n_M = n_{LM} = \hat{n}_M = \hat{n}_{CM} = 0$, each memristor introduces a new state variable and we thus have: $m = (b_L + b_C + b_M) - n_{CE} - \hat{n}_{LJ}$.

Observe next that a constraint among state variables occurs whenever an independent loop consisting of elements corresponding to those specified in the definition of n_M and n_{LM} is present in the network. This is because we assume the algebraic sum of flux linkages around any loop (charges flowing into any node, recall equivalence of KCL node to cut sets, Theorem 3.1) is zero. We now have: $m = (b_L + b_C + b_M) - (n_M + n_{CE} + n_{LM}) - \hat{n}_{LJ}$.

Finally, by duality, a constraint among state variables again occurs whenever an independent cut set consisting of elements corresponding to those specified in the definition of $\hat{n}_M$ and $\hat{n}_{CM}$ is present in the network. We thus have: $m = (b_L + b_C + b_M) - (n_M + n_{CE} + n_{LM}) - (\hat{n}_M + \hat{n}_{LJ} + \hat{n}_{CM})$.   □

---

[2]With respect to $\mathscr{D}$ with memristors, the concept of using (ϕ_M, q_M) to determine the degree of complexity and write network equations is further explored in Sect. 4.4.1.

4.1.2 Principles of Duality

In light of the enormous solution space of dynamic nonlinear networks, it would be instructive to check if there are techniques that help us reduce this solution space. One such powerful technique is duality (alluded to in earlier chapters), and since duality is particularly useful in the analysis of dynamic networks, we have deferred a rigorous discussion of duality till this chapter.

A significant fact about dual networks is that once we know the solution of one network, the solution of the dual network can be obtained immediately by simply interchanging the symbols. This means that as soon as we know the behavior and properties of one network, we immediately know the behavior of the properties of dual network. Hence a lot of redundancy is avoided if we can recognize dual networks.

Generally speaking, we say two systems or phenomena are **duals** of each other if we can exhibit some kind of one-to-one correspondence between various quantities or attributes of the two systems. For example, in physics, for each translational system or problem there exists a corresponding rotational system or problem, and they are usually referred to as dual systems. In mathematics, two equations which differ only in symbols but are otherwise identical in form are said to be dual equations. In electrical engineering, besides circuits, duality is widely used in digital design because of dual Boolean relationships. The recognition of dual quantities, attributes, phenomena, properties, or concepts often leads to the discovery and invention of new ideas.

Before we render the concept of duality more precise, it is instructive to consider first the two nonlinear networks shown in Fig. 4.4a and b. The laws of elements and the laws of interconnection for these two networks are readily obtained and tabulated in Table 4.1. A careful comparison of the expressions in the two columns of this table reveals a one-to-one correspondence between the equations. As a matter of fact, except for the symbols, the equations in the two columns are in identical form. Observe that, had we replaced v_j by i'_j, i_j by v'_j, ϕ_j by q'_j, q_j by ϕ'_j for the variables in the left column, the result would be identical with that in the right column, and therefore the two networks are said to be dual networks. We are now ready to precisely define the concept of duality.

Definition 4.5 (Duality) Let $\mathscr{D}$ and $\mathscr{D}'$ be a pair of networks each containing b two-terminal network elements which are not controlled sources. Then $\mathscr{D}$ and $\mathscr{D}'$ are **dual networks** if the elements in $\mathscr{D}$ and $\mathscr{D}'$ can be labeled, respectively, as $b_1, b_2, \cdots, b_b$ and $b'_1, b'_2, \cdots, b'_b$ such that the circuit equations for the two networks are identical.

A few points to note from Definition 4.5:

1. It is possible to generalize the definition of dual networks to include controlled sources. However, the procedure for constructing such networks is much more complicated and will not be discussed in this book.

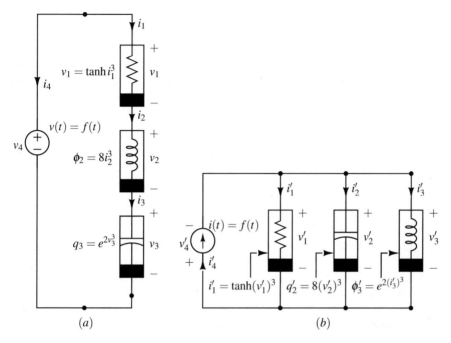

Fig. 4.4 The dual of a series nonlinear network is a parallel nonlinear network

Table 4.1 Circuit equations for the networks in Fig. 4.4

Network of Fig. 4.4a	Network of Fig. 4.4b
Laws of elements	*Laws of elements*
$v_1 = \tanh i_1^3$	$i_1' = \tanh(v_1')^3$
$\dfrac{di_2}{dt} = \dfrac{v_2}{24i_2^2}$	$\dfrac{dv_2'}{dt} = \dfrac{i_2'}{24v_2'}$
$\dfrac{dv_3}{dt} = \dfrac{i_3}{2e^{2v_3}}$	$\dfrac{di_3'}{dt} = \dfrac{v_3'}{2e^{2i_3'}}$
$v_4 = f(t)$	$i_4' = f(t)$
Laws of interconnection	*Laws of interconnection*
KVL: $v_1 + v_2 + v_3 - v_4 = 0$	KCL: $i_1' + i_2' + i_3' - i_4' = 0$
KCL: $i_1 + i_4 = 0$	KVL: $v_1' + v_4' = 0$
$i_1 - i_2 = 0$	$v_1' - v_2' = 0$
$i_2 - i_3 = 0$	$v_2' - v_3' = 0$

Note the derivative relationships in the laws of elements have been written in normal form

2. We have defined duality for dynamic networks, $\mathscr{D}$, but it should be obvious that the definition is also applicable to (nonlinear) resistive networks $\mathscr{N}$.
3. In order to find $\mathscr{D}'$, we need to uncover the duality relationships that must be satisfied by the laws of elements **and** the laws of interconnections. Due to space limitations, we will only cover the laws of elements. With respect to duality and the laws of interconnections, we will restrict our discussion to memristive networks. For a general graph theoretic approach to duality relationships from the laws of interconnections, the reader is referred to [3].

Definition 4.6 (Dual Resistor) If element b_j is a two-terminal resistor in $\mathscr{D}$ characterized by a curve Γ in the $v - i$ plane, then the corresponding dual element b'_j in $\mathscr{D}'$ must also be a two-terminal resistor characterized by the same curve Γ in the $i' - v'$ plane.

For example, if element b_j of $\mathscr{D}$ is a resistor characterized by $i_j = v_j^3 - 3v_j$, then the dual resistor in $\mathscr{D}'$ is a resistor characterized by $v'_j = i'^3_j - 3i'_j$. Observe that the dual of a given resistor is a new resistor, which may need a new name and a new symbol. However, there are some two-terminal elements which have the interesting property that the dual of the element is the same element with its two terminals interchanged. For such elements, a new symbol is obviously not needed. The simplest example of this type of element is the ideal diode.

Definition 4.7 (Dual Inductor) If element b_j in $\mathscr{D}$ is a two-terminal inductor characterized by a curve Γ in the $i - \phi$ plane, then the corresponding dual element b'_j in $\mathscr{D}'$ must be a capacitor characterized by the same curve Γ in the $v' - q'$ plane.

For example, the dual of an inductor characterized by $\phi = \log i$ is a capacitor characterized by $q' = \log v'$.

Definition 4.8 (Dual Capacitor) If element b_j in $\mathscr{D}$ is a two-terminal capacitor characterized by a curve Γ in the $v - q$ plane, then the corresponding dual element b'_j in $\mathscr{D}'$ must be an inductor characterized by the same curve Γ in the $i' - \phi'$ plane.

For example, the dual of a capacitor characterized by $q = \tanh v$ is an inductor characterized by $q' = \log v'$.

Definition 4.9 (Dual Ideal Memristor) If element b_j in $\mathscr{D}$ is a two-terminal **ideal memristor** characterized by a curve Γ in the $\phi - q$ plane, the corresponding dual element b'_j in $\mathscr{D}'$ must be an **ideal menductor** characterized by the same curve Γ in the $q' - \phi'$ plane.

Note that by mutatis mutandis, we can define the dual of an ideal menductor.

Definition 4.10 (Dual Memristive Device) If element b_j in $\mathscr{D}$ is a two-terminal **current-controlled (voltage-controlled) memristive device**, the corresponding dual element b'_j in $\mathscr{D}'$ must be a **voltage-controlled (current-controlled) memristive device**.

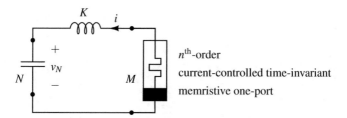

Fig. 4.5 Circuit for Example 4.1.3

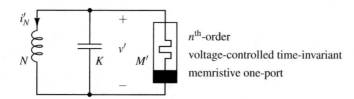

Fig. 4.6 Dual network $\mathscr{D}'$ for the circuit in Fig. 4.5

Example 4.1.3 Determine the dual of the memristive circuit in Fig. 4.5.

Solution Based on Definitions 4.8, 4.7, and 4.10, the dual of the circuit is shown in Fig. 4.6.

In other words, the dual of a linear capacitor with capacitance N F ($q - v$ relationship: $q = Nv$) is a linear inductor with inductance N H ($\phi' - i'$ relationship: $\phi' = Ni'$). Analogously, the dual of a linear inductor with inductance K H ($\phi - i$ relationship: $\phi = Ki$) is a linear capacitor with capacitance K F ($q' - v'$ relationship: $q' = Kv'$).

For the memristive device in $\mathscr{D}$, we have (recall Eq. (1.86)):

$$v = R(\mathbf{x}, i)i$$

$$\dot{\mathbf{x}} = f(\mathbf{x}, i) \tag{4.18}$$

Hence the dual voltage-controlled equations are:

$$i' = G(\mathbf{x}', v')v'$$

$$\frac{d\mathbf{x}'}{dt} = f(\mathbf{x}', v') \tag{4.19}$$

(continued)

Example 4.1.3 (continued)
Since we have a series network for $\mathscr{D}$, simple application of KVL and the element laws gives:

$$\frac{dv_N}{dt} = \frac{i}{N}$$

$$\frac{di}{dt} = \frac{1}{K} \left(v_N + R(\mathbf{x}, i)i \right)$$

$$\frac{d\mathbf{x}}{dt} = f(\mathbf{x}, i) \qquad\qquad (4.20)$$

Notice we have normal form equations for the $\mathscr{D}$. Using duality, we get:

$$\frac{di'_N}{dt} = \frac{v'}{N}$$

$$\frac{dv'}{dt} = \frac{1}{K} \left(i'_N + G(\mathbf{x}', v')v' \right)$$

$$\frac{d\mathbf{x}'}{dt} = f(\mathbf{x}', v') \qquad\qquad (4.21)$$

Table 4.2 summarizes the dual relationships that we have discussed.

On a brief note, the question of existence and uniqueness theorems for dynamic nonlinear networks does not carry much meaning [6], unlike linear dynamic networks. Two reasons are: the solution of the normal form Eq. (4.1) can exhibit many qualitatively different behaviors, depending only on the choice of the initial state. The second reason is that some steady state behavior can be extremely complicated (chaos in Chap. 5) **precluding** the existence of a closed form solution.

So, the correct **approach** is to study the qualitative behavior of dynamic nonlinear networks. There are a variety of techniques, in the context of the scope of this book, we will discuss impasse points later in Sect. 4.2.1.6. Other advanced concepts can be found in [6].

4.2   Time Domain Analysis of nth-Order Nonlinear Networks

In this section, we will analyze nth-order dynamic nonlinear networks **in the time domain**. That is, we will write differential equations as functions of time for the dynamic networks in question. We will start with first-order networks because a variety of important results can be easily understood using first-order networks [12].

Table 4.2 Common dual quantities

Network $\mathscr{D}$	Network $\mathscr{D}'$
Current i_j	Voltage v'_j
Voltage v_j	Current i'_j
Flux linkage ϕ_j	Charge q'_j
Charge q_j	Flux linkage ϕ'_j
Nonlinear resistor characterized by a curve Γ in the $v - i$ plane	Nonlinear resistor characterized by a curve Γ in the $i' - v'$ plane
Linear resistor with a resistance R Ω	Linear resistor with a conductance R S
Nonlinear inductor characterized by a curve Γ in the $i - \phi$ plane	Nonlinear capacitor characterized by a curve Γ in the $v' - q'$ plane
Linear inductor with an inductance K H	Linear capacitor with a capacitance K F
Nonlinear capacitor characterized by a curve Γ in the $v - q$ plane	Nonlinear inductor characterized by a curve Γ in the $i' - \phi'$ plane
Linear capacitor with a capacitance N F	Linear inductor with an inductance N H
Voltage source, $v_j = f(t)$	Current source, $i'_j = f(t)$
Current source, $i_j = g(t)$	Voltage source, $v'_j = g(t)$
Short circuit	Open circuit
Open circuit	Short circuit
Series branches	Parallel branches
Ideal memristor	Ideal menductor
Current-controlled memristive device	Voltage-controlled memristive device

Since circuit analysis techniques for memristor networks are still a topic of active research, we will postpone discussion of such networks till Sect. 4.4. Hence until then, our circuits will contain only capacitors and inductors as the dynamic element(s).

4.2.1   First-Order Circuits

Circuits made of one capacitor,[3] resistors, and independent sources are called **first-order circuits** [8]. Note that "resistor" is understood in the broad sense: it includes controlled sources, gyrators, ideal transformers, etc.

In this section,[4] we study first-order circuits made of linear time-invariant elements and independent sources. Any such circuit can be redrawn as shown in

---

[3] We will primarily focus on capacitor circuits in this section since the corresponding dual inductor circuit(s) can be easily derived using the ideas of duality discussed in Sect. 4.1.2. The reader is encouraged to derive the results for the dual inductor case as they read this section, to enhance their conceptual understanding.

[4] It would be helpful to review Sect. 1.9.3, specifically the memory and continuity properties.

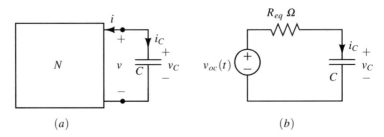

Fig. 4.7 (a) First-order RC circuit. (b) Thévenin equivalent

Fig. 4.7a, where the one-port N is assumed to include all other elements (e.g., independent sources, resistors, controlled sources, gyrators, ideal transformers, etc.). Applying the Thévenin equivalent one-port Theorem 3.6 from Chap. 3, we can, in most instances, replace N by the equivalent circuit shown in Fig. 4.7b.

Applying KVL we obtain

$$R_{eq}i_C + v_C = v_{oc}(t) \tag{4.22}$$

Substituting $i_C = C\dot{v}_C$ and solving for $\dot{v}_C$, we obtain:

$$\dot{v}_C = -\frac{v_C}{R_{eq}C} + \frac{v_{oc}(t)}{R_{eq}C} \tag{4.23}$$

Since the **first**-order linear differential equation above is in normal form, $v_C(t)$ is the state variable. Recall from our discussion of initial conditions in Sect. 4.1.1 that $v_C(t)$ depends only on the initial condition $v_C(t_0)$ and the waveform $v_{oc}(\cdot)$ over $[t_0, t]$.

In Sect. 4.2.1.1 we show that the solution of any first-order linear circuit can be found by inspection, provided N contains only DC sources. By repeated application of this "inspection method," Sect. 4.2.1.2 shows how the solution can be easily found if N contains only piecewise-constant sources. This method is then applied in Sect. 4.2.1.3 for finding the solution—called the **impulse response**—when the circuit is driven by an **impulse** $\delta(t)$. Finally, Sect. 4.2.1.4 gives an explicit integration formula for finding solutions under arbitrary excitations, which is then applied in Sects. 4.2.1.5 and 4.2.1.6.

4.2.1.1 Circuits Driven by DC Sources

When N contains only DC sources, $v_{oc}(t) = v_{oc}$ is a constant in Fig. 4.7b and in Eq. (4.23). Let us rewrite the equation as follows:

$$\overset{\bullet}{x} = \frac{x}{\tau} + \frac{x(t_\infty)}{\tau} \qquad (4.24)$$

where

$$x \overset{\triangle}{=} v_C$$

$$x(t_\infty) \overset{\triangle}{=} v_{oc}$$

$$\tau \overset{\triangle}{=} R_{eq}C \qquad (4.25)$$

Given any initial condition $x = x(t_0)$ at $t = t_0$, Eq. (4.24) has the unique solution:

$$x(t) = x(t_\infty) + [x(t_0) - x(t_\infty)]e^{\frac{-(t-t_0)}{\tau}} \qquad (4.26)$$

which holds for all times t, i.e., $-\infty < t < \infty$. To verify that this is indeed the solution, simply substitute Eq. (4.26) into Eq. (4.24) and show that both sides are identical. Observe that at $t = t_0$, Eq. (4.26) reduces to $x(t) = x(t_0)$, which makes physical sense. Note also that the solution given by Eq. (4.26) is valid whether τ is positive or negative.

The solution in Eq. (4.26) is determined only by three parameters $x(t_0), x(t_\infty)$ and τ. We call them **initial state**, **equilibrium state**, and **time constant**, respectively. To see why $x(t_\infty)$ is called the equilibrium state, note that if $x(t_0) = x(t_\infty)$, then Eq. (4.24) gives $\overset{\bullet}{x}(t_0) = 0$ and thus $x(t) = x(t_\infty)$ for all t. Hence the circuit remains "motionless" or in equilibrium.

Since the inspection method to be developed in this section depends crucially on the ability to sketch the exponential waveform quickly, the following properties are extremely useful. These properties in turn depend on whether τ is positive or negative. For $\tau > 0$, the exponential waveform in Eq. (4.26) tends to a constant as $t \to \infty$. For $\tau < 0$, the exponential waveform in Eq. (4.26) tends to $\pm\infty$, as $t \to \infty$. Hence it is convenient to consider the two cases separately.

Case 1: $\tau > 0$ In this case Eq. (4.26) shows that $x(t) - x(t_\infty)$, i.e., the distance between the present state and the equilibrium state $x(t_\infty)$ decreases exponentially. For all initial states, the solution $x(t)$ approaches equilibrium and $|x(t) - x(t_\infty)|$ decreases exponentially with a time constant τ. The solution in Eq. (4.26) for $\tau > 0$ is sketched in Fig. 4.8 for two different initial states $x(t_0)$ and $\tilde{x}(t_0)$. Observe that because τ is positive, $x(t) \to x(t_\infty)$ as $t \to \infty$.

Thus when $\tau > 0$ we say the system in Eq. (4.24) is **stable**,[5] because any initial deviation $x(t_0) - x(t_\infty)$ decays exponentially and $x(t) \to x(t_\infty)$ as $t \to \infty$.

---

[5]Stability is a system property, not a signal property. We say the signals associated with a stable system are bounded. In system terminology, we are using the concept of bounded-input bounded-output (BIBO) stability.

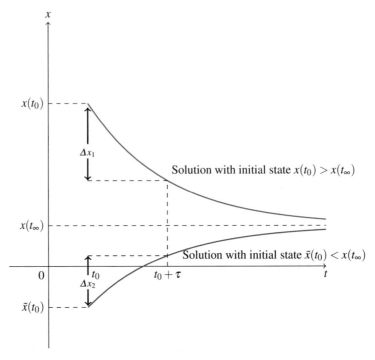

Fig. 4.8 The solution tends to the equilibrium state $x(t_\infty)$ as $t \to \infty$ when the time constant τ is positive. $\Delta x_1 = 0.63[x(t_0) - x(t_\infty)]$, $\Delta x_2 = 0.63[x(t_\infty) - \tilde{x}(t_0)]$

The **exponential waveforms** in Fig. 4.8 can be accurately sketched using the following observations:

1. After one time constant τ, the distance between $x(t)$ and $x(t_\infty)$ decreases approximately by 63% of the initial distance $|x(t_0) - x(t_\infty)|$.
2. After five time constants, $x(t)$ practically attains the equilibrium state (or **steady-state**) value $x(t_\infty)$ ($e^{-5} \approx 0.007$).

Example 4.2.1 Recall the opamp voltage follower from Example 2.5.3, but now we have a switch closing at $t = 0$ as shown in Fig. 4.9. Sketch $v_o(t)$ for $t \geq 0$.

Solution The switch shown models the fact that in practice, the output is observed to reach the 10 V solution after a small but finite time. In order to predict this **transient** behavior before equilibrium is reached, we will use the

(continued)

Example 4.2.1 (continued)
finite gain opamp model from Exercise 2.5, augmented with a capacitor, to
obtain the **dynamic model** shown in Fig. 4.10a.

To analyze this first-order circuit, we extract the capacitor and replace the
remaining circuit by its Thévenin equivalent as shown in Fig. 4.10b, where:

$$R_{eq} = \frac{R}{A+1} \approx \frac{R}{A} \quad \text{since } A \gg 1$$

$$v_{oc} = \frac{10A}{A+1} \approx 10 \quad \text{since } A \gg 1 \tag{4.27}$$

Assuming $A = 10^5, R = 100\,\Omega, C = 3\,\mathrm{F}$, we obtain $R_{eq} \approx 10^{-3}\,\Omega$
and $v_{oc} \approx 10\,\mathrm{V}$. Consequently, the time constant and equilibrium state are
given, respectively, by $\tau = R_{eq}C = 3\,\mathrm{ms}$ and $v_o(t_\infty) \approx 10\,\mathrm{V}$. Assuming
the capacitor is initially uncharged, the resulting output voltage can be easily
sketched as shown in Fig. 4.11. Note that after five time constants or 15 ms,
the output is practically equal to 10 V.

Case 2: $\tau < 0$ In this case Eq. (4.26) shows that the quantity $x(t) - x(t_\infty)$ increases
exponentially for all initial states, i.e., the solution $x(t)$ diverges from equilibrium
and hence the corresponding system is **unstable**. The solution for Eq. (4.26) is
sketched in Fig. 4.12 for two different initial states $x(t_0)$ and $\tilde{x}(t_0)$. Observe that
since the time constant τ is negative, as $t \to \infty$, $x(t) \to \infty$ if $x(t_0) > x(t_\infty)$ and
$x(t) \to -\infty$ if $x(t_0) < x(t_\infty)$.

However, if we run time "backward," then $x(t) \to x(t_\infty)$ as $t \to -\infty$.
Consequently, $x(t_\infty)$ can be interpreted as a **virtual equilibrium state**.

Fig. 4.9 Circuit for
Example 4.2.1

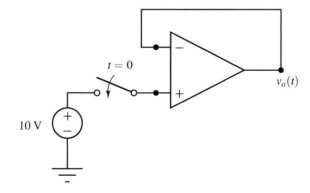

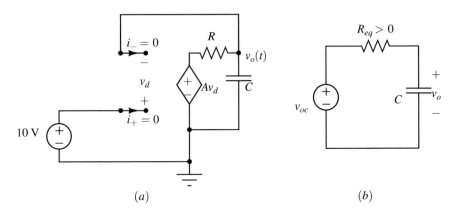

Fig. 4.10 (a) Dynamic opamp model (b) Thévenin equivalent, notice R_{eq} is positive

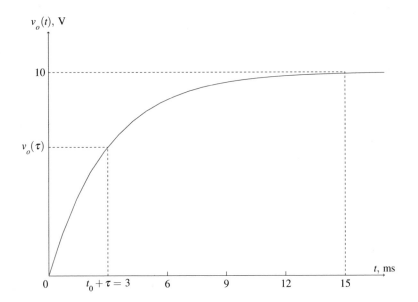

Fig. 4.11 Exponential voltage waveform for Example 4.2.1

Analogous to the stable case, the exponential waveform can be accurately sketched using the observation that at $t = t_0 + |\tau|$, the distance $|x(t_0 + |\tau|) - x(t_\infty)|$ is approximately 1.72 times the initial distance $|x(t_0) - x(t_\infty)|$.

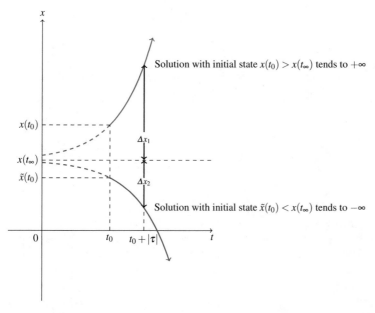

Fig. 4.12 The solution tends to the "virtual" equilibrium state $x(t_\infty)$ as $t \to -\infty$ when the time constant τ is negative. $\Delta x_1 = 1.72[x(t_0) - x(t_\infty)]$, $\Delta x_2 = 1.72[x(t_\infty) - \tilde{x}(t_0)]$

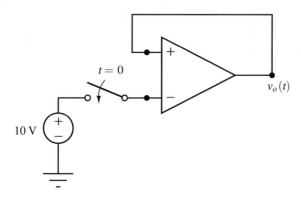

Fig. 4.13 Circuit for Example 4.2.2

Example 4.2.2 Consider the positive feedback opamp circuit shown in Fig. 4.13. Determine $v_o(t)$ for $t \geq 0$.

Solution The opamp circuit in Fig. 4.13 is identical to that of Fig. 4.9 except for an interchange between the inverting $(-)$ and noninverting $(+)$ terminals.

(continued)

Example 4.2.2 (continued)
Using the ideal opamp model in the linear region, we would obtain exactly the same answer as before, namely $v_o = 10$ V for $t \geq 0$, provided $E_{\text{sat}} > 10$ V.

But, let us see what happens if the opamp is replaced by the dynamic model adopted earlier, as shown in Fig. 4.14a. Note now the polarity of v_d is reversed. The parameters in the Thévenin equivalent circuit now become:

$$R_{\text{eq}} = -\frac{R}{A-1} \approx -\frac{R}{A} \text{ since } A \gg 1$$

$$v_{\text{oc}} = \frac{10A}{A-1} \approx 10 \text{ since } A \gg 1 \qquad (4.28)$$

Notice R_{eq} is now negative. Assuming the same parameter values as in Example 4.2.1, we obtain $R_{\text{eq}} = -10^{-3}\,\Omega$ and $v_{\text{oc}} \approx 10$ V. Consequently, the time constant and **virtual** equilibrium state are now given by $\tau \approx -3$ ms and $v_{\text{oc}}(t_\infty) \approx 10$ V, respectively. Hence the solution drastically differs from that of Example 4.2.1:

$$v_o(t) = 10 \left(1 - e^{\frac{t}{3\,\text{ms}}}\right) \qquad (4.29)$$

$v_o(t) \to -\infty$ as $t \to \infty$. Of course, in practice, when $v_o(t)$ decreases to $-E_{\text{sat}}$, the opamp saturates and the solution would remain constant at $-E_{\text{sat}}$. The sketch of $v_o(t)$ is trivial and is left as an exercise for the reader.

Example 4.2.2 shows us why the "middle" segment in the positive feedback circuit (Fig. 2.36) and Schmitt trigger VTCs from Sect. 2.5.3 are physically absent.

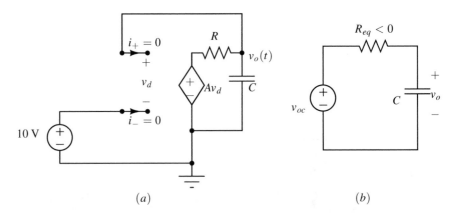

(a)   *(b)*

Fig. 4.14 (a) Dynamic opamp model (b) Thévenin equivalent, notice R_{eq} is negative

Parasitic elements such as capacitors result in the opamp circuit model corresponding to the "middle" segment to display unstable behavior. In other words, the R_{eq} seen by the parasitic capacitor turns out to be negative. A detailed analysis is given in [20].

We will often need to calculate the time interval between two prescribed points on an exponential waveform. Given any two points $[t_j, x(t_j)]$ and $[t_k, x(t_k)]$ on an exponential waveform (see for example Figs. 4.8 and 4.12), the time it takes to go from $x(t_j)$ to $x(t_k)$ is given by the **elapsed time formula**:

$$t_k - t_j = \tau \ln \frac{x(t_j) - x(t_\infty)}{x(t_k) - x(t_\infty)} \tag{4.30}$$

To derive Eq. (4.30), let $t = t_j$ and $t = t_k$ in Eq. (4.26), respectively:

$$x(t_j) - x(t_\infty) = [x(t_0) - x(t_\infty)]e^{\frac{-(t_j - t_0)}{\tau}} \tag{4.31}$$

$$x(t_k) - x(t_\infty) = [x(t_0) - x(t_\infty)]e^{\frac{-(t_k - t_0)}{\tau}} \tag{4.32}$$

Dividing Eq. (4.31) by (4.32) and taking the logarithm on both sides, we obtain Eq. (4.30). Notice the derivation does not depend on whether τ is positive or negative.

We are now ready to formally state the inspection method. Consider again the first-order RC circuit from Fig. 4.7a where all independent sources inside N are DC sources. Equation (4.26) gives us the voltage across the capacitor:

$$v_C(t) = v_C(t_\infty) + [v_C(t_0) - v_C(t_\infty)]e^{-\frac{(t-t_0)}{\tau}} \tag{4.33}$$

Suppose we replace the capacitor with a voltage source defined by Eq. (4.33). Let v_{jk} denote the voltage across any pair of nodes j and k in N. Assume that N contains α independent DC voltage sources $V_{s1}, V_{s2}, \cdots, V_{s\alpha}$ and β independent DC current sources $I_{s1}, I_{s2}, \cdots, I_{s\beta}$. Applying the superposition theorem 3.5, we know that the solution $v_{jk}(t)$ is given by an expression of the form:

$$v_{jk}(t) = H_0 v_C(t) + \sum_{j=1}^{\alpha} H_j V_{sj} + \sum_{k=1}^{\beta} K_j I_{sj} \tag{4.34}$$

where H_0, H_j, and K_j are constants (which depend on element values and circuit configuration). Substituting for $v_C(t)$ in Eq. (4.34) from (4.33) and rearranging the terms, we obtain:

$$v_{jk}(t) = v_{jk}(t_\infty) + [v_{jk}(t_0) - v_{jk}(t_\infty)]e^{-\frac{(t-t_0)}{\tau}} \tag{4.35}$$

where

$$v_{jk}(t_\infty) \triangleq H_0 v_C(t_\infty) + \sum_{j=1}^{\alpha} H_j V_{sj} + \sum_{j=1}^{\beta} K_j I_{sj} \qquad (4.36)$$

and

$$v_{jk}(t_0) \triangleq H_0 v_C(t_0) + \sum_{j=1}^{\alpha} H_j V_{sj} + \sum_{j=1}^{\beta} K_j I_{sj} \qquad (4.37)$$

Since Eq. (4.35) has the exact same form as Eq. (4.26), and since nodes j and k are arbitrary, we conclude that: **the voltage $v_{jk}(t)$ across any pair of nodes in a first-order RC circuit driven by DC sources is an exponential waveform having the same time constant τ as** $v_C(t)$. By the same reasoning, we can also conclude that the current $i_j(t)$ in any branch j of a first-order RC circuit driven by DC sources is an exponential waveform having the same constant τ as that of $v_C(t)$.

The above "exponential solution waveform" property of course assumes that the first-order circuit is not degenerate, i.e., that it is uniquely solvable and $0 < |\tau| < \infty$. Also note that as we approach equilibrium, i.e., when $t \to +\infty$ (if $\tau > 0$) or $t \to -\infty$ (if $\tau < 0$), the capacitor current tends to zero. This follows from Figs. 4.8 and 4.12, $i_C = C\dot{v}_C$.

Since an exponential waveform is uniquely determined by only three parameters (initial state $x(t_0)$, equilibrium state $x(t_\infty)$ and time constant τ), we can now formally state the inspection method for first-order RC circuits driven by DC sources:

First-order Circuit Inspection Method:

1. Replace the capacitor by a DC voltage source with a terminal voltage equal to $v_C(t_0)$. Label the voltage across node-pair j, k as $v_{jk}(t_0)$ and the current i_j as $i_j(t_0)$. Solve the resulting resistive circuit for $v_{jk}(t_0)$ and $i_j(t_0)$. In other words, we are solving for the initial state.
2. Replace the capacitor by an open circuit. Label the voltage across node-pair j, k as $v_{jk}(t_\infty)$ and the current i_j as $i_j(t_\infty)$. Solve the resulting resistive circuit for $v_{jk}(t_\infty)$ and $i_j(t_\infty)$. In other words, we are solving for the equilibrium state.
3. Find the Thévenin equivalent circuit of N, so that the time constant can be computed as $\tau = R_{eq}C$.

The reader should use the above three parameters to make a quick sketch of the exponential waveform, as a sanity check.

4.2.1.2 Circuits Driven by Piecewise-Constant Signals

Consider next the case where the independent sources in N of Fig. 4.7a are piecewise-constant for $t > t_0$. This means that the semi-infinite time interval $t_0 \leq t < \infty$ can be partitioned into subintervals $[t_j, t_{j+1}), j = 1, 2, \cdots$ such that all sources assume a constant value during each subinterval. Hence we can analyze the circuit as a sequence of first-order circuits driven by DC sources, each one analyzed separately by the inspection method. Since the circuit remains unchanged except for the sources, the time constant τ remains unchanged throughout the analysis.

The initial state $x(t_0)$ and equilibrium state $x(t_\infty)$ will of course vary from one subinterval to another. Although the inspection method holds in the determination of $x(t_\infty)$, one must be careful in calculating the initial value at the beginning of each subinterval t_j because at least once source changes its value discontinuously at each boundary time t_j. In general, $x(t_j^-) \neq x(t_j^+)$, where the $-$ and $+$ denote the usual limit of $x(t)$ as $t \to t_j$, from the left and from the right, respectively. The initial value to be used in the calculation during the subinterval $[t_j, t_{j_1})$ is $x(t_j^+)$.

Although in general both $v_{jk}(t)$ and $i_j(t)$ can jump, the continuity property from Sect. 1.9.3 guarantees that in the usual case where the capacitor current (inductor voltage) waveform is bounded, the capacitor voltage (inductor current) waveform is a continuous function of time and therefore cannot jump. This property is the key to finding the solution by inspection, as Example 4.2.3 illustrates.

Example 4.2.3 Find and sketch $v_C(t), i_C(t)$ and $v_R(t)$ in Fig. 4.15 by inspection, for $t \geq 0$. Assume $v_C(0) = 0$ V (capacitor is initially discharged).

Solution Since $v_C(0) = 0$ and $v_s(t) = 0$ for $t \leq 0$, it follows that $i_C(t) = 0, v_C(t) = 0, v_R(t) = 0$ for $t \leq 0$.

The solution waveforms for $t > 0$ obviously consists of exponentials with a time constant $t = RC$. At $t = 0^+$, using the continuity property, we have $v_C(0^+) = v_C(0^-) = 0$. Therefore, by KVL, $v_R(0^+) = v_s(0^+) - v_C(0^+) = E$ and $i_C(0^+) = E/R$, by Ohm's law. To find the equilibrium state, we open the capacitor and hence find that $i_C(t_\infty) = 0, v_C(t_\infty) = E, v_R(t_\infty) = 0$. We now have enough information to determine the expressions ($t \geq 0$) as:

$$v_C(t) = E \left(1 - e^{-\frac{t}{RC}} \right)$$

$$i_C(t) = \frac{E}{R} \left(e^{-\frac{t}{RC}} \right)$$

$$v_R(t) = E \left(e^{-\frac{t}{RC}} \right) \tag{4.38}$$

The waveforms are sketched in Fig. 4.16. Note that $i_C(t) = C dv_C(t)/dt$ and $v_R(t) + v_C(t) = E$ for $t \geq 0$, as they should. Also observe that whereas $v_R(t)$ and $i_C(t)$ are discontinuous at $t = 0$, $v_C(t)$ is continuous for all t as expected.

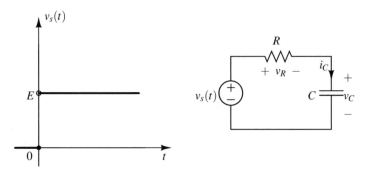

Fig. 4.15 Circuit for Example 4.2.3

The circuit in Fig. 4.15 is often used to model the situation where a DC voltage source is suddenly connected across a resistive circuit which normally draws a zero-input current. The linear capacitor in this case is used to model the small parasitic capacitance between the connecting wires. Without this capacitor, the input voltage would be identical to $v_s(t)$. However, in practice, a "transient" is always observed and the circuit in Fig. 4.15 represents a more realistic situation. In this case, the time constant τ gives a measure of how "fast" the circuit can respond to a step input. Such a measure is of crucial importance in the design of high-speed circuits.

Since the term time constant is meaningful only for first-order circuits, a more general measure of such "response speed" called the **rise time** is used. The rise time t_r is defined as the time it takes the output waveform to rise from 10% to 90% of the steady-state value after application of a step input. For first-order circuits, t_r is easily calculated from the elapsed time formula in Eq. (4.30):

$$t_r = \tau \ln \frac{0.1E - E}{0.9E - E}$$

$$= \tau \ln 9$$

$$\approx 2.2\tau \tag{4.39}$$

4.2.1.3   Linear Time-Invariant Circuits Driven by an Impulse

Consider the RC circuit shown in Fig. 4.17. Let the input voltage source $v_s(t)$ be a square pulse $p_\Delta(t)$ of width Δ and height $1/\Delta$, as shown in Fig. 4.18a. Assuming zero initial state (i.e., $v_C(0^-) = 0$), the response voltage $v_C(t)$ is shown in Fig. 4.18b. We define:

$$h_\Delta(\Delta) \triangleq \frac{1 - e^{\frac{-\Delta}{\tau}}}{\Delta} \triangleq \frac{f(\Delta)}{g(\Delta)} \tag{4.40}$$

Fig. 4.16 Exponential
waveforms for Example 4.2.3

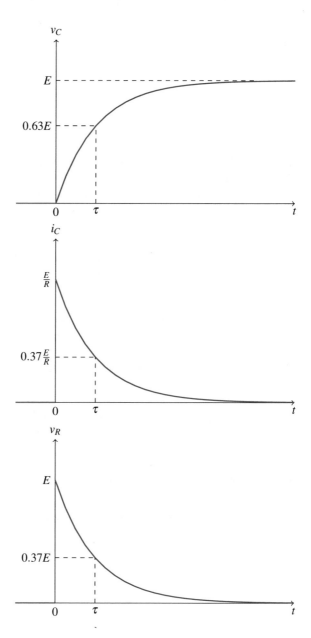

Fig. 4.17 Various $v_s(t)$
inputs are shown in Fig. 4.18

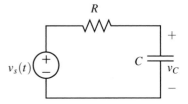

The input and response corresponding to $\Delta = 1, \frac{1}{2}, \frac{1}{3}$ are shown in Fig. 4.18c and
d, respectively. Note that as $\Delta \rightarrow 0$, $p_\Delta(t)$ tends to the **unit impulse** shown in
Fig. 4.18e. The unit impulse or the **Dirac delta function**[6] tends to infinity at $t = 0$
and to zero elsewhere, while the area under the pulse is unity. More precisely, the
unit impulse is defined such that the following two properties are satisfied:

$$1.\ \delta(t) \triangleq \begin{cases} \text{singular} & t = 0 \\ 0 & t \neq 0 \end{cases} \tag{4.41}$$

$$2.\ \int_{-\epsilon_1}^{\epsilon_2} \delta(t)dt = 1 \quad \text{for any } \epsilon_1 > 0, \epsilon_2 > 0 \tag{4.42}$$

The derivative "in the distribution sense"[7] of $\delta(t)$ is the **unit step function** defined
as:

$$u(t) \triangleq \begin{cases} 0 & t < 0 \\ 1 & t \geq 0 \end{cases} \tag{4.43}$$

Note that the "peak" value $h_\Delta(\Delta)$ of the response waveform in Fig. 4.18b increases
as Δ increases. To obtain the limiting value of $h_\Delta(\Delta)$ as $\Delta \rightarrow 0$, we apply
L'Hospital's rule:

$$\lim_{\Delta \to 0} h_\Delta(\Delta) = \lim_{\Delta \to 0} \frac{f'(\Delta)}{g'(\Delta)}$$

$$= \lim_{\Delta \to 0} \frac{(1/\tau)e^{(-\Delta/\tau)}}{1}$$

$$= \frac{1}{\tau} \tag{4.44}$$

---

[6]The delta function is used to model point charges in physics. Using the theory of distributions
from advanced mathematics, the unit impulse can be rigorously defined as a "generalized" function
imbued with most of the standard properties of a function. In particular, most of the time, $\delta(t)$ can
be manipulated like an ordinary function.

[7]We say "differentiating in the distribution sense" to emphasize that whenever we differentiate a
function which has a jump discontinuity at $t = t_0$, i.e., $f(t)$ jumps from $f(t_0^-)$ to $f(t_0^+)$, we must
include the corresponding impulse in the derivative: $f'(t_0) = [f(t_0^+)] - f(t_0^-)]\delta(t - t_0)$.

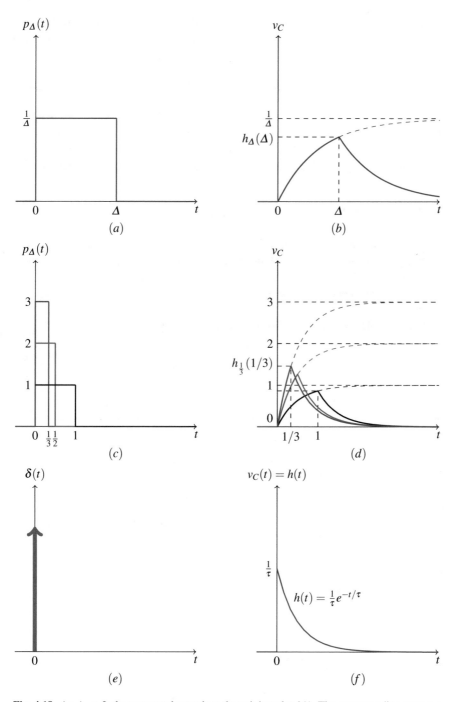

Fig. 4.18 As $\Delta \to 0$, the square pulse tends to the unit impulse $\delta(\cdot)$. The corresponding response tends to the impulse response h

Hence the response waveforms in Fig. 4.18d tend to the exponential waveform for $t \geq 0$ shown in Fig. 4.18f, compactly written using the unit step function defined earlier as:

$$h(t) = \frac{1}{\tau} e^{-t/\tau} u(t) \tag{4.45}$$

Because $h(t)$ is the response of the circuit when driven by a unit impulse under zero initial conditions, it is called the **impulse response**. In Sect. 4.3.3, we will show that given the impulse response of any linear time-invariant circuit, we can use it to calculate the response when the circuit is driven by any other input waveform.

4.2.1.4 Circuits Driven by Arbitrary Signals

Let us consider now the general case where the one-port N in Fig. 4.7a contains arbitrary independent sources. This means that the Thévenin equivalent voltage source $v_{oc}(t)$ in Fig. 4.7b can be any function of time, say, a PWL function, a sine wave, etc. Our objective is to derive an explicit solution and draw conclusions from our result.

Consider the RC circuit in Fig. 4.7b whose state equation is:

$$\dot{v}_C(t) = -\frac{v_C(t)}{\tau} + \frac{v_{oc}(t)}{\tau} \tag{4.46}$$

where $\tau \overset{\triangle}{=} R_{eq}C$.

Theorem 4.2 (Explicit Solution for First-Order Linear Time-Invariant RC Circuits) *Given any prescribed waveform $v_{oc}(t)$, the solution of Eq. (4.46) corresponding to any initial state $v_C(t_0)$ at $t = t_0$ is given by*

$$v_C(t) = \underbrace{v_C(t_0)e^{-\frac{(t-t_0)}{\tau}}}_{\text{zero-input response}} + \underbrace{\int_{t_0}^{t} \frac{1}{\tau} e^{-\frac{(t-t')}{\tau}} v_{oc}(t')dt'}_{\text{zero-state response}} \tag{4.47}$$

$\forall t \geq t_0$. *Here $\tau = R_{eq}C$.*

Proof

(a) At $t = t_0$, Eq. (4.47) reduces to

$$v_C(t)\Big|_{t=t_0} = v_C(t_0) \tag{4.48}$$

Hence Eq. (4.47) has the correct initial condition.

(b) To prove that Eq. (4.47) is a solution of Eq. (4.46), differentiate both sides of Eq. (4.47) with respect to t. First, we rewrite Eq. (4.47) as:

$$v_C(t) = v_C(t_0)e^{-\frac{(t-t_0)}{\tau}} + \frac{1}{\tau}e^{-t/\tau}\int_{t_0}^{t} e^{t'/\tau}v_{oc}(t')dt' \qquad (4.49)$$

Then upon differentiating with respect to t, we obtain for $t > 0$:

$$\dot{v}_C(t) = -\frac{1}{\tau}v_C(t_0)e^{-\frac{(t-t_0)}{\tau}} + \left(-\frac{1}{\tau^2}e^{\frac{-t}{\tau}}\right)\int_{t_0}^{t} e^{t'/\tau}v_{oc}(t')dt'$$

$$+ \left(\frac{1}{\tau}e^{\frac{-t}{\tau}}\right)\left[e^{\frac{t}{\tau}}v_{oc}(t)\right] \qquad (4.50)$$

where we used the second fundamental theorem of calculus [29]:

$$\frac{d}{dt}\int_{a}^{t} f(t')dt = f(t) \qquad (4.51)$$

Simplifying Eq. (4.50), we obtain:

$$\dot{v}_C(t) = -\frac{1}{\tau}v_C(t_0)e^{\frac{-(t-t_0)}{\tau}} - \frac{1}{\tau}\left[\int_{t_0}^{t} \frac{1}{\tau}e^{\frac{-(t-t')}{\tau}}v_{oc}(t')dt'\right] + \frac{1}{\tau}v_{oc}(t)$$

$$= -\frac{v_C(t)}{\tau} + \frac{v_{oc}(t)}{\tau} \qquad (4.52)$$

Hence Eq. (4.47) is a solution of Eq. (4.46).

(c) From our basic calculus courses, we know that the differential equation (4.46) has a unique solution. Hence Eq. (4.47) is indeed the solution.   □

The solution Eq. (4.47) consists of two terms. The first term is called the **zero-input response** because when all independent sources in N are set to zero, we have $v_{oc}(t) = 0$ for all times and $v_C(t)$ reduces to the first term only. The second term is called the **zero-state response** because when the initial state $v_C(t_0) = 0$, $v_C(t)$ reduces to the second term only.

Example 4.2.4 Find the solution $v_C(t)$ in Fig. 4.15, using Eq. (4.47).

Solution In this case we have: $v_C(t_0) = 0, t_0 = 0, \tau = RC$ and $v_{oc}(t) = E$, $t \geq 0$. Substituting these parameters in Eq. (4.47) and simplifying, we get:

$$v_C(t) = E\left(1 - e^{\frac{-t}{RC}}\right) \tag{4.53}$$

which coincides with the solutions in Example 4.2.3, as it should.

Note that in Eq. (4.47), the total response can be interpreted as the superposition of two terms, one due to the initial condition acting alone (with all independent sources set to zero) and the other due to the input acting alone (with the initial condition set to zero). Also, the equation is valid for both $\tau > 0$ and $\tau < 0$. Consider the case $\tau > 0$. For all values t' such that $t - t' \gg \tau$, the factor $e^{\frac{-(t-t')}{\tau}}$ is very small; consequently, the values of $v_{oc}(t)$ for such times contribute almost nothing to the integral in Eq. (4.47). In other words, the stable RC circuit has a fading memory. Inputs that have occurred many time constants ago have practically no effect at the present time. Thus we may say that the time constant τ is a measure of the memory time of the circuit.

4.2.1.5 First-Order Linear Switching Circuits

Suppose now that the one-port N in Fig. 4.7a contains one or more switches, where the state (open or closed) of each switch is specified for all $t \geq t_0$. Typically, a switch may be open over several disjoint time intervals, and closed during the remaining times. Although a switch is a time-varying linear resistor, such a linear switching circuit may be analyzed as a sequence of first-order linear time-invariant circuits, each one valid over a time interval where all switches remain in a given state. This class of circuits can therefore be analyzed by the procedures given in the previous sections. The only difference here is that the time constant τ will generally vary whenever a switch changes, as demonstrated in Example 4.2.5.

Example 4.2.5 Determine $v_o(t)$, $t \geq 0$ for the circuit in Fig. 4.19. Assume that the switch S has been open for a long time prior to $t = 0$.

Solution Given that the switch is closed at $t = 1\,\mathrm{s}$ and then reopened at $t = 2\,\mathrm{s}$, our objective is to first find $v_C(t)$ (since voltage across a capacitor should be a continuous function of time) and then find $v_o(t)$.

(continued)

Example 4.2.5 (continued)

Since we are only interested in $v_C(t)$ and $v_o(t)$, let us replace the remaining part of the circuit by its Thévenin equivalent circuit. The result is shown in Fig. 4.20a and b, corresponding to the case when S is open or closed, respectively. The corresponding τs are $\tau_2 = 1$ s and $\tau_1 = 0.9$ s, respectively.

Since the switch is initially open and the capacitor is initially in equilibrium, it follows from Fig. 4.20a that $v_C(t) = 6$ V and $v_o(t) = 0$ V for $t \le 1$ s. At $t = 1^+$, we change the equivalent circuit to Fig. 4.20b. Since by continuity, $v_C(1^+) = v_C(1^-) = 6$ V, we have $i_C(1^+) = (10 - 6)\,\text{V}/(2 + 1.6)\,\text{k}\Omega \approx 1.11$ mA and hence $v_o(1^+) = (1.6\,\text{k}\Omega)(1.11\,\text{mA}) \approx 1.78$ V. Note that we have obviously used the passive sign convention when computing the currents.

To determine $v_C(t_\infty)$ and $v_o(t_\infty)$ for the equivalent circuit in Fig. 4.20b, we replace the capacitor with an open circuit and obtain $v_C(t_\infty) = 10$ V and $v_o(t_\infty) = 0$ V. The waveforms of v_C and v_o during $[1, 2)$ are drawn as solid lines in Fig. 4.21a and b, respectively. The dashed portion shows the respective waveform if S had been left closed $\forall\, t \ge 1$ s.

Since S is closed at $t = 2$ s, we must write the equation of these two waveforms to calculate $v_C(2^-) \approx 8.68$ V and $v_o(2^-) \approx 0.59$ V (we leave the verification of these calculations to the reader). At $t = 2^+$, we return to the equivalent circuit in Fig. 4.20a. Since $v_C(2^-) = v_C(2^+) \approx 8.68$ V, we have $i_C(2^+) = (6 - 8.68)/(2.4 + 1.6)$ mA ≈ -0.67 mA and $v_o(2^+) = (1.6\,\text{k}\Omega)(-0.67\,\text{mA}) \approx -1.07$ V. Note that v_o has a discontinuous jump at $t = 2$ s.

To determine $v_C(t_\infty)$ and $v_o(t_\infty)$ for the circuit in Fig. 4.20a, we again replace the capacitor with an open circuit to obtain $v_C(t_\infty) = 6$ V and $v_o(t_\infty) = 0$ V. We have completed the waveform plots in Fig. 4.21.

Fig. 4.19 An *RC* switching circuit, where S is open during $t < 1$ s and $t \ge 2$ s, and closed during $1 \le t < 2$

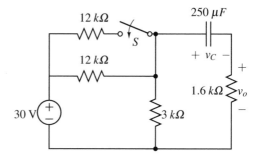

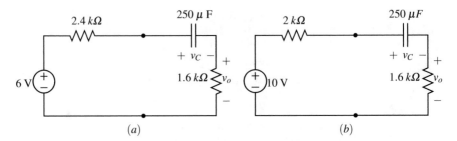

Fig. 4.20   Equivalent circuits from Fig. 4.19 when (**a**) switch is open, (**b**) switch is closed

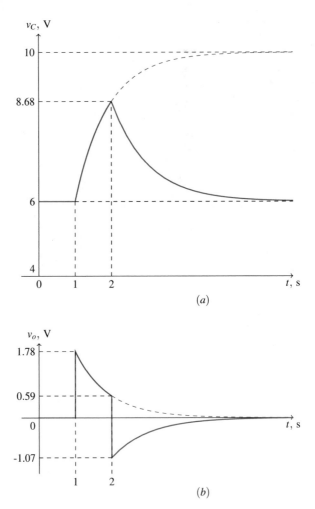

Fig. 4.21   (**a**) $v_C(t)$ and (**b**) $v_o(t)$ plots for Example 4.2.5

Fig. 4.22 A PWL RC circuit

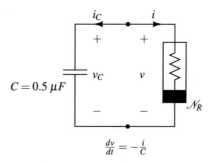

$$\frac{dv}{dt} = -\frac{i}{C}$$

4.2.1.6 First-Order PWL Circuits: Dynamic Route, Jump Phenomenon, and Relaxation Oscillations

Consider the first-order circuit in Fig. 4.22 where the nonlinear resistive one-port $\mathcal{N}_R$ may now contain nonlinear resistors (in addition to linear resistors and DC sources). As before, all resistors and the capacitor are time-invariant. This class of circuits includes many important nonlinear electronic circuits such as multivibrators, relaxation oscillators, etc. In this section, we assume that all nonlinear elements inside $\mathcal{N}_R$ are PWL so that the one-port is described by a PWL DP characteristic.

Our main problem is to find the solution $v_C(t)$ for the RC circuit, subject to any given initial state. Since the corresponding port variables of $\mathcal{N}_R$, namely $[v(t), i(t)]$, must fall on the DP characteristic of $\mathcal{N}_R$, the evolution of $[v(t), i(t)]$ can be visualized as the motion of a point on the characteristic starting from a given initial point.

Since the DP characteristic is PWL, the solution $[v(t), i(t)]$ can thus be found by determining first the specific "route" and "direction," henceforth called as the **dynamic route**, along the characteristic where the motion actually takes place. Once this route is identified, we can apply the "inspection method" developed in Sect. 4.2.1.1 to obtain the solution traversing along each segment separately, as illustrated in Example 4.2.6.

Example 4.2.6 Given the circuit in Fig. 4.22 and the associated DP characteristic for $\mathcal{N}_R$ in Fig. 4.23, determine $v_C(t)$ $\forall t \geq 0$. Let $v_C(0) = 2.5$ V.

Solution Step 1: Identify the initial point. Since $v(t) = v_C(t)$, for all t, initially $v(0) = v_C(0) = 2.5$ V. Hence the initial point on the DP characteristic is P_0, as shown in Fig. 4.23.

Step 2: Determine the dynamic route. The dynamic route starting from P_0 contains two pieces of information: (a) the route traversed and (b) the direction of motion. They are determined from the following information:

Key to dynamic route for RC circuit:

(continued)

Example 4.2.6 (continued)
1. The DP characteristic of $\mathcal{N}_R$.
2. $\dot{v}(t) = -\frac{i(t)}{C}$.

Since $\dot{v} = -i/C < 0$ whenever $i > 0$, the voltage $v(t)$ decreases as long as the associated current $i(t)$ is positive. Hence for $i(t) > 0$, the dynamic route starting at P_0 must always move along the $v - i$ curve **toward the left**, as indicated by the red directed (red) line segments $P_0 \to P_1$ and $P_1 \to P_2$ in Fig. 4.23. The dynamic route for this circuit ends at P_2 because at P_2, $i = 0$, so $\dot{v} = 0$. Hence the capacitor is in equilibrium at P_2.

Step 3: Obtain the solution for each straight line segment. Replace $\mathcal{N}_R$ by a sequence of Thévenin equivalent circuits corresponding to each line segment in the dynamic route. Using the method from Sect. 4.2.1.1, find a sequence of solutions $v_C(t)$. For this example, the dynamic route $P_0 \to P_1 \to P_2$ consists of only two segments. The corresponding equivalent circuits are shown in Fig. 4.24a and b, respectively.

To obtain $v_C(t)$ for segment $P_0 \to P_1$, we calculate $\tau = -62.5\,\mu\text{s}$. $v_C(0) = 2.5\,\text{V}$ and $v_C(t_\infty) = 3.25\,\text{V}$. Since the time constant in this case is negative, the corresponding circuit is unstable and hence the exponential is unbounded. We leave it to the reader to verify that $v_C(t)$ for $P_0 \to P_1$ is given by:

$$v_C(t) = 3.25 - 0.75 e^{\frac{t}{62.5\,\mu\text{s}}} \tag{4.54}$$

Since $v_C(t) = 2\,\text{V}$ at P_1, we can use the expression above to find the time $t \approx 31.9\,\mu\text{s}$ when $v_C(t) = 2\,\text{V}$. We hence use $v_C(0) = 2\,\text{V}$ for the following bounded exponential from $P_1 \to P_2$:

$$v_C(t) = 2 e^{\frac{-t}{100\,\mu\text{s}}} \tag{4.55}$$

A plot of $v_C(t)$ is given in Fig. 4.25.

After some practice, one can obtain the solution in Fig. 4.25 directly from the dynamic route, i.e., without drawing the Thévenin equivalent. Note that in the RC case, the dynamic route always terminates upon intersecting the v axis ($i = 0$).

We will now discuss a very important application of the dynamic route technique—the **opamp relaxation oscillator**. Oscillation is one of the most important and exciting phenomena that occurs in physical systems (e.g., electronic watch) and in nature (e.g., planetary motions). In this section, we will focus on

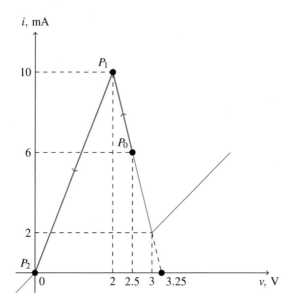

Fig. 4.23 DP characteristic of $\mathcal{N}_R$, with dynamic route (red) indicated, for Example 4.2.6

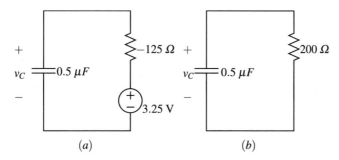

Fig. 4.24 (a) Equivalent circuit corresponding to $P_0 \to P_1$ (b) Equivalent circuit corresponding to $P_1 \to P_2$

a particular type of oscillator, the relaxation oscillator.[8] Section 4.6.3 will further explore the ideas behind nonlinear oscillators.

Consider the RC opamp circuit shown in Fig. 4.26a. The DP characteristic of the resistive one-port $\mathcal{N}_R$ was derived in Sect. 2.5.3.2, and is reproduced in Fig. 4.26b for convenience.

---

[8]Historically, relaxation oscillators were designed using only two vacuum tubes, or two transistors, such that one device is operating in a "cut-off" or relaxing mode, while the other device is operating in an "active" or "saturation" mode.

Fig. 4.25 $v_C(t)$ for
Example 4.2.6, unbounded
(red) and bounded (blue)
exponential functions
corresponding to the unstable
and stable circuits in
Fig. 4.24a,b respectively

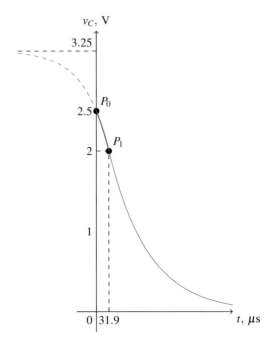

Consider the four different initial points Q_1, Q_2, Q_3, Q_4 (corresponding to four different initial capacitor voltages at $t = 0$) on this characteristic. Since $\dot{v}(t) = \dot{v}_C(t) = -i/C$ and $C > 0$, we have:

$$\dot{v}(t) > 0 \quad \text{for all } t \text{ such that } i(t) < 0 \tag{4.56}$$

and

$$\dot{v}(t) < 0 \quad \text{for all } t \text{ such that } i(t) > 0 \tag{4.57}$$

Hence the dynamic route from any initial point must move toward the left in the upper half plane, and towards the right in the lower half plane, as indicated by the arrowheads in Fig. 4.26b.

Since $i \neq 0$ at the two breakpoints Q_A and Q_B, they are not equilibrium points of the circuit. It follows from Eq. (4.30) that the amount of time T it takes to go from any initial point to Q_A or Q_B is finite because $x(t_k) \neq x(t_\infty)$.

Since the arrowheads from Q_1 and Q_2 (or from Q_3 and Q_4) are oppositely directed, it is impossible to continue drawing the dynamic route beyond Q_A or Q_B. In other words, an **impasse** is reached whenever the solution reaches Q_A or Q_B.

Any circuit which exhibits an impasse is the result of poor modeling. For the circuit of Fig. 4.26a, the impasse can be resolved by inserting a small linear inductor

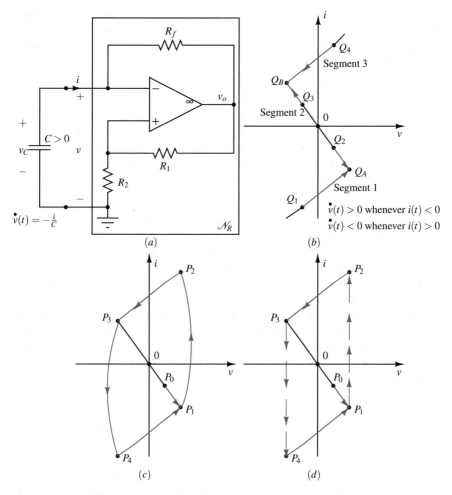

Fig. 4.26 (a) RC opamp circuit. (b) DP characteristic of $\mathcal{N}_R$. (c) Solution locus of $(v(t), i(t))$ for the remodeled circuit. (d) Dynamic route for the limiting case

in series with the capacitor; this inductor models the inductance L of the connecting wires. As will be shown in Sect. 4.6.3, the remodeled circuit has a well-defined solution $\forall\ t \geq 0$, so long as $L > 0$. A typical solution locus of $(v(t), i(t))$ corresponding to the initial condition at P_0 is shown in Fig. 4.26c. Our analysis in Sect. 4.6.3 will show that the transition time from P_1 to P_2, or from P_3 to P_4, decreases with L. In the limit $L \to 0$, the solution locus tends to the limiting case shown in Fig. 4.26d with a zero transition time. In other words in the limit where L decreases to zero, the solution **jumps** from the impasse point P_1 to P_2, and from the impasse point P_3 to P_4. We have used arrows to emphasize the instantaneous transition.

Both analytical and experimental studies [20] support the existence of a **jump phenomenon**, such as the one depicted in Fig. 4.26d, whenever a solution reaches an impasse point. This observation allows us to state the following rule which greatly simplifies the solution procedure.

Jump Rule

Let Q be an impasse point of any first-order RC circuit (respectively, RL circuit). Upon reaching Q at $t = T$, the dynamic route can be continued by jumping (instantaneously) to another point Q' on the DP characteristic of $\mathcal{N}_R$ such that $v_C(T^+) = v_C(T^-)$ [respectively, $i_L(T^+) = i_L(T^-)$] provided Q' is the only point satisfying the continuity property.

Note that the jump rule is also consistent with the continuity property of v_C or i_L. Also, the concepts of an impasse point and the jump rule are applicable regardless of whether the DP characteristic of $\mathcal{N}_R$ is PWL or not. A first-order RC circuit has at least one impasse point if $\mathcal{N}_R$ is described by a continuous nonmonotonic current-controlled DP characteristic. The instantaneous transition in this case consists of a vertical jump in the $v - i$ plane, assuming i is the vertical axis. A dual argument is applicable to a first-order RL circuit. Once the dynamic route is determined, with the help of the jump rule, for all $t > t_0$, the solution waveforms of $v(t)$ and $i(t)$ can be determined by inspection, refer to Exercise 4.7. This exercise should also enlighten the reader as to why the circuit in Fig. 4.26a is a prototypical[9] relaxation oscillator.

4.2.2 General Dynamic Circuits

So far we have analyzed first-order capacitor networks. As we transition from first to second (and higher) order nonlinear circuits, the complexity of steady-state behavior increases tremendously. In fact, third (and higher) order nonlinear continuous-time circuits exhibit the fascinating phenomenon of chaos (to be studied in Chap. 5).

Since it is impossible to cover all the techniques for general dynamic circuits in one section, we will instead present techniques that will help the reader formulate the equations governing such circuits. This is tremendously helpful because:

1. The reader will notice that we will extend the primary techniques from Chap. 3, nodal and tableau analysis, to cover dynamic networks.
2. Formulating the dynamic equations is the first (and probably most important) step in using a computer to simulate the associated network. Due to the complex behavior of third (and higher order) networks, computer simulations play an

---

[9]In fact, Fig. 4.26a could model the classic 555 timer, since the nonlinear DP characteristic can also be obtained by simply using two BJTs.

important role in studying such networks. Hence it is vital that the reader understand how to obtain the associated circuit equations.

4.2.2.1   Modified Nodal Analysis (MNA)

In Sect. 3.4, we studied node analysis for resistive circuits. For any resistive circuit made up of voltage-controlled resistors, we can write the node equations by inspection. MNA is based on node analysis but is suitably modified so that it can be used on **any** dynamic circuit. The goal of MNA is to obtain a set of coupled algebraic and differential equations. Consequently to specify a linear time-invariant inductor we use the differential equation

$$v(t) = L\frac{di}{dt} \tag{4.58}$$

rather than the integral equation

$$i(t) = i(t_0) + \frac{1}{L}\int_{t_0}^{t} v(t')dt' \tag{4.59}$$

The underlying ideas of MNA are:

1. Write node equations using node voltages as variables.
2. Whenever an element is encountered that is not voltage-controlled, introduce in the node equation the corresponding branch current as a new variable and add, as a new equation, the branch equation of that element.

The result is a system of equations where the unknowns are node voltages and some selected branch currents.

The equations of MNA can be written down by inspection. The number of equations is always smaller than that of tableau analysis (Sect. 4.2.2.2). But since MNA equations contain information about the interconnection as well as the nature of the branches, the equations of MNA do not have the conceptual clarity of the tableau equations. Many circuit analysis programs use MNA, SPICE in particular. As in the case of Chap. 3 we will first use example(s) to illustrate the ideas and then detail the algorithm.

Fig. 4.27 Circuit for
Example 4.2.7

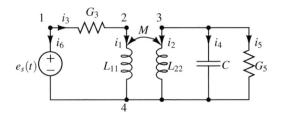

Example 4.2.7 Write MNA equations for the circuit in Fig. 4.27.

Solution The circuit shown in Fig. 4.27 includes an independent voltage source, a pair of coupled inductors (with mutual inductance M, self-inductances L_{11}, L_{12}), two resistors and a capacitor. We have $b = 6$ and $n = 4$. In writing the node equation for node 1, since the independent source is not voltage-controlled, we have inserted the branch current i_6. In considering nodes 2 and 3, we introduce inductor currents i_1 and i_2. We append these three suitably modified node equations with the branch equations of the voltage source and of the two (coupled) inductors. The result is:

Node Equations:
$$\begin{cases} G_3 e_1 - G_3 e_2 \qquad\qquad\quad + i_6 = 0 \\ -G_3 e_1 + G_3 e_2 \;\; + i_1 \qquad\qquad = 0 \\ \qquad\qquad\; C\dot{e}_3 \;\; + G_5 e_3 + i_2 \qquad = 0 \end{cases}$$

Coupled inductors:
$$\begin{cases} -e_2 + L_{11}\dot{i}_1 \;\; + M\dot{i}_2 \;\; = 0 \\ -e_3 + M\dot{i}_1 \;\; + L_{22}\dot{i}_2 \;\; = 0 \end{cases} \qquad (4.60)$$

Voltage Source: $\quad e_1 = e_s(t)$

MNA gives six equations in the node voltages e_1, e_2, and e_3 and in the selected currents i_1, i_2, and i_6. Eq. (4.60) forms the required set of coupled algebraic and differential equations.

Example 4.2.8 shows that the basic idea of MNA works quite easily for nonlinear circuits.

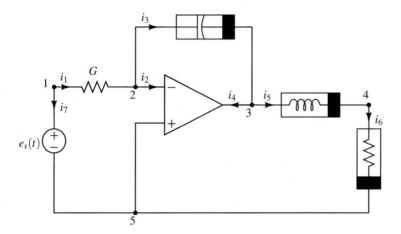

Fig. 4.28 Circuit for Example 4.2.8

Example 4.2.8 Write MNA equations for the circuit in Fig. 4.28. For the opamp, we will use the finite-gain model from Exercise 2.5, the nonlinear capacitor is specified by its small-signal capacitance $C(\cdot)$, the nonlinear inductor by its small-signal inductance $L(\cdot)$, and the current-controlled nonlinear resistor is specified by its characteristic $\hat{v}_6(\cdot)$.

Solution   Recall that the finite-gain model of an opamp:

$$v_o(v_d) = \frac{A}{2}|v_d + \epsilon| - \frac{A}{2}|v_d - \epsilon| \tag{4.61}$$

where v_o is the output voltage of the opamp and $v_d \triangleq v_+ - v_-$. The MNA equations can be easily written by inspection:

Node Eqs.:
$$\begin{cases} Ge_1 - Ge_2 & + i_7 = 0 \\ -Ge_1 + Ge_2 + C(e_2 - e_3)\dot{e}_2 - C(e_2 - e_3)\dot{e}_3 + i_2 & = 0 \\ \qquad\qquad -C(e_2 - e_3)\dot{e}_2 + C(e_2 - e_3)\dot{e}_3 + i_4 + i_5 & = 0 \\ \qquad\qquad\qquad\qquad\qquad\qquad -i_5 + i_6 = 0 \end{cases}$$

Opamp:
$$\begin{cases} i_2 & = 0 \\ -v_o(-e_2) + e_3 = 0 \end{cases} \tag{4.62}$$

$\mathcal{N}_L$:
$$\begin{cases} -e_3 + e_4 + L(i_5)\dot{i}_5 = 0 \end{cases}$$

(continued)

Example 4.2.8 (continued)

$$\mathscr{N}_R: \qquad \left\{ e_4 - \hat{v}_6(i_6) = 0 \right.$$

Voltage source: $e_1 = e_s$

Equation (4.62) constitutes a set of nine coupled algebraic and differential equations in nine unknown functions: the four node voltages $e_1(\cdot), e_2(\cdot), e_3(\cdot), e_4(\cdot)$ and the five selected currents $i_2(\cdot), i_4(\cdot), i_5(\cdot), i_6(\cdot), i_7(\cdot)$. Note that the variable i_4, the opamp output current, appears only in the third node equation. This node equation is thus a recipe for calculating i_4, once e_2, e_3, and i_5 are known. If i_4 is not required, the third node equation can be dropped.

Examples 4.2.7 and 4.2.8 have shown how easy it is to write MNA equations for any circuit, the algorithm is summarized below.

> **MNA Algorithm:**
> **Data:**
>
> - Circuit diagram with assigned node numbers and assigned current reference directions
> - Branch equation(s) for each element of the circuit
>
> **Steps:**
>
> 1. Choose a ground node, say n and draw a connected digraph (may require hinging some nodes).
> 2. For $k = 1, 2, \cdots, n-1$, write KCL for node k using the node-to-ground voltages as variables, keeping in mind that a if one or more inductors are connected to node k, then the branch currents of that inductor is entered in the node equation and the branch equation of the inductor is appended to the $n-1$ node equation; b if one or more branches which are not voltage-controlled are connected to node k, then the corresponding branch current is entered in the node equation and the corresponding branch equation is appended to the $n-1$ node equations.

4.2.2.2   Tableau Analysis

Tableau analysis is the second method for writing dynamic circuit equations. The method essentially mirrors the technique in Sect. 3.5, hence we will simply show a nonlinear example (very similar to Example 4.2.8) and then discuss the general technique.

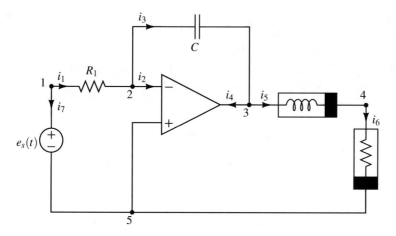

Fig. 4.29   Circuit for Example 4.2.9

Example 4.2.9   Write tableau equations for the circuit in Fig. 4.29.

Solution   We will assume that $\mathcal{N}_R$ is voltage-controlled. We will assume the same characteristics for $\mathcal{N}_L$ and the opamp as Example 4.28.
   By inspection we can write KCL and KVL for the circuit:

$$\mathbf{A}i(t) = \mathbf{0}$$

$$\mathbf{v}(t) - \mathbf{A}^T\mathbf{e}(t) = \mathbf{0} \qquad (4.63)$$

using the suitable reduced incidence matrix $\mathbf{A}$. Using the branch equations from the circuit, we get:

$$v_1 - R_1 i_1 = 0$$

$$i_2 = 0$$

$$C\dot{v}_3 - i_3 = 0$$

$$v_4 - v_o(-v_2) = 0$$

$$L(i_5)\dot{i}_5 - v_5 = 0$$

$$i_6 - \dot{i}_6(v_6) = 0$$

$$v_7 = e_s(t) \qquad (4.64)$$

(continued)

Example 4.2.9 (continued)

In Eq. (4.64), we know the constants R_1, C and the functions $v_o(\cdot)$, $L(\cdot)$, $\hat{i}_6(\cdot)$ and $e_s(\cdot)$. The unknown functions are $\mathbf{e}(\cdot)$, $\mathbf{v}(\cdot)$, and $\mathbf{i}(\cdot)$. Equations (4.63) and (4.64) are the tableau equations for the given circuit.

In general, the tableau equations for a nonlinear dynamic circuit are:

$$\text{KCL:} \quad \mathbf{A}\mathbf{i}(t) = \mathbf{0}$$

$$\text{KVL:} \quad \mathbf{v}(t) - \mathbf{A}^T \mathbf{e}(t) = \mathbf{0}$$

$$\text{Branch eqs.:} \quad \mathbf{h}(\dot{\mathbf{v}}(t), \mathbf{v}(t), \dot{\mathbf{i}}(t), , \mathbf{i}(t), t) = \mathbf{0} \tag{4.65}$$

Comparing Eqs. (3.121) and (4.65), we see that for the dynamic case we have derivatives in the branch equation.

For a connected digraph of b branches and n nodes, the tableau equations (4.65) constitute a system of $2b + n - 1$ scalar equations in $2b + n - 1$ unknown functions $e_j(\cdot)$, $j = 1, 2, \cdots, n - 1$, $v_k(\cdot)$, $k = 1, 2, \cdots, b$ and $i_l(\cdot)$, $l = 1, 2, \cdots, b$.

In the derivation of the tableau equations (4.65) we considered only a nonlinear inductor specified by its small-signal inductance $L(i)$. The dual case would be a nonlinear capacitor specified by its small-signal capacitance $C(v)$.

Suppose, however, we have a capacitor that is charge-controlled ($v_C = \hat{v}(q)$) and an inductor that is flux-controlled ($i_L = \hat{i}(\phi)$). If we use the chain rule as before, we are stuck because q and ϕ appear as arguments in $\hat{v}'(q)$ and $\hat{i}'(\phi)$. The remedy is to use q and ϕ as additional variables and to describe the capacitor by:

$$v_C = \hat{v}(q)$$

$$\dot{q} = i_C \tag{4.66}$$

and the inductor by:

$$i_L = \hat{i}(\phi)$$

$$\dot{\phi} = -v_L \tag{4.67}$$

4.2.2.3 Small Signal Analysis Revisited

We have already encountered the concept of small-signal analysis with respect to $\mathcal{N}_R$ in Sect. 3.1.1. We will see in this section that the method of small-signal

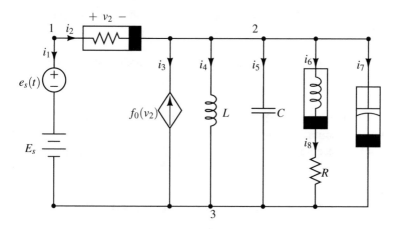

Fig. 4.30 Nonlinear time-invariant circuit $\mathscr{D}$ driven by the DC source E_s and the AC source $e_s(\cdot)$

analysis, when applied to dynamic circuits[10] helps reduce the analysis of a nonlinear dynamic circuit to that of a nonlinear resistive circuit, then to that of a linear dynamic circuit. The goal of this section is to state and justify the algorithm which delivers the small-signal equivalent circuit of any nonlinear time-invariant dynamic circuit about a fixed operating point.

In order to avoid complicated notations, this section studies the circuit shown in Fig. 4.30. We have chosen this circuit so that it includes most of the analyses required for obtaining a small-signal equivalent circuit. The aim of small-signal analysis is to take advantage of the fact that $e_s(\cdot)$ is small (in the sense that, for all $t \geq 0$, the values of $|e_s(t)|$ are small: higher order terms of any nonlinear expression are negligible). The circuit $\mathscr{D}$ includes a linear resistor R, a linear capacitor C, a linear inductor L, a nonlinear VCCS specified by its characteristic $f_0(\cdot)$, a nonlinear current-controlled inductor specified by $\hat{\phi}_6(\cdot)$, a nonlinear voltage-controlled capacitor specified by $\hat{q}_7(\cdot)$, and a nonlinear voltage-controlled resistor specified by $\hat{i}_2$.

The tableau equations of $\mathscr{D}$ can be written as:

$$\text{KCL:} \quad \mathbf{Ai}(t) = \mathbf{0}$$

$$\text{KVL:} \quad \mathbf{v}(t) - \mathbf{A}^T \mathbf{e}(t) = \mathbf{0}$$

$$\text{Branch eqs.:} \quad \mathbf{f}(\dot{\mathbf{v}}(t), \mathbf{v}(t), \dot{\mathbf{i}}(t), \mathbf{i}(t)) = \mathbf{u}_s(t) \tag{4.68}$$

---

[10]We will utilize the ideas from this section and Sect. 4.3 to derive important small-signal AC characteristics of memristors in Sect. 4.4.2.

Notice that Eq. (4.68) is slightly different from Eq. (4.65): we emphasize the fact that $\mathbf{f}$ does not depend explicitly on time. Column vector $\mathbf{u}_s(t)$ bookkeeps the contribution of the independent sources: $E_s + e_s(t)$.

In order to derive approximate equations representing $\mathscr{D}$ we proceed in three steps:

Step 1. Calculate the DC Operating Point   Q, i.e., $\mathbf{E_Q}$, $\mathbf{V_Q}$, $\mathbf{I_Q}$

Set the AC source $e_s(\cdot)$ to zero, turn on the DC source, and call $\mathbf{E_Q}$, $\mathbf{V_Q}$, $\mathbf{I_Q}$ the resulting DC steady-state. The corresponding tableau equations read:

$$\text{KCL:} \quad \mathbf{AI_Q} = 0$$

$$\text{KVL:} \quad \mathbf{V_Q} - \mathbf{A}^T\mathbf{E_Q} = 0$$

$$\text{Branch eqs.:} \quad \mathbf{f}(0, \mathbf{V_Q}, 0, \mathbf{I_Q}) = \mathbf{U}_s \qquad (4.69)$$

where $\mathbf{U}_s$ denotes the contribution of the DC source E_s. Since $\mathbf{V_Q}$ and $\mathbf{I_Q}$ are constant vectors, $\dot{\mathbf{V}}_Q = 0$ and $\dot{\mathbf{I}}_Q = 0$. For this particular circuit, the branch equations read:

$$V_1 = E_s$$

$$\hat{i}_2(V_2) - I_2 = 0$$

$$-f_0(V_2) + I_3 = 0$$

$$V_4 = 0 \quad \left(\text{because } \frac{dI_4}{dt} = 0\right)$$

$$I_5 = 0 \quad \left(\text{because } \frac{dV_5}{dt} = 0\right)$$

$$V_6 = 0 \quad \left(\text{because } \frac{dI_6}{dt} = 0\right)$$

$$I_7 = 0 \quad \left(\text{because } \frac{dV_7}{dt} = 0\right)$$

$$V_8 - RI_8 = 0 \qquad (4.70)$$

From Eq. (4.70), we see that to calculate the DC operating point, (a) we replace each inductor by a short circuit and (b) we replace each capacitor by an open circuit; (c) we solve the resulting nonlinear resistive circuit, shown in Fig. 4.31. In the next step, we assume that $\mathbf{E_Q}$, $\mathbf{V_Q}$, $\mathbf{I_Q}$ are known.[11]

---

[11] If Eq. (4.69) have several solutions, we choose one and stick to it.

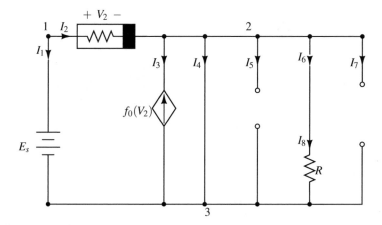

Fig. 4.31 Nonlinear resistive circuit whose solution $V_1, V_2, \cdots, I_1, I_2, \cdots$ specifies the operating point Q. Note the inductors have been replaced by short circuits and the capacitors by open circuits

Step 2. Change of Variables

The idea is to use the fact that the AC source is small, and consequently[12] the actual node voltages $\mathbf{e}(t)$ will be close to $\mathbf{E_Q}$, $\mathbf{v}(t)$ will be close to $\mathbf{V_Q}$ and $\mathbf{i}(t)$ will be close to $\mathbf{I_Q}$. So we write:

$$\mathbf{e}(t) = \mathbf{E_Q} + \tilde{\mathbf{e}}(t)$$
$$\mathbf{v}(t) = \mathbf{V_Q} + \tilde{\mathbf{v}}(t)$$
$$\mathbf{i}(t) = \mathbf{I_Q} + \tilde{\mathbf{i}}(t) \tag{4.71}$$

The point is that $\tilde{\mathbf{e}}(t), \tilde{\mathbf{v}}(t), \tilde{\mathbf{i}}(t)$ are small deviations from the operating point $\mathbf{E_Q}, \mathbf{V_Q}, \mathbf{I_Q}$, respectively. If we substitute the expressions for $\mathbf{e}, \mathbf{v}, \mathbf{i}$ from Eq. (4.71) into the KCL Eq. (4.68) and the KVL Eq. (4.68) while taking into account the corresponding tableau KCL, KVL Eq. (4.69) about the DC operating point, we obtain:

$$\mathbf{A}\tilde{\mathbf{i}}(t) = 0$$
$$\tilde{\mathbf{v}}(t) - \mathbf{A}^T \tilde{\mathbf{e}}(t) = 0 \tag{4.72}$$

Note that the equations above are exact, no approximation is involved.

---

[12]We will implicitly assume that the system is stable in the neighborhood of Q. For details, please refer to [12].

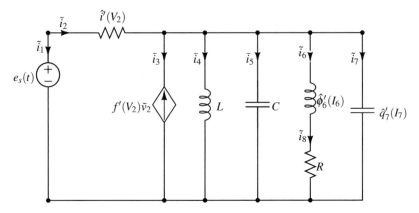

Fig. 4.32 The small-signal linear time-invariant circuit of $\mathscr{D}$ about the operating point $(\mathbf{V_Q}, \mathbf{I_Q})$

We could perform the same substitution in the branch Eq. (4.68) and use Eq. (4.69) to obtain:

$$\mathbf{f}\left(\dot{\tilde{\mathbf{v}}}(t), \mathbf{V_Q} + \tilde{\mathbf{v}}(t), \dot{\tilde{\mathbf{i}}}(t), \mathbf{I_Q} + \tilde{\mathbf{i}}(t)\right) - \mathbf{f}(\mathbf{0}, \mathbf{V_Q}, \mathbf{0}, \mathbf{I_Q}) = \mathbf{u}_s(t) - \mathbf{U}_s \qquad (4.73)$$

However it is more instructive to proceed by considering one branch at a time, because Eq. (4.73) is still a nonlinear equation and we would like to linearize it by using Taylor series (since $\tilde{\mathbf{e}}(t), \tilde{\mathbf{v}}(t), \tilde{\mathbf{i}}(t)$ are small).

Step 3. Obtain Approximate Branch Equations

We consider successively resistors, controlled sources, capacitors, and independent sources. Since inductors are the dual of capacitors, the corresponding derivation is trivial and is left as an exercise for the reader.

The final result will be obtained by using a Taylor series expansion and dropping the higher-order terms. The result is a set of approximate linear time-invariant equations relating $\tilde{\mathbf{v}}(t), \tilde{\mathbf{i}}(t)$ and the AC source. The linear small-signal circuit corresponding to $\mathscr{D}$ is shown in Fig. 4.32.

For the nonlinear resistor, we have:

$$i_2(t) = \hat{i}_2(v_2(t)) \qquad (4.74)$$

Substituting for $i_2(t)$ and $v_2(t)$, we get:

$$I_2 + \tilde{i}_2(t) = \hat{i}_2(V_2 + \tilde{v}_2(t)) \qquad (4.75)$$

Expanding the RHS using Taylor series, we get:

$$I_2 + \tilde{i}_2(t) = \hat{i}_2(V_2) + \left.\frac{d\hat{i}_2}{dv}\right|_{V_2} \tilde{v}_2(t) + \text{higher order terms} \tag{4.76}$$

Now if $\tilde{v}_2(t)$ is small, we may neglect the higher order terms and since $I_2 = \hat{i}_2(V_2)$, we get:

$$\tilde{i}_2(t) = \left.\frac{d\hat{i}_2}{dv}\right|_{V_2} \tilde{v}_2(t) \tag{4.77}$$

Equation (4.77) is the equation of a linear time-invariant resistor with conductance $\left.\frac{d\hat{i}_2}{dv}\right|_{V_2}$, the slope of the resistor characteristic at its operating point. Note that for a linear resistor, $\left.\frac{d\hat{i}_2}{dv}\right|_{V_2} = \frac{1}{R}$. Hence, comparing Figs. 4.30 and 4.32, we see that the linear resistor remains.[13]

For the controlled source, we have:

$$i_3(t) = f_0(v_2(t)) \tag{4.78}$$

Substituting for $i_3(t)$ and $v_2(t)$, we get:

$$I_3 + \tilde{i}_3(t) = f_0(V_2 + \tilde{v}_2(t)) \tag{4.79}$$

Expanding the RHS using Taylor series, we get:

$$I_3 + \tilde{i}_3(t) = f_0(V_2) + \left.\frac{df_0}{dv}\right|_{V_2} \tilde{v}_2(t) + \text{higher order terms} \tag{4.80}$$

Now if $\tilde{v}_2(t)$ is small, we may neglect the higher order terms to get:

$$\tilde{i}_3(t) = \left.\frac{df_0}{dv}\right|_{V_2} \tilde{v}_2(t) \tag{4.81}$$

Equation (4.81) is the equation of a linear time-invariant VCCS.
For the nonlinear capacitor, we have:

$$q_7(t) = \hat{q}_7(v_7(t)) \tag{4.82}$$

Using the chain rule and substituting for $v_7(t)$, we get:

$$\tilde{i}_7(t) = \hat{q}_7'(V_7 + \tilde{v}_7(t)) \cdot \dot{\tilde{v}}_7(t) \tag{4.83}$$

---

[13] In fact, this is also true for linear capacitors and inductors.

Expanding the RHS using Taylor series, dropping the higher order terms and using the fact that the DC equivalent of a capacitor is an open circuit, we get:

$$\tilde{i}_7(t) = \frac{d\hat{q}_7}{dv}\bigg|_{V_7} \cdot \dot{\tilde{v}}_7(t) \tag{4.84}$$

Equation (4.84) is the slope of the nonlinear capacitor at the operating point V_7.

For the independent AC source, we trivially get: $\tilde{v}_1(t) = e_s(t)$.

Hence the resulting branch equations for the small-signal linear time-invariant circuit in Fig. 4.32:

$$\tilde{v}_1(t) = e_s(t)$$

$$\tilde{i}_2(t) - \hat{i}_2'(V_2)\tilde{v}_2(t) = 0$$

$$\tilde{i}_3(t) - f_0'(V_2)\tilde{v}_2(t) = 0$$

$$\tilde{v}_4(t) - L\dot{\tilde{i}}_4(t) = 0$$

$$\tilde{i}_5(t) - C\dot{\tilde{v}}_5(t) = 0$$

$$\tilde{v}_6(t) - \hat{\phi}_6'(I_6)\dot{\tilde{i}}_6(t) = 0$$

$$\tilde{i}_7(t) - \hat{q}_7'(V_7)\dot{\tilde{v}}_7(t) = 0$$

$$\tilde{v}_8(t) - R\tilde{i}_8(t) = 0 \tag{4.85}$$

Let us abbreviate these equations in the form:

$$(\mathbf{M}_{0Q}D + \mathbf{M}_{1Q})\tilde{\mathbf{v}}(t) + (\mathbf{N}_{0Q}D + \mathbf{N}_{1Q})\tilde{\mathbf{i}}(t) = \tilde{\mathbf{u}}_s(t) \tag{4.86}$$

where the constant matrices $\mathbf{M}_{0Q}, \mathbf{M}_{1Q}, \mathbf{N}_{0Q}, \mathbf{N}_{1Q}$ are directly read from Eq. (4.85) and $\tilde{\mathbf{u}}_s(t)$ is the column vector of AC sources in Eq. (4.85).

Conclusion

If we collect KCL, KVL from Eqs. (4.72) and (4.86), we get the **tableau equation of a small-signal equivalent circuit**:

$$\mathbf{A}\tilde{\mathbf{i}}(t) = 0$$

$$\tilde{\mathbf{v}}(t) - \mathbf{A}^T\tilde{\mathbf{e}}(t) = 0$$

$$(\mathbf{M}_{0Q}D + \mathbf{M}_{1Q})\tilde{\mathbf{v}}(t) + (\mathbf{N}_{0Q}D + \mathbf{N}_{1Q})\tilde{\mathbf{i}}(t) = \tilde{\mathbf{u}}_s(t) \tag{4.87}$$

We will denote the small-signal equivalent circuit as $\mathscr{L}_Q$. Since the concept of small-signal equivalent circuits is very important, we summarize the procedure in detail by the following algorithm.

Algorithm to obtain the small-signal equivalent circuit $\mathscr{L}_Q$ of $\mathscr{D}$
Data

- Circuit diagram of the nonlinear time-invariant circuit $\mathscr{D}$ driven by DC and AC sources, with nodes numbered and with current reference directions
- Branch equations for each element in $\mathscr{D}$

First we determine the operating point Q

1. In $\mathscr{D}$, set all AC independent sources to zero.
2. Replace all inductors by short circuits and all capacitors by open circuits.
3. Solve the resulting resistive circuit, which is now driven by DC sources only. Call Q the resulting operating point specified by the solution $(\mathbf{V_Q}, \mathbf{I_Q})$. If there are multiple operating points, we choose the one of interest and study the dynamics of the circuit about that operating point.

Second, we determine $\mathscr{L}_Q$

1. In $\mathscr{D}$, set all DC independent sources to zero.
2. Leave all linear elements.
3. Replace every nonlinear element by its (linear) small-signal equivalent circuit about the operating point found in step 3. The resulting linear time-invariant circuit is $\mathscr{L}_Q$, the small-signal equivalent circuit of $\mathscr{D}$ about the operating point Q.

4.3 Frequency Domain Analysis of Linear Time-Invariant Circuits

In this section, we consider exclusively linear time-invariant circuits and we concentrate on their sinusoidal steady-state behavior, that is, their behavior when they are driven by one or more sinusoidal sources at some frequency ω and when, after all "transients" have died down, all currents and voltages are sinusoidal at frequency ω.

This section has a somewhat narrow focus in the sense that we do not discuss nonlinear circuits. However, the concepts and techniques this section covers are fundamental to science, in the sense that frequency domain analysis helps transform the analysis of differential equations from the time domain (Sect. 4.2), into analysis of algebraic (albeit complex) equations in the frequency domain. Also, we will see later in this chapter that a variety of small-signal AC analysis techniques (for example, with higher-order circuit elements in Sect. 4.6.2) will make use of frequency response concepts.

Moreover, discussing frequency domain techniques for nonlinear circuits is beyond the scope of this book, as we need to develop the mathematical machinery (such as describing functions) first. We plan to add this topic as part of our follow-up advanced volume on nonlinear circuits and networks.

The analysis technique when sinusoidal inputs are applied to linear time-invariant circuits is called **AC analysis** or **sinusoidal steady-state analysis**. Our first task would be to systematically develop the concept of a **phasor**: to each sine wave (of voltage or current) we associate a complex number, to encode both the magnitude and the phase.

4.3.1 Complex Numbers and Phasors

We will first discuss some important ideas regarding complex numbers. We would like to emphasize that our approach to deriving the phasor concept from complex numbers is probably unique because we use a historical approach [18], covering important concepts along the way. Hence we encourage readers who are familiar with complex numbers to at least glance through this section to make sure that they do not miss out some on fascinating facts. Many texts seek to introduce complex numbers with a convenient historical **fiction** based on solving quadratic equations[14] [25]:

$$x^2 = mx + c \tag{4.88}$$

Two thousand years BC, it was already known that such equations could be solved using a method that is equivalent to the modern formula:

$$x_{1,2} = \frac{m \pm \sqrt{m^2 + 4c}}{2} \tag{4.89}$$

But what if $m^2 + 4c$ (discriminant) is negative? This is where many textbooks are historically inaccurate in the sense that they state: the need for Eq. (4.88) to always have a solution forced mathematicians to take complex numbers seriously for negative discriminants.

But that is simply false. For the ancient Greeks mathematics was synonymous with geometry. Thus an algebraic relation such as Eq. (4.88) was not so much thought of as a problem in its own right, but rather as a mere vehicle for solving a genuine problem in geometry. In other words, Eq. (4.88) was simply seen to

---

[14]Dr. Muthuswamy thanks Dr. Jevtic for valuable discussions over the years, including suggesting Needham's excellent text on "Visual Complex Analysis."

represent the problem of finding the intersection points of the parabola $y = x^2$ with the line $y = mx + c$. Thus, depending on the sign of the discriminant, the equation either had two, one, or no real solutions. So, if the solution was absent, then it was correctly manifested by the occurrence of "impossible" (now known as complex) numbers in the formula.

It was **not** the quadratic that forced complex numbers to be taken seriously, it was the **cubic**:

$$x^3 = 3px + 2q \tag{4.90}$$

Exercise 4.9 shows that any cubic equation can be reduced to the form above. This equation represents the analogous problem of finding the intersection points of the cubic $y = x^3$ with the line $y = 3px + 2q$. Girolamo Cardano in his Ars Magna (which appeared in 1545) showed that this equation could be solved by means of the elegant formula (see Exercise 4.10):

$$x = s + t$$

$$\text{where: } s^3 = q + \sqrt{q^2 - p^3} \quad t^3 = q - \sqrt{q^2 - p^3} \tag{4.91}$$

Some 30 years after this formula appeared, Rafael Bombelli in L'Algebra recognized that there was something strange and paradoxical about it. First note that if the line $y = 3px + 2q$ is such that $p^3 > q^2$ then the formula involves complex numbers. For example, Bombelli considered $x^3 = 15x + 4$ which yields as one of the solutions:

$$x = \sqrt[3]{2 + 11j} + \sqrt[3]{2 - 11j} \tag{4.92}$$

In the previous case of the quadratic, this merely signaled that the geometric problem had no solution but in the case of the cubic, the line will **always**[15] hit the curve. In fact, we can (graphically) show that Bombelli's example yields the solution $x = 4$.

As he struggled to resolve this paradox, Bombelli had what he called a "wild thought": perhaps the solution $x = 4$ could be recovered from the above expression if $\sqrt[3]{2 + 11j} = 2 + nj$ and $\sqrt[3]{2 + 11j} = 2 - nj$. Of course for this to work he would have to assume that the addition of two complex numbers $A = a + j\tilde{a}$ and $B = b + j\tilde{b}$ obeyed the plausible rule,

$$A + B = (a + j\tilde{a}) + (b + j\tilde{b})$$

$$= (a + b) + j(\tilde{a} + \tilde{b}) \tag{4.93}$$

---

[15]This is a consequence of the Fundamental Theorem of Algebra: a cubic will have at least one real root.

Table 4.3 Complex numbers terminology

Name	Meaning	Notation
Modulus of z	Length r of z	$|z|$
Argument of z	Angle θ of z	$\arg(z)$
Real part of z	x coordinate of z	$\mathrm{Re}(z)$
Imaginary part of z	y coordinate of z	$\mathrm{Im}(z)$
Imaginary number	Real multiple of j	
Real axis	Set of real numbers	
Imaginary axis	Set of imaginary numbers	
Complex conjugate of z	Reflection of z in the real axis	$\bar{z}$

Next, to see if there was indeed a value of n for which $\sqrt[3]{2 + 11j} = 2 + nj$, he needed to calculate $(2 + jn)^3$. To do so he assumed that he could multiply out the brackets as in ordinary algebra and assuming $j^2 = -1$:

$$(a + j\tilde{a})(b + j\tilde{b}) = ab + j(a\tilde{b} + \tilde{a}b) + j^2\tilde{a}\tilde{b}$$

$$= (ab - \tilde{a}\tilde{b}) + j(a\tilde{b} + \tilde{a}b) \tag{4.94}$$

This rule vindicated his "wild thought," for he was now able to show that $(2 \pm j)^3 = 2 \pm 11j$.

While complex numbers themselves remained mysterious, Bombelli's[16] work on cubic equations thus established that **perfectly real problems requires complex arithmetic for their solution**. This justifies our use of complex arithmetic in AC circuit analysis: complex numbers provide an elegant way to encode both the magnitude and phase of a sinusoid. In fact, the subsequent development of the theory of complex numbers was bound with progress in other areas of physics and mathematics. That discussion is beyond the scope of this book, the interested reader is referred to [25].

We will now introduce the modern terminology and notation for complex numbers. Throughout this discussion, refer to Table 4.3 and Fig. 4.33.

It is valuable to grasp from the outset that (according to the geometric view) a complex number is a single, indivisible entity—a point in the plane. Only when we choose to describe such a point with numerical coordinates does a complex number appear to be compounded or "complex." More precisely, $\mathbb{C}$ is said to be **two dimensional**, meaning that **two** real numbers (coordinates) are needed to label a point within it, but exactly **how** the labeling is done is entirely up to us.

One way is to label the points with Cartesian coordinates (the real part x and imaginary part y), the complex number being written as $z = x + jy$. This form, called the **standard form** (encountered earlier via Bombelli's work), is the "natural"

---

[16]Bombelli is generally regarded as the father of complex numbers.

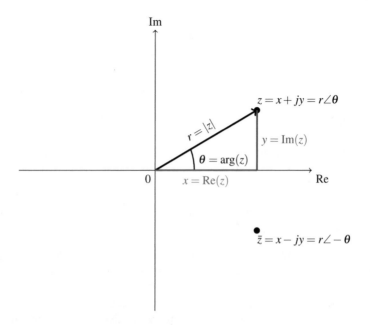

Fig. 4.33 Complex numbers terminology (contd.)

labeling when dealing with addition (or subtraction) of two complex numbers $z_1 = a + jb, z_2 = c + jd$:

$$z_1 + z_2 = (a + c) + j(b + d) \tag{4.95}$$

We simply add the real parts to get the real part for the sum, and add the imaginary parts to get the imaginary part for the sum.

But, when multiplying (or dividing) two complex numbers, the standard form is cumbersome. To emphasize this point, let us again multiply two complex numbers in standard form:

$$z_1 * z_2 = (a + jb) * (c + jd)$$
$$= (ac - bd) + j(ad + bc) \quad (j^2 = -1) \tag{4.96}$$

There is a more elegant way to multiply (divide) complex numbers. We will simply state the rule since a detailed explanation is beyond the scope of this book: labeling z with its **polar** coordinates, $r = |z|$, $\theta = \arg(z)$, we can now write $z = r \angle \theta$ where the symbol $\angle$ serves to remind us that θ is the **angle** of z.

The geometry multiplication rule takes the simple form:

$$(R \angle \phi)(r \angle \theta) = (Rr) \angle (\phi + \theta) \tag{4.97}$$

In words: **The length of z_1z_2 is the product of the lengths of z_1 and z_2, and the angle of z_1z_2 is the sum of the angles of z_1 and z_2.**

Complex division can now be defined in a simple manner:

$$\frac{R\angle\phi}{r\angle\theta} = \frac{R}{r}\angle(\phi - \theta) \tag{4.98}$$

One important concept is that: in common with the Cartesian label $x + jy$, a given polar label $r\angle\theta$ specifies a unique point, but (unlike the Cartesian case) **a given point does not have a unique polar label!** Since any two angles that differ by a multiple of 2π correspond to the same direction, a given point has infinitely many labels:

$$\cdots = r\angle(\theta - 4\pi) = r\angle(\theta - 2\pi) = r\angle\theta = r\angle(\theta + 2\pi) = r\angle(\theta + 4\pi) = \cdots \tag{4.99}$$

This simple fact about angles is one of the most important concepts in complex numbers that is encountered many times in science and engineering. Before proceeding, you should solve Exercise 4.11 so that you thoroughly understand and are comfortable with the concepts, terminology and notation for complex numbers.

We are now in a position to look at probably the most elegant formula in mathematics, called **Euler's formula**:

$$e^{j\theta} = \cos(\theta) + j\sin(\theta) \tag{4.100}$$

Simply stated, "Euler's formula relates polar form to standard form." But this **does not** help us understand what the formula means. Simply stating "Euler's formula relates polar form to standard form" reduces one of Euler's greatest achievements to a mere tautology. Perhaps the best approach to understanding Euler's formula is to go visualize $e^{j\theta}$ in the complex plane, as shown in Fig. 4.34. Given the fact that the complex number $e^{j\theta}$ in standard form is $x + jy$, we can see from Fig. 4.34 that since the magnitude of $e^{j\theta}$ is 1, $x = \cos(\theta)$, $y = \sin(\theta)$ by the definition of the trigonometric functions from a right-angled triangle. Obviously, if we scale the magnitude of a complex number by r, Fig. 4.34 shows $re^{j\theta} = r\cos(\theta) + jr\sin(\theta)$.

Now, we are ready to discuss the concept of a phasor.

Definition 4.11 A **sinusoid** of angular frequency ω (rad/s) is by definition a function of the form $A_m\cos(\omega t + \theta)$ where the amplitude A_m, phase θ, and the frequency ω are real constants. The amplitude A_m is always taken to be positive. The period $T = 2\pi/\omega$ is in seconds. Also note that given a frequency f in Hz, $\omega = 2\pi f$.

Definition 4.12 To the sinusoid in Definition 4.11, we associate a complex number A called the **phasor**[17] (of that sinusoid) according to the rule: $A \triangleq A_m e^{j\theta}$.

---

[17] A phasor is essentially a complex number written in exponential or Euler form.

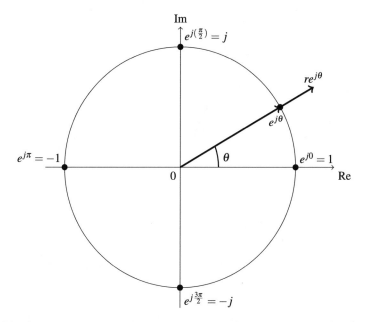

Fig. 4.34 Interpreting Euler's formula via the complex plane

It is crucial to note that the phasor **does not explicitly** involve ωt! The best way to understand this is visually, refer to Fig. 4.35, called the **phasor diagram**. We plot the phasor A in the complex plane as a vector from the origin to the point $A = A_m e^{j\theta}$. We now imagine the vector rotating **counterclockwise** at angular velocity of ω rad/s, namely, we consider $Ae^{j\omega t}$ as t increases. Whenever we want $x(t)$, we project orthogonally on the x-axis the tip of the vector.

In other words, **knowing the frequency** ω, the phasor A specifies **uniquely** the sinusoid by the formula:

$$\text{Re}[Ae^{j\omega t}] = \text{Re}[A_m e^{j(\omega t + \theta)}]$$

$$= A_m \cos(\omega t + \theta) \tag{4.101}$$

In summary, there is a one-to-one correspondence between sinusoids (at frequency ω) and phasors:

Sinusoid   Phasor

$$\left. \begin{array}{l} A_m \cos(\omega t + \theta) \\ = (A_m \cos\theta)\cos\omega t \\ +(-A_m \sin\theta)\sin\omega t \end{array} \right\} \Leftrightarrow \begin{cases} A = A_m e^{j\theta} \\ = (A_m \cos\theta) + j(A_m \sin\theta) \end{cases} \tag{4.102}$$

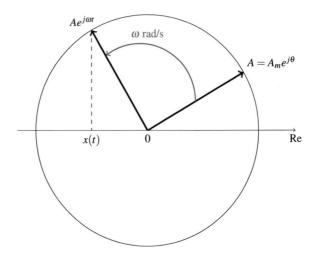

Fig. 4.35 The sinusoid $x(t) = A_m \cos(\omega t + \theta)$ is viewed as being generated by the projection of the tip of the "rotating phasor" $Ae^{j\omega t}$

Equivalent (4.102) states that

$$A_m \cos(\omega t + \theta) = \text{Re}[A] \cos \omega t - \text{Im}[A] \sin \omega t \qquad (4.103)$$

4.3.2 Sinusoidal Steady-State Analysis Using Phasors

The use of phasors in the analysis of linear time-invariant circuits in sinusoidal steady-state becomes completely obvious once the following lemmas are thoroughly understood.

Lemma 4.1 (Uniqueness) *Two sinusoids are equal iff they are represented by the same phasor; symbolically for all t,*

$$Re(Ae^{j\omega t}) = Re(Be^{j\omega t}) \Leftrightarrow A = B \qquad (4.104)$$

Proof

(a) Assume $A = B$. Consequently, for all t,

$$Ae^{j\omega t} = Be^{j\omega t} \quad \text{and} \quad \text{Re}(Ae^{j\omega t}) = \text{Re}(Be^{j\omega t})$$

(b) Assume, for all t:

$$\text{Re}(Ae^{j\omega t}) = \text{Re}(Be^{j\omega t}) \qquad (4.105)$$

In particular, for $t = 0$, we get: $\text{Re}(A) = \text{Re}(B)$. Similarly for $t_0 = \pi/(2\omega)$, $e^{j\omega t_0} = e^{j\pi/2} = j$. Thus $\text{Re}(Aj) = -\text{Im}(A)$ and hence Eq. (4.105) gives $\text{Im}(A) = \text{Im}(B)$. Therefore:

$$A = \text{Re}(A) + j\text{Im}(A)$$

$$= \text{Re}(B) + j\text{Im}(B)$$

$$= B \tag{4.106}$$

$$\square$$

Lemma 4.2 (Linearity) *The phasor representing a linear combination of sinusoids (with **real** coefficients) is equal to the same linear combination of the phasors representing the individual sinusoids. Symbolically, let the sinusoids be*

$$x_1(t) = Re[A_1 e^{j\omega t}] \quad and \quad x_2(t) = Re[A_2 e^{j\omega t}]$$

Thus the phasor A_1 represents sinusoid $x_1(t)$ and the phasor A_2 represents $x_2(t)$. Let $a_1, a_2 \in \Re$; then the sinusoid $a_1 x_1(t) + a_2 x_2(t)$ is represented by the phasor $a_1 A_1 + a_2 A_2$.

Proof We verify the assertion by computation:

$$a_1 x_1(t) + a_2 x_2(t) = a_1 \text{Re}[A_1 e^{j\omega t}] + a_2 \text{Re}[A_2 e^{j\omega t}] \tag{4.107}$$

Now a_1 and a_2 are real numbers, hence for any complex numbers z_1 and z_2,

$$a_i \text{Re}[z_i] = \text{Re}[a_i z_i] \quad i = 1, 2$$

$$\text{and } a_1 \text{Re}[z_1] + a_2 \text{Re}[z_2] = \text{Re}[a_1 z_1 + a_2 z_2] \tag{4.108}$$

Now applying this fact to Eq. (4.107) we have:

$$a_1 \text{Re}[A_1 e^{j\omega t}] + a_2 \text{Re}[A_2 e^{j\omega t}] = \text{Re}[(a_1 A_1 + a_2 A_2) e^{j\omega t}] \tag{4.109}$$

Combining the equation above with Eq. (4.107) we get:

$$a_1 x_1(t) + a_2 x_2(t) = \text{Re}[(a_1 A_1 + a_2 A_2) e^{j\omega t}] \tag{4.110}$$

$$\square$$

The proof is easily extended to a linear combination (with real coefficients) of n sinusoids.

Lemma 4.3 (Phasor Differentiation) *A is the phasor of a given sinusoid $A_m \cos(\omega t + \theta)$ iff $j\omega A$ is the phasor of its derivative, $\frac{d}{dt}[A_m \cos(\omega t + \theta)]$. Symbolically,*

$$Re[j\omega A e^{j\omega t}] = \frac{d}{dt}[Re(A e^{j\omega t})] \tag{4.111}$$

Proof Note that it is convenient to think of Eq. (4.111) as stating that the linear operators Re and $\frac{d}{dt}$ commute:

$$Re\left[\frac{d}{dt}(A e^{j\omega t})\right] = Re[j\omega A e^{j\omega t}] = \frac{d}{dt}[Re(A e^{j\omega t})]$$

Now:

$$\frac{d}{dt}[Re(A e^{j\omega t})] = \frac{d}{dt}[Re(A_m e^{j(\omega t + \theta)})]$$

$$= \frac{d}{dt}[A_m \cos(\omega t + \theta)]$$

$$= -A_m \omega \sin(\omega t + \theta)$$

$$= Re[j\omega A_m e^{j(\omega t + \theta)}]$$

$$= Re[A e^{j\omega t}] \tag{4.112}$$

$\square$

Example 4.3.1 Simplify: $12\cos(\omega t + 23°) + 7\cos(\omega t - 57°) + \frac{d}{dt}(0.2\cos(\omega t + 71°))$

Solution We could combine all the functions using trigonometric formulae, however, this approach gets very complicated. Instead let us use the phasor rules we just learned, **understanding that ω for all the functions is the same.** Let $\omega = 377$ rad/s. Hence, the phasor formulation for each function is:

$$A_1 = 12 e^{j23°}$$

$$A_2 = 7 e^{-j57°}$$

$$A_3 = j\omega 0.2 e^{j71°} = 75.4 e^{j161°} \tag{4.113}$$

(continued)

Example 4.3.1 (continued)

For A_3, we used the differentiation rule. Since we are going to be adding complex numbers, let us write each of the phasor in standard form:

$$A_1 \approx 11.05 + j4.69$$

$$A_2 \approx 3.81 - j5.87$$

$$A_3 \approx -71.29 + j24.55 \tag{4.114}$$

We will add and convert back to phasor form, so we can interpret the result as a sinusoid:

$$A_1 + A_2 + A_3 = -56.43 + j23.34$$

$$= 61.08e^{j157.51°}$$

$$\stackrel{\triangle}{=} A \tag{4.115}$$

Thus the resulting sinusoid is: $\text{Re}[Ae^{j\omega t}] = 61.08\cos(377t + 157.51°)$.

We will now solve a differential equation using phasor formulation.

Example 4.3.2 Given the circuit in Fig. 4.36, determine $i_L(t)$ for all t. $i_s(t) = I_{sm}\cos(\omega t + \angle I_s)$. Assume $L > 0, R > 0, C > 0$.

Solution The time domain equation for $i_L(t)$ can be easily found via inspection as:

$$\frac{d^2}{dt^2}i_L(t) + 2\alpha \dot{i}_L(t) + \omega_0^2 i_L(t) = \omega_0^2 i_s(t) \tag{4.116}$$

$w_0^2 = 1/LC, 2\alpha = 1/RC$. Let the phasor representation of the sinusoidal current source be $I_s = I_{sm}e^{j\angle I_s}$. Since a phasor is an exponential function and Eq. (4.116) is a linear ODE, we try the solution $\text{Re}(I_L e^{j\omega t})$ where the complex number I_L is the yet-undetermined phasor which specifies this particular sinusoidal solution. Substituting into Eq. (4.116) we obtain for all t:

$$\frac{d^2}{dt^2}[\text{Re}(I_L e^{j\omega t})] + 2\alpha\frac{d}{dt}[\text{Re}(I_L e^{j\omega t})] + \omega_0^2\text{Re}(I_L e^{j\omega t}) = \omega_0^2\text{Re}(I_s e^{j\omega t}) \tag{4.117}$$

(continued)

Example 4.3.2 (continued)

1. Using the differentiation lemma three times, we get:

$$\mathrm{Re}[(j\omega)^2 I_L e^{j\omega t}] + 2\alpha\mathrm{Re}[(j\omega)I_L e^{j\omega t}] + \omega_0^2\mathrm{Re}(I_L e^{j\omega t}) = \omega_0^2\mathrm{Re}(I_s e^{j\omega t})$$

2. Using the linearity lemma we obtain (since α and ω_0^2 are real):

$$\mathrm{Re}[(j\omega)^2 + 2\alpha(j\omega) + w_0^2]I_L e^{j\omega t} = \omega_0^2\mathrm{Re}(I_s e^{j\omega t})$$

3. Using uniqueness lemma, we obtain an algebraic equation for I_L:

$$[(j\omega)^2 + 2\alpha(j\omega) + \omega_0^2]I_L = \omega_0^2 I_s \qquad (4.118)$$

Hence

$$I_L = \frac{\omega_0^2 I_s}{(\omega_0^2 - \omega^2) + 2\alpha j\omega} \triangleq I_{\mathrm{Lm}} e^{j(\theta_L + \angle I_s)} \qquad (4.119)$$

with

$$I_{\mathrm{Lm}} = \frac{\omega_0^2}{\sqrt{(\omega_0^2 - \omega^2)^2 + (2\alpha\omega)^2}} I_{\mathrm{sm}} \quad \theta_L = -\tan^{-1}\frac{2\alpha\omega}{\omega_0^2 - \omega^2}$$

The sinusoidal solution is then:

$$i_{\mathrm{Lp}}(t) = \frac{\omega_0^2 I_{\mathrm{sm}}}{\sqrt{(\omega_0^2 - \omega^2)^2 + (2\alpha\omega)^2}} \cos(\omega t + \angle I_s + \theta_L) \qquad (4.120)$$

where the subscript p reminds us that i_{Lp} is the sinusoidal particular solution. The physical meaning of this particular solution is the following: since R, L, C are positive constants, it follows that $\alpha > 0$ and $\omega_0^2 > 0$. Consequently, the two natural frequencies s_1, s_2 of the circuit, i.e., the zeros of its characteristic polynomial $C(s) = s^2 + 2\alpha s + \omega_0^2$ have negative real parts. Therefore, any solution of Eq. (4.116) starting at any t_0 from any initial condition has the form:

$$i_L(t) = k_1 e^{s_1(t-t_0)} + k_2 e^{s_2(t-t_0)} + i_{\mathrm{Lp}}(t) \qquad (4.121)$$

Note that we have assumed $s_1 \neq s_2$.

Fig. 4.36 Circuit for
Example 4.3.2

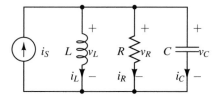

Example 4.3.2 illustrates the following ideas:

1. Since $\mathrm{Re}(s_1) < 0, \mathrm{Re}(s_2) < 0$, as $t \to x, i_L(t) \to i_{\mathrm{Lp}}(t)$. This particular solution is called the **sinusoidal steady-state solution of the circuit**. The difference between the **total response** $i_L(t)$ given by Eq. (4.121) and the particular solution given by Eq. (4.120) is called the **transient response**.
2. Note that the frequency of the output is the same as the frequency of the input. This property is true in general for any linear time-invariant circuit: if all its natural frequencies have negative real parts, then for any initial conditions and for any set of independent sources, each one sinusoidal at the same frequency ω, all currents and all voltages will tend exponentially as $t \to \infty$ to sinusoidal waveforms at frequency ω. When that situation occurs the circuit is said to be in the **sinusoidal steady-state**. Note that sinusoidal steady-state does not depend on the initial conditions. A general proof is beyond the scope of this book, the interested reader is referred to [12].
3. Comparing Eqs. (4.116) and (4.118) we see that a differential equation in the time domain has been converted to a complex algebraic equation in the phasor domain. So a natural question is: can we obtain the algebraic equation in the phasor domain directly from the circuit, instead of writing the time domain differential equation?

The answer is yes, and simply involves reformulating the laws of interconnections (KCL, KVL) and laws of elements in the phasor domain.

For example, in Fig. 4.36, KCL reads for all t:

$$i_L(t) + i_R(t) + i_C(t) = 0 \qquad (4.122)$$

For $k = L, R, C$, let I_k be the phasor representing the sinusoid $i_k(t)$. Thus, Eq. (4.122) gives, for all t:

$$\mathrm{Re}(I_L e^{j\omega t}) + \mathrm{Re}(I_R e^{j\omega t}) + \mathrm{Re}(I_C e^{j\omega t}) = 0 \qquad (4.123)$$

Using the linearity and uniqueness lemmas, we obtain:

$$I_L + I_R + I_C = 0 \qquad (4.124)$$

Since the reasoning is quite general, we can state the following conclusion.

Theorem 4.3 (KCL in the Phasor Domain)   *In the sinusoidal steady-state, for any connected circuit $\mathscr{D}$, KCL reads:*

$$\mathbf{A}\bar{\mathbf{I}} = 0 \tag{4.125}$$

where $\mathbf{A}$ is the $(n-1) \times b$ reduced incidence matrix of real numbers and $\bar{\mathbf{I}}$ is an b-vector current phasor. We use $\bar{\mathbf{I}}$ to avoid confusion with the identity matrix.

We can make a similar argument for KVL and hence we get:

Theorem 4.4 (KVL in the Phasor Domain)   *In the sinusoidal steady-state, for any connected circuit $\mathscr{D}$, KVL reads:*

$$\mathbf{V} = \mathbf{A}^T \mathbf{E} \tag{4.126}$$

where $\mathbf{A}$ is the $(n-1) \times b$ reduced incidence matrix of real numbers and $\mathbf{E}$ is a $(n-1)$-vector voltage phasor. Notice that $\mathbf{V}$ is a matrix with complex components.

The laws of elements in the phasor domain can also be derived in a straightforward manner by application of the three lemmas to the time domain element laws. Table 4.4 has the results. The expressions R, $j\omega L$ and $\frac{1}{j\omega C}$, are the **impedances** at frequency ω of the circuit elements R, L, and C, respectively; $\frac{1}{R}$, $\frac{1}{j\omega L}$, $j\omega C$ are the corresponding **admittances**; μ is a voltage gain; α is a current gain; g_m is a transconductance, and r_m is a transresistance. The crucial point again is that in terms of phasors, the branch equations become algebraic equations with complex coefficients in the phasor domain.

Also, as shown in Fig. 4.35, it is common to visualize phasors as rotating counterclockwise. Hence, referring to the phasor domain constitutive relations for the inductor and capacitor, we say **the inductor current phasor I_L lags the inductor voltage phasor V_L by** $90°$ and **the capacitor current phasor I_C leads the capacitor voltage phasor V_C by** $90°$.[18] We will see in Sect. 4.4.2 that capacitive and inductive parasitic effects in physical memristors lead to "unpinching" of memristor hysteresis loops, due to the leading (or lagging) behavior of current and voltage variables (under sinusoidal excitation).

Thus we have in essence "resistive" circuits in the frequency domain, except now our resistances are in the form of frequency-dependent impedances. Therefore,

---

[18]The convention is to say "current leads/lags voltage,'" **not** "voltage lags/leads current."

Table 4.4 Laws of elements in the time domain and phasor domain

Element	Time domain Constitutive relation	Phasor domain Constitutive relation
Resistor	$v(t) = Ri(t)$	$V = RI$
Inductor	$v(t) = L\frac{di}{dt}$	$V = j\omega LI$
Capacitor	$i(t) = C\frac{dv}{dt}$	$I = j\omega CV$
VCVS	$v_3(t) = \mu v_1(t)$	$V_3 = \mu V_1$
VCCS	$i_4(t) = g_m v_5(t)$	$I_4 = g_m V_5$
CCVS	$v_6(t) = r_m i_5(t)$	$V_6 = r_m I_5$
CCCS	$i_8(t) = \alpha i_7(t)$	$I_8 = \alpha I_7$
Gyrator	$i_9(t) = G v_{10}(t)$	$I_9 = G V_{10}$
	$i_{10}(t) = -G v_9(t)$	$I_{10} = -G V_9$
Ideal transformer	$v_1(t) = \frac{1}{n} v_2(t)$	$V_1 = \frac{1}{n} V_2$
	$i_1(t) = -n i_2(t)$	$I_1 = -n I_2$

techniques such as tableau analysis are applicable and to avoid repeating the concepts from Chap. 3, we will simply summarize the main ideas, by drawing an analogy with tableau analysis for resistive circuits.

Let $\mathcal{N}_R$ be a linear time-invariant resistive circuit with a connected graph having n nodes and b branches. Suppose that we first replace a number of resistors of $\mathcal{N}_R$ by inductors or capacitors, and second, drive the resulting circuit by sinusoidal sources **all operating at the same frequency** ω. Assume that the resulting circuit is in the sinusoidal steady-state and call the circuit N_ω. We have chosen this label to emphasize that we consider its sinusoidal steady at frequency ω.

Linear time-invariant resistive circuit N_R (see Eq. (3.115))

$$\begin{bmatrix} \mathbf{0} & \mathbf{0} & \mathbf{A} \\ -\mathbf{A}^T & \mathbf{I} & \mathbf{0} \\ \mathbf{0} & \mathbf{M}(t) & \mathbf{N}(t) \end{bmatrix} \begin{bmatrix} \mathbf{e}(t) \\ \mathbf{v}(t) \\ \mathbf{i}(t) \end{bmatrix} = \begin{bmatrix} \mathbf{0} \\ \mathbf{0} \\ \mathbf{u}_s(t) \end{bmatrix} \tag{4.127}$$

1. $\mathbf{e}(\cdot), \mathbf{v}(\cdot), \mathbf{i}(\cdot), \mathbf{u}_s(\cdot)$ are vector-valued functions of time.
2. The tableau matrix $\mathbf{T}$ has real entries.
3. N_R is completely described by Eq. (4.127), i.e., a set of linear algebraic equations with real coefficients.

Linear time-invariant circuit N_ω operating in the sinusoidal steady-state

$$\begin{bmatrix} 0 & 0 & A \\ -A^T & I & 0 \\ 0 & M(j\omega) & N(j\omega) \end{bmatrix} \begin{bmatrix} E \\ V \\ \bar{I} \end{bmatrix}$$

$$= \begin{bmatrix} 0 \\ 0 \\ U_s \end{bmatrix} \qquad (4.128)$$

1. $E, V, \bar{I}, U_s$ are vectors whose components are phasors.
2. The tableau matrix $T(j\omega)$ has complex entries in its bottom b rows.
3. N_ω is completely described by Eq. (4.128), i.e., a set of linear algebraic equations with complex coefficients.

Moreover:

---

1. The superposition theorem holds for N_ω: provided $\det[T(j\omega)] \neq 0$, the sinusoidal steady-state (at frequencies ω) due to several independent sources (at frequency ω) is equal to the sum of the sinusoidal steady-states due to each independent source acting alone (see Sect. 3.6.1).
2. Thévenin-Norton equivalent: For example, if the DP characteristic of N_ω at a pair of terminals $1, 1'$ is current-controlled, then the resulting one-port may be replaced by a Thévenin equivalent, but with a V_{oc} that is the phasor representing the open-circuit voltage at $1, 1'$ and Z_{eq} is the impedance of $N_{\omega 0}$ seen at $1, 1'$, $\omega 0$ is the particularly forcing frequency at which the impedance is determined (see Sect. 3.6.2).

4.3.3 Laplace Transforms

In the preceding section, we studied linear time invariant circuits in the sinusoidal steady-state, and our main tool was phasor analysis. In this section, we continue to study linear time-invariant circuits, but we do it now under general excitation. We will again encounter a number of basic concepts and properties that are indispensable to the solution of many scientific problems.

Since the Laplace transform is a **generalization of the phasor concept**, we will avoid repetition and discuss the main differences in this section between the Laplace transform and phasors through examples. Particularly:

1. The Laplace transform can be utilized to obtain both the transient and steady-state response.

2. Inverse Laplace transforms (usually by partial fraction expansion) are needed to obtain the corresponding time-response.

Throughout this section, the variable s will be a **complex** variable expressed in standard form: $s = \sigma + j\omega$, $\sigma, \omega \in \Re$. We view s as a point in the complex plane: σ is its abscissa and ω is its ordinate. The (one-sided) Laplace transform of a time domain function $f(t)$ is defined as:

$$F(s) \stackrel{\triangle}{=} \int_{0^-}^{\infty} f(t)e^{-st}dt \qquad (4.129)$$

In the integral above, t is the integration variable and hence the integral depends only on the time function $f(\cdot)$ and on a **particular** value of s, the **complex frequency**. Few remarks:

1. The lower limit of integration is chosen to be $0-$ so that whenever $f(t)$ includes an impulse at the origin, it is included in the interval of integration (see Example 4.3.5).
2. The operation of taking the Laplace transform is denoted by $\mathscr{L}$, thus we write: $F(s) = \mathscr{L}\{f\}(s)$.
3. The operation of taking the inverse Laplace transform is denoted by $\mathscr{L}^{-1}$: $f(t) = \mathscr{L}^{-1}\{F\}(t)$.
4. If we take the Laplace transform of a voltage $v(t)$ or current $i(t)$, we denote them by $V(s)$ and $I(s)$. Thus we use uppercase letters to denote Laplace transforms.

Example 4.3.3 Show that the Laplace transform of the impulse function $\delta(t)$ is $\mathscr{L}(\delta) = 1$.

Solution Let us approximate $\delta(t)$ by using the procedure from Sect. 4.2.1.3. Consider the unit area rectangular pulse $p_\Delta(t)$:

$$p_\Delta(t) = \begin{cases} \frac{1}{\Delta} & \text{for } 0 \leq t \leq \Delta \\ 0 & \text{elsewhere} \end{cases}$$

Using p_Δ in the definition of the Laplace transform in Eq. (4.129) and simplifying:

$$\int_0^{\infty} p_\Delta(t)e^{-st}dt = \int_0^{\Delta} \frac{1}{\Delta}e^{-st}dt$$

$$= \frac{e^{-st}}{-s\Delta}\Big|_0^{\Delta}$$

(continued)

Example 4.3.3 (continued)

$$= \frac{1 - e^{-s\Delta}}{\Delta}$$

Now let $\Delta \to 0$, then $p_\Delta(t) \to \delta(t)$ and $\mathscr{L}\{p_\Delta\} \to \mathscr{L}\{\delta\}$. Thus we have:

$$\mathscr{L}\{\delta\} = \lim_{\Delta \to 0} \frac{1 - e^{-s\Delta}}{s\Delta}$$

$$= \lim_{\Delta \to 0} \frac{1 - (1 - s\Delta + s^2\Delta^2/2 - \cdots)}{s\Delta}$$

$$= 1$$

Example 4.3.3 shows the significance of the impulse response: since the Laplace transform of δ is unity, from a (complex) frequency standpoint, we say $\delta(t)$ contains "all frequencies." Hence the impulse response of a linear time-invariant circuit (system) contains **all** information about the system.

There are also a variety of properties of Laplace transforms that follow from phasors: linearity, etc. But the **uniqueness** property of Laplace transforms is **general** in the sense Eq. (4.129) establishes a **one-to-one correspondence** between f and F. This is a deep theorem of mathematical analysis, whose proof is beyond the scope of this text. But it is extremely useful and justifies the fact that we can transform a time-domain problem into a frequency-domain problem, solve it in the frequency domain, and then go back to the time-domain solution. The uniqueness of Laplace transforms guarantees that the procedure gives **the** solution of the original problem.

The important difference of Laplace transforms being able to "handle" initial conditions (as opposed to phasors) is illustrated by Example 4.3.4.

Example 4.3.4 Show that: $\mathscr{L}\{\frac{d}{dt} f(t)\} = sF(s) - f(0^-)$.

Solution Using integration by parts in the definition of the Laplace transform:

$$\int_{0^-}^{\infty} \underbrace{e^{-st}dt}_{u} \underbrace{\dot{f}(t)}_{dv} = \underbrace{e^{-st}}_{u} \underbrace{f(t)}_{v} \Big|_{0^-}^{\infty} - \int_{0^-}^{\infty} \underbrace{f(t)}_{v} \underbrace{(-se^{-st})\,dt}_{du}$$

$$= -f(0^-) + s\int_{0^-}^{\infty} f(t)e^{-st}dt$$

$$= sF(s) - f(0^-) \qquad\qquad (4.130)$$

(continued)

Example 4.3.4 (continued)
To obtain the final result note that we have used the fact that Re(s) is
sufficiently large so that $f(t)e^{-st} \to 0$ at $t \to \infty$. This is true for all non-
pathological physical functions $f(t)$.

Exercise 4.14 generalizes Example 4.3.4 to nth-order.

The analysis of a circuit by Laplace transforms yields the transform of the output
variable. The next step is to go from the Laplace transform back to the time function,
or as engineers say, from the frequency domain to the time domain. An extremely
useful technique is the partial fraction expansion.

Suppose we are given a Laplace transform $F_0(s)$ which is a rational function
$n_0(s)/d_0(s)$, where $n_0(s)$ and $d_0(s)$ are polynomials with real coefficients. We
further assume that $n_0(s)$ and $d_0(s)$ are coprime, that is, any nontrivial common
factor has been canceled out.

If the degree of n_0 is greater than or equal to the degree of d_0, we first divide the
polynomial $n_0(s)$ by $d_0(s)$ to obtain the quotient polynomial $q(s)$ and the remainder
polynomial $r(s)$. For example:

$$\frac{2s^2 + 8s + 7}{(s+1)(s+3)} = 2 + \frac{1}{(s+1)(s+3)}$$

with $q(s) = 2$, $r(s) = 1$. Since the property of linearity carries over to the Laplace
transform from phasors:

$$\mathcal{L}^{-1}\left(\frac{2s^2 + 8s + 7}{(s+1)(s+3)}\right) = \mathcal{L}^{-1}(2) + \mathcal{L}^{-1}\left(\frac{1}{(s+1)(s+3)}\right)$$

The inverse Laplace transform can be looked up from tables, but we know from
Example 4.3.3 that:

$$\mathcal{L}^{-1}(2) = 2\delta(t)$$

To determine the inverse Laplace transform of $\frac{1}{(s+1)(s+3)}$, we know from basic
algebra that:

$$\frac{1}{(s+1)(s+3)} = \frac{A}{s+1} + \frac{B}{s+3}$$

We can solve for A and B by any convenient technique. We thus have:

$$\frac{1}{(s+1)(s+3)} = \frac{0.5}{s+1} - \frac{0.5}{s+3}$$

From Laplace transform tables (or the reader can easily derive the expression below from the Laplace transform definition), we get:

$$\mathscr{L}^{-1}\left(\frac{k}{s+a}\right) = ke^{-at}u(t)$$

We insert the unit step function to remind ourselves that $f(t) < 0$ for $t < 0$ (we have not defined the double-sided Laplace transform). Thus:

$$\mathscr{L}^{-1}\left(\frac{2s^2 + 8s + 7}{(s+1)(s+3)}\right) = 2\delta(t) + (0.5e^{-t} - 0.5e^{-3t})u(t) \tag{4.131}$$

The subject of partial fraction expansion as applied to Laplace transforms can be found in any text on electrical engineering. Hence, we will not discuss the topic further and instead we will now illustrate how to reformulate a linear time-invariant circuit in the frequency domain using Laplace transforms, with Example 4.3.5.

Example 4.3.5 Reconsider the series RC circuit from Sect. 4.2.1.3. Derive the impulse response.

Solution Consider the element law (following the passive sign convention) for the linear capacitor:

$$i = C\frac{dv}{dt} \tag{4.132}$$

Assuming zero initial conditions, taking Laplace transforms on both sides and using the differentiation rule, we get:

$$I(s) = sCV(s) \tag{4.133}$$

For a linear resistor, the $V(s) - I(s)$ relationship is trivial: $V(s) = RI(s)$. Therefore, the circuit in Sect. 4.2.1.3 can be transformed to the Laplace domain as shown in Fig. 4.37. As stated earlier, since the Laplace transform is a generalization of the phasor technique, KCL, KVL, etc. are all valid in the Laplace domain. Therefore, using voltage divider and simplifying:

$$\begin{aligned}
V_C(s) &= \frac{1/sC}{R + 1/sC} \\
&= \frac{1}{1 + sRC} \\
&= \frac{1/RC}{s + 1/RC}
\end{aligned} \tag{4.134}$$

(continued)

Example 4.3.5 (continued)
Using inverse Laplace transform:

$$v_C(t) = \frac{1}{RC}e^{-t/RC}u(t) \tag{4.135}$$

which is exactly Eq. (4.45), since $\tau = RC$.

Note Example 4.3.5 shows the Laplace transform is applicable even when the input is nonsinusoidal. Example 4.3.5 also shows that **provided all time functions are 0 at $t = 0^-$** (equivalently, all initial conditions are zero at $t = 0^-$) **the rules for manipulating phasors and the rules for manipulating Laplace transforms are identical**, except for replacing $j\omega$ by s. Example 4.3.6 further illustrates this point.

Example 4.3.6 Reconsider the *RLC* circuit from Example 4.3.2. Determine $I_L(s)$.

Solution We can redraw the *RLC* circuit in the Laplace domain and solve for $I_L(s)$. But, let us simply take the differential equation from Example 4.3.2:

$$\frac{d^2}{dt^2}i_L(t) + 2\alpha\dot{i}_L(t) + \omega_0^2 i_L(t) = \omega_0^2 i_s(t)$$

and take its Laplace transform (assuming zero initial conditions):

$$(s^2 + 2\alpha s + \omega_0^2)I_L(s) = \omega_0^2 I_s(s) \tag{4.136}$$

We have used Exercise 4.14 for the Laplace transform of the second derivative. Simplifying:

$$I_L(s) = \frac{\omega_0^2 I_s(s)}{(s^2 + 2\alpha s + \omega_0^2)} \tag{4.137}$$

Fig. 4.37 Circuit for
Example 4.3.5

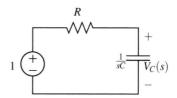

The phasor Eq. (4.118) and the Laplace Eq. (4.136) have the **exact same form** except for $j\omega$ being replaced by s. But, we would like to again emphasize that the two equations have **different meanings**: Eq. (4.118) is only valid for sinusoidal inputs at **steady-state**. The Laplace Eq. (4.136) is valid for arbitrary inputs. Moreover, Exercise 4.15 generalizes Example 4.3.6 to the case when the initial conditions are not zero.

Example 4.3.6 also shows an example of a **network function**. A detailed discussion is beyond the scope of this book but can be found in excellent references such as [12].

However, one can understand the concept by considering $H(s) \overset{\triangle}{=} I_L(s)/I_s(s)$ in Eq. (4.137). Notice $H(s)$ (or the **current transfer function**) depends only on the circuit parameters, it does **not** depend on $I_s(s)$ (the input). Thus, we will adopt the following **general definition** of a **network function**, which basically describes the properties of the circuit:

$$\text{Network Function} \overset{\triangle}{=} \frac{\mathscr{L}(\text{zero-state response})}{\mathscr{L}(\text{input})} \tag{4.138}$$

For example, Exercise 4.16 asks you to derive the input impedance of a gyrator, which is a network function.

4.4   Memristive Networks

We will now discuss memristive networks. We will split the discussion into two parts—discussion of ideal memristors and memristive devices. For the ideal memristors, we will introduce the Flux-Charge Analysis Method (FCAM) developed by Fernando Corinto and Mauro Forti [13], that helps us write minimal number of ODEs for ideal memristive networks. For memristive devices, we will study some very fundamental properties related to sinusoidal excitation. We will also use only linear L, R, C in memristor networks.

4.4.1   Flux-Charge Analysis Method (FCAM)

Example 4.4.1 illustrates the main concept behind FCAM: the idea of an incremental flux (charge).

Fig. 4.38 Circuit for
Example 4.4.1

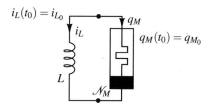

$i_L(t_0) = i_{L_0}$

$q_M(t_0) = q_{M_0}$

Example 4.4.1 Derive circuit equations for the $L - M$ network in Fig. 4.38.
Notice that we have a charge-controlled memristor.

Solution We can easily derive the normal form circuit equations by inspection:

$$\frac{di_L}{dt} = \frac{-R(q_M)i_L}{L} \tag{4.139}$$

$$\frac{dq_M}{dt} \triangleq i_L \tag{4.140}$$

with the given initial conditions. Notice, however, that Eq. (4.139) can be
rewritten using the fact that $\phi = s(q_M)$ (Sect. 1.9.4) for $\mathcal{N}_M$:

$$\frac{di_L}{dt} = \frac{1}{L}\frac{d}{dt}s(q_M(t)) \tag{4.141}$$

Note from the passive sign convention: $\frac{dq_M}{dt} = -i_L$. Integrating both sides
from t_0 to t and applying the first fundamental theorem of calculus, we get:

$$i_L(t) - i_L(t_0) = \frac{1}{L}(s(q_M(t)) - s(q_M(t_0))) \tag{4.142}$$

In other words, we have the following **first-order ODE** (with **two** initial
conditions):

$$\frac{dq_M(t)}{dt} = -\frac{s(q_M(t))}{L} + \frac{s(q_{M_0})}{L} - i_{L_0}$$

$$q_M(t_0) = q_{M_0} \tag{4.143}$$

Example 4.4.1 shows that for **ideal memristor networks**, an **nth-order ODE
in the (v, i) domain can be reduced to an $(n - 1)$th-order ODE in the (ϕ, q)
domain**. But, the **order of complexity is still n, because we still need n initial
conditions**.

The example also shows that the fundamental step in reducing the number of ODES by one is the **integral of KVL** in (t_0, t), referred to as KϕL [13]. Formally:

Definition 4.13 (KϕL)   The algebraic sum of incremental flux around any closed circuit is zero.

With respect to Example 4.4.1, Eq. (4.142) can be written as:

$$Li_L(t) - Li_L(t_0) - [s(q_M(t)) - s(q_M(t_0))] = 0$$

$$\phi_L(t; t_0) - \phi_M(t; t_0) = 0 \qquad (4.144)$$

where we have used the notation: $\phi_L(t; t_0) \overset{\triangle}{=} Li_L(t) - Li_L(t_0)$ (similar notation for $\phi_M(t; t_0)$). Notice as expected KϕL is simply the equivalent of KVL in the flux domain: there is only one flux in the circuit of Fig. 4.38 since the voltage across both elements is equal. By duality, we have KqL:

Definition 4.14 (KqL)   The algebraic sum of incremental charge in a closed surface is zero.

Now that we have the laws of interconnections for ideal memristor networks, we can easily reformulate the fundamental circuit elements in the (ϕ, q) domain [13] as shown in Fig. 4.39. In Fig. 4.39, we have:

(a) Ideal voltage source: $\phi(t; t_0) = \phi_e(t; t_0)$, $\forall q_e(t; t_0)$
(b) Ideal current source: $q(t; t_0) = q_a(t; t_0)$, $\forall \phi_a(t; t_0)$
(c) R: $\phi_R(t; t_0) = Rq_R(t; t_0)$
(d) L: $\phi_L(t) = L\frac{d}{dt}(q_L(t))$ $\phi_L(t; t_0) = -\phi_{L_0} + L\frac{d}{dt}(q_L(t; t_0))$
(e) C: $q_C(t) = C\frac{d}{dt}(\phi_C(t))$ $q_C(t; t_0) = -q_{C_0} + C\frac{d}{dt}(\phi_C(t; t_0))$
(f) Flux-controlled $\mathscr{N}_M$: $q_M(t; t_0) = f(\phi_M(t; t_0) + \phi_{M_0}) - q_{M_0}$
(g) Charge-controlled $\mathscr{N}_M$: $\phi_M(t; t_0) = h(q_M(t; t_0) + q_{M_0}) - \phi_{M_0}$

Although we could have reduced the number of relationships above by invoking duality, we would like for the reader to have a complete reference for FCAM.

4.4.2   Memristive Devices

It is possible to systematically derive differential-algebraic equations for memristive devices, based on tableau analysis, see [27]. But since this topic is beyond the scope of this book, we will simply obtain the circuit equations for networks with memristive devices in the (v, i) domain by inspection.

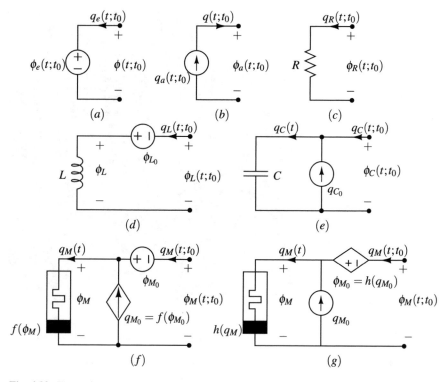

Fig. 4.39 The various two-terminal circuit element equivalents in the (ϕ, q) domain

We will also focus on **passivity** and **frequency-characteristics**[19] theorems the memristor. These theorems will help us identify physical memristors. We will not rigorously prove these theorems as all the proofs can be found in [9]. Rather, we will give examples from physical memristors. We will state all theorems for current-controlled (recall Eq. (1.86)) memristive devices:

$$\dot{\mathbf{x}} = f(\mathbf{x}, i, t)$$

$$v = R(\mathbf{x}, i, t)i \qquad (4.145)$$

The theorems are valid for voltage-controlled memristive devices, by duality.

---

[19]It is important to note that we **do not say** frequency response since that is a term reserved for linear systems.

Fig. 4.40 Measured
discharge tube characteristics

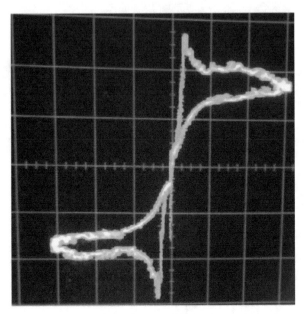

Theorem 4.5 (Passivity Criterion) *Let a current-controlled memristive one-port be time-invariant and let its nonlinear memristance function $R(\cdot)$ satisfy the constraint $R(x, i) = 0$ only if $i = 0$. Then the one-port is **passive** iff $R(x, i) \geq 0$ for any admissible input current $i(t)$, for all $t \geq t_0$ where t_0 is chosen such that $x(t_0) = x^*$, where x^* is the state of minimum energy storage.*

This theorem essentially says that for a memristor to be passive, its (v, i) characteristic should lie in the first and third quadrant. For example, consider the discharge tube $v - i$ from Chap. 1, reproduced in Fig. 4.40. Notice how the Lissajous figure is only present in the first and third quadrants, hence the discharge tube is a passive memristor. **However**, in each quadrant, the curve is **passive** but **not strictly passive**.

Theorem 4.6 (DC Characteristics) *A time-invariant current-controlled memristive one-port under DC operation is equivalent to a time-invariant current-controlled nonlinear resistor if $f(x, I) \approx 0$ has a unique solution $x = X(I)$ such that for each value of $I \in \Re$, the equilibrium point $x = X(I)$ is globally asymptotically stable.*

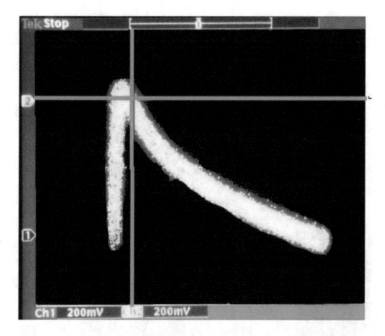

Fig. 4.41 Measured DC memristor characteristics. Experimental oscilloscope picture has been offset for clarity, the x axis is current mapped to voltage. We have marked axes in blue

An example of DC characteristics is shown in Fig. 4.41 for an emulated memristor [23] that is used in the Muthuswamy-Chua (Sect. 5.4.1) chaotic circuit.

Theorem 4.7 (Double-Valued Lissajous Figure) *A current-controlled memristive one-port under periodic operation (i.e., response is periodic with same period as input) with $i(t) = I\cos(\omega t)$ always gives rise to a $v - i$ Lissajous figure whose voltage v is at most a double-valued function of i.*

Figure 4.40 shows the classic **pinched-hysteresis fingerprint** of a memristor.

Theorem 4.8 (Limiting Linear Characteristics) *If a time-invariant current-controlled memristive one-port described by Eq. (4.145) is BIBO stable, then under periodic operation it degenerates into a **linear time-invariant** resistor as the excitation frequency increases towards infinity.*

The effect of limiting linear characteristics is shown in Fig. 4.42. Notice from Fig. 4.42 that we have "lost" the limiting linear characteristics as the frequency

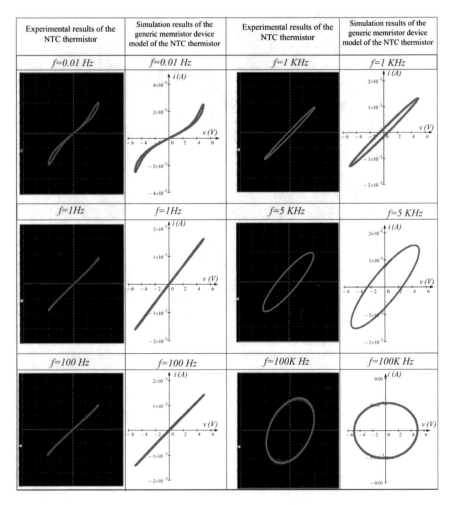

Fig. 4.42 Experimental measurements and corresponding simulated results (red) [28] of Thermometric's NTC diode thermistors (NTC-3.896KGJG), illustrating Theorem 4.8. The input is a sinusoidal source with amplitude $A = 5\,\text{V}$. The experiment on the NTC thermistor was conducted at room temperature. The parameters used for simulations of the generic memristor device model of the NTC thermistor are: $T_{0N} = 300$ K, $R_{0N} = 3.89$ KΩ, $H_{CN} = 0.14$ J/K, $\delta_N = 0.1$ W/K, $\beta_N = 5 \times 10^5$ K. For parasitic effects (Fig. 4.43), $C_P = 5\,\text{nF}$, $L_p = 2\,\text{mH}$, $E_P = 0\,\text{V}$, $I_P = 0\,\text{A}$

increases. This **does not** imply Theorem 4.8 is invalid. Rather, a physical memristor is not exactly modeled by Eq. (4.145).

Recall from Sect. 1.7.1 about the essence of modeling: we extract the essential factors of the device based on the circuit in question. In the case of physical memristive devices, we need a generic device model since measured pinched-hysteresis loops need not pass through the origin due to parasitics. This generic device model is shown in Fig. 4.43 [28] and has been used in the simulation results

Fig. 4.43 Generic memristor
device model

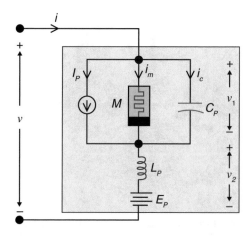

for Fig. 4.42. The NTC thermistor model used in the circuit of Fig. 4.43 for obtaining
Fig. 4.42 is given by Eqs. (4.146) and (4.147).

$$W(T_N) \triangleq \left(R_{0N} e^{-\beta_N \left(\frac{1}{T_N} - \frac{1}{T_{0N}} \right)} \right)^{-1}$$

$$i_N = W(T_N) v_N \tag{4.146}$$

$$\frac{dT_N}{dt} = \frac{\delta_N}{H_{CN}} (T_{0N} - T_N) + \frac{W(T_N)}{H_{CN}} v_N^2 \tag{4.147}$$

We will now examine another example of Fig. 4.43, detailed analysis can be found
in [28]. Consider two simulated pinched-hysteresis loops for the discharge tube
memristor, shown in Fig. 4.44a and b [24]. Experimental confirmation can be found
in [24].

If we have a parasitic inductor in series with a memristor, as in Fig. 4.44a, we
know that an inductor causes current to start lagging voltage. Hence when $i_M =
0$, if $v_M > 0$, then v_M should be increasing because current is lagging voltage.
Thus $\dot{v}_M > 0$. Similarly, when $i_M = 0$ and $v_M < 0$, then v_M should continue
to decrease and thus $\dot{v}_M < 0$. Hence the parasitic pinched-hysteresis loop ends up
having no "crossings." A dual argument applies to Fig. 4.44b but in this case we get
two "crossings."

Another very important point about modeling: it is **irrelevant with respect
to terminal behavior** how the internal state of a memristor is represented. For
example, there are two known internal state variables for the memristive model of a
discharge tube: the number of conduction electrons n [11]:

$$v = M(n)i \tag{4.148}$$

$$\frac{dn}{dt} = -\beta n + \alpha M(n) i^2 \tag{4.149}$$

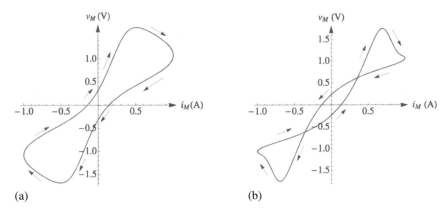

Fig. 4.44 Simulation parameters for the discharge tube model in Eqs. (4.148) and (4.149) are: $\beta = 0.1$, $\alpha = 0.1$, $F = 1$, $\omega = 0.063$. (a) Simulated $i_M - v_M$ curve for an inductor $L_p = 5\,\mathrm{H}$ in series with a memristive discharge tube. The arrows indicate the trajectory of $(i_M(t), v_M(t))$ at $t \to \infty$. We have assumed $E_p = 0$, $I_p = 0$ and $|Z_{C_p}(j\omega)| \to \infty$. (b) Simulated $i_M - v_M$ curve for a capacitor $C_p = 1\,\mathrm{F}$ in parallel with a memristive discharge tube. The arrows indicate the trajectory of $(i_M(t), v_M(t))$ at $t \to \infty$. We have assumed $E_p = 0$, $I_p = 0$ and $|Z_{L_p}| \to 0$

or tube temperature T [21]:

$$R(T) \overset{\triangle}{=} a_5 T^{-3/4} \exp(ea_6/2kT) \tag{4.150}$$

$$v(t) = R(T)i(t) \tag{4.151}$$

$$\frac{dT}{dt} = a_1[i^2 R(T) - a_2 \exp(-ea_3/kT) - a_4 \exp(T - T_0)] \tag{4.152}$$

The values of the constants and the physical meaning of the variables in Eqs. (4.151) and (4.152) depend on whether the discharge tube being modeled is either a high-pressure lamp or a low-pressure lamp. For instance, for high pressure lamps, T is the gas temperature T_g and T_0 is the tube-wall temperature. $a_1 = 20976.1$, $a_2 = 54350.4$, $a_3 = 0.986$, $a_4 = 0.128$, $a_5 = 2012.0$, $a_6 = 0.375$, $T_0 = 1000\,\mathrm{K}$. $e = 1.6 \times 10^{-19}\,\mathrm{C}$ is the electron charge and $k = 1.38 \times 10^{-23}\,\mathrm{J/K}$ is Boltzmann's constant.

Irrespective of the choice of the internal state variable for the memristive model of the discharge tube, the $v - i$ terminal behavior **still** shows pinched-hysteresis. For investigating the parasitic behavior, we chose the **simpler of the two models**: the internal state being a function of the number of conduction electrons [24]. This point bolsters our theme of modeling throughout the book, which is summarized by a quote from Einstein: "It can scarcely be denied that the **supreme goal of theory is to make irreducible basic elements as simple and as few as possible without having to surrender the adequate representation of a single datum of experience**" [2].

Fig. 4.45 The small-signal
AC equivalent circuit for
Eq. (4.145)

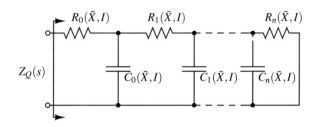

Fig. 4.46 The small-signal
AC equivalent circuit of a
thermistor [9]

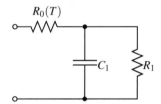

Theorem 4.9 (Small-Signal AC Characteristics) *If a time-invariant current-controlled memristive one-port is globally asymptotically stable for all DC input current* I*, then its small-signal equivalent circuit about the DC operating point is shown in Fig. 4.45, with a **small-signal impedance** given by:*

$$Z_Q(s) \triangleq \frac{\Delta V(s)}{\Delta I(s)} = \frac{\partial R(X, I)I}{\partial i} + \frac{\beta_1 s^{n-1} + \beta_2 s^{n-2} + \cdots + \beta_{n-1}s + \beta_n}{s^{n-1} + \alpha_1 s^{n-1} + \cdots + \alpha_{n-1}s + \alpha_n}$$

$$(4.153)$$

A small-signal equivalent for the thermistor is shown in Fig. 4.46, where:

$$C_1 = \frac{C}{2\alpha P R(T)} \triangleq \hat{C}_1(T, I)$$

$$R_1 = \frac{2\alpha P R(T)}{\delta_N - \alpha P} \triangleq \hat{R}_1(T, I)$$

$$\alpha \triangleq = -\frac{\beta_N}{T^2} < 0, \quad P \triangleq VI = R(T)I^2$$

Since C_1 is negative, the thermistor is inductive under small-signal operation.

The reader should hence realize from this section that a memristor is described by **two** concepts: **memory** and **resistance**. Memory occurs in the form of hysteresis in a $v - i$ plot, resistance in the form of **pinching** behavior at the origin in the $v - i$ plot. Note that memory **need not** imply "storage" in the sense of a capacitor or inductor. Rather, a memristor's resistance (conductance) depends on past history of a particular state variable.

Therefore, in conclusion to this section, we have the following working hypothesis for memristors:

Since a memristor is described by two concepts: memory and resistance, memristor physics **cannot be fully explained by electromagnetic field theory**. Specifically, the memristor state equation requires another branch of science. This is in sharp contrast to resistors, capacitors, and inductors, whose material behavior is the subject of electromagnetic fields in matter (conductive, dielectric, and ferromagnetic media respectively).

For example:

1. The Josephson junction ideal menductance is described using superconductivity (and hence quantum mechanics).
2. Discharge tube state equation is described using plasma physics.
3. *pn*-junction diode memristance requires junction physics. In fact, the memristance arises because the semiconductor bulk resistance is not a constant, but a function of the charge flowing through it [11, 26].

We encourage readers to rigorously investigate and prove or disprove the hypothesis above.

4.5   Energy Approach: Lagrangian and Hamiltonian

In this section,[20] we will start out by discussing energy expressions for two-terminal resistor, capacitor, and inductor.[21] As examples, we will obtain system equations for a circuit using the Lagrangian and Hamiltonian. The purpose of doing so is to provide the reader with a third (in addition to time and frequency) approach to writing circuit equations.

A key aspect of the Lagrangian and Hamiltonian frameworks is that they bring to forefront one of the most fundamental concepts in physics—energy. A second motivation is that the energy based approach helps us to view a circuit as a (usually simpler) set of subsystems that exchange energy among themselves and the environment. Unfortunately, we can only scratch the surface of this fascinating topic in this section. The interested reader is referred to [19] and [17] as starting points.

---

[20]Many thanks to Dr. Jevtic and Dr. Thomas for reviewing and correcting errors in this section.

[21]We will only focus on linear elements, for brevity. Specifically with respect to action and coaction definitions for the memristor, please see [4, 22].

We know from basic physics that energy is defined as the integral of power:

$$w(t_1, t_2) = \int_{t_1}^{t_2} p(\tau)d\tau \tag{4.154}$$

From the definition of power for a two-terminal element, we get:

$$w(t_1, t_2) = \int_{t_1}^{t_2} v(\tau)i(\tau)d\tau \tag{4.155}$$

With respect to a resistor, Eq. (4.155) would imply that no energy is stored. For example, for a linear resistor, we get:

$$w_R(t_1, t_2) = \int_{t_1}^{t_2} [i(\tau)R]i(\tau)d\tau$$

$$= R \int_{t_1}^{t_2} i^2(\tau)d\tau$$

$$= \frac{1}{R} \int_{t_1}^{t_2} v^2(\tau)d\tau \tag{4.156}$$

If $R > 0$, the energy is dissipated usually in the form of heat and is lost as far as the circuit is concerned. Such an element is therefore said to be **lossy**.

In contrast, capacitors and inductors store energy. The energy $w_c(t_1, t_2)$ entering a charge-controlled capacitor during any time interval $[t_1, t_2]$ is **independent of the capacitor voltage or current waveforms**: It is uniquely determined by the capacitor charge at the end points, namely, $q(t_1)$ and $q(t_2)$:

$$w_c(t_1, t_2) = \int_{t_1}^{t_2} \hat{v}(q(t))\frac{dq}{dt}dt$$

$$= \int_{q(t_1)}^{q(t_2)} \hat{v}(q)dq \tag{4.157}$$

Suppose we have a C-F linear capacitor having an initial voltage $v(t_1) = V$ and initial charge $q(t_1) = Q = CV$ at $t = t_1$. Let the capacitor be connected to an external circuit at $t = t_1$. The energy entering the capacitor during $[t_1, t_2]$ is given by Eq. (4.157):

$$w_C(t_1, t_2) = \frac{1}{2C}[q^2(t_2) - Q^2]$$

$$= \frac{1}{2}C[v^2(t_2) - V^2] \tag{4.158}$$

Note that whenever $q(t_2) < Q$ or $v(t_2) < V$, then $w_C(t_1, t_2) < 0$. This means energy is actually being sent out of the capacitor and returned to the external circuit. It follows from Eq. (4.158) that $w_C(t_1, t_2)$ is **most negative** when $q(t_2) = v(t_2) = 0$, whereupon $w_c(t_1, t_2) = -\frac{Q^2}{2C} = -\frac{1}{2}CV^2$. Since this represents the maximum amount of energy that could be **extracted** from the capacitor, it is natural to say that **an energy equal to**

$$\mathscr{E}_C(Q) = \frac{Q^2}{2C}$$

$$= \frac{1}{2}CV^2 \tag{4.159}$$

is stored in a linear capacitor C having an initial voltage $v(t_1) = V$ or initial charge $q(t_1) = Q = CV$.

By duality, an energy equal to:

$$\mathscr{E}_L(\phi) = \frac{1}{2L}\phi^2$$

$$= \frac{1}{2}LI^2 \tag{4.160}$$

is stored in a linear inductor L having an initial current $i(t_1) = I$ or initial flux $\phi(t_1) = \phi = LI$.

Now that we have expressions for the energy stored in a (linear) capacitor or inductor, we need to understand the meaning of "kinetic" and "potential" energy in electric circuits, before we can discuss how to obtain circuit equations via the Lagrangian and the Hamiltonian. To do this, we will appeal to the reader's "natural intelligence" with respect to (translational) mechanical systems. Consider the following (we will again assume all mechanical elements are linear and we will not worry about relativistic effects):

- m (mass)—Characteristic Equation: $p = mv$, p: linear momentum, v: velocity
- k (Spring constant)—Characteristic Equation: $F = kx$, F : force, x: displacement

We know the energy expressions for a mass and spring as:

- m: $\mathscr{E}_m = \frac{1}{2}mv^2$
- k: $\mathscr{E}_k = \frac{1}{2}kx^2$

Obviously, a **moving mass** has a **kinetic energy** $\mathscr{E}_m$ and a **compressed spring** has a **potential energy** $\mathscr{E}_k$. Now consider the energy expressions for L and C:

- L: $\mathscr{E}_L = \frac{1}{2}LI^2$
- C: $\mathscr{E}_C = \frac{1}{2}CV^2$

It should now be clear that since an inductor's stored energy is due to **current or moving charge**, our **mechanical analog of an inductor is the mass**. Since a capacitor's stored energy is due to **electrostatic potential**, our **mechanical analog of a capacitor is the spring**. Hence ($\longleftrightarrow$ stands for analog):

$$p \text{ (momentum)} \longleftrightarrow \phi \text{ (flux)} \tag{4.161}$$

$$v \text{ (velocity)} \longleftrightarrow i \text{ (current)} \tag{4.162}$$

$$x \text{ (displacement)} \longleftrightarrow q \text{ (charge)} \tag{4.163}$$

$$F \text{ (force)} \longleftrightarrow v \text{ (voltage)} \tag{4.164}$$

Next, let $x = [x_1, \cdots, x_n]^T$ denote a column vector and $V(\mathbf{x})$ denote a scalar function $V : \mathfrak{R}^n \to R$. The gradient of $V(\mathbf{x})$ with respect to x is denoted by:

$$\nabla V_x(\mathbf{x}) \overset{\triangle}{=} \begin{bmatrix} \frac{\partial V}{\partial x_1} \\ \frac{\partial V}{\partial x_2} \\ . \\ . \\ . \\ \frac{\partial V}{\partial x_n} \end{bmatrix} \tag{4.165}$$

Mechanically, the Lagrangian and Hamiltonian are described in terms of generalized coordinates. In the case of electric circuits, we will use $\mathbf{q}$ (capacitor charge(s)) and consequently $\dot{\mathbf{q}}$ as our generalized coordinates. The formalism requires that the Lagrangian L be expressed in terms of $\mathbf{q}$ and $\dot{\mathbf{q}}$:

$$L(\mathbf{q}, \dot{\mathbf{q}}) = \mathscr{E}_L - \mathscr{E}_C \tag{4.166}$$

where $\mathscr{E}_L$ represents the **total** energy stored in inductor(s) and $\mathscr{E}_C$ represents the **total** energy stored in capacitor(s). Notice this is equivalent to the definition from mechanics, **if** we consider energy stored in inductor(s) as "kinetic" energy and energy stored in capacitor(s) as "potential" energy.

The total energy of the capacitors $\mathscr{E}_C$ can be readily expressed in terms of charge:

$$\mathscr{E}_C(\mathbf{q}) = \sum_{n=1}^{N_C} \frac{q_n^2}{2C_n} \tag{4.167}$$

Fig. 4.47 Circuit for
Example 4.5.1

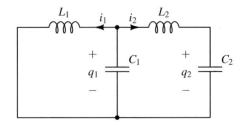

where N_C is the total number of capacitors in the circuit. However the total energy
of the inductors $\mathscr{E}_L$ is usually expressed in terms of inductor currents $\mathbf{i}$:

$$\mathscr{E}_L(\mathbf{i}) = \sum_{n=1}^{N_L} \frac{1}{2} L_n i_n^2 \tag{4.168}$$

where N_L is the total number of inductors in the circuit. We must therefore first
express the inductor currents $\mathbf{i}$ in terms of $\dot{\mathbf{q}}$:

$$\mathbf{i} = \mathbf{A}\dot{\mathbf{q}} \tag{4.169}$$

where $\mathbf{A}$ is an $N_L \times N_C$ matrix. This can be done using KCL (as shown in Example 4.5.1). Now, we can write Lagrange's equations in terms of the Lagrangian:

$$\frac{d}{dt} \nabla L_{\dot{q}}(\mathbf{q}, \dot{\mathbf{q}}) - \nabla L_q(\mathbf{q}, \dot{\mathbf{q}}) = \mathbf{0} \tag{4.170}$$

where $L(\mathbf{q}, \dot{\mathbf{q}}) = \mathscr{E}_L(\mathbf{A}\dot{\mathbf{q}}) - \mathscr{E}_C(\mathbf{q})$.

Example 4.5.1 Write system equations for the circuit in Fig. 4.47 using the
Lagrangian, for $t \geq 0$. Assume the inductors have initial current $i_1(0), i_2(0)$
and capacitors have initial charge $q_1(0), q_2(0)$ at $t = 0$.

Solution For the circuit, we have:

$$\mathscr{E}_L - \mathscr{E}_C = \frac{1}{2} L_1 i_1^2 + \frac{1}{2} L_2 i_2^2 - \frac{q_1^2}{2C_1} - \frac{q_2^2}{2C_2} \tag{4.171}$$

(continued)

Example 4.5.1 (continued)

We need to rewrite the energy expression above in terms of $(\mathbf{q}, \dot{\mathbf{q}})$ for the Lagrangian, where $\mathbf{q} = \begin{bmatrix} q_1 \\ q_2 \end{bmatrix}$. From KCL:

$$i_1 = -\dot{q}_1 - \dot{q}_2$$

$$i_2 = \dot{q}_2 \tag{4.172}$$

Thus the Lagrangian for the circuit is:

$$L(\mathbf{q}, \dot{\mathbf{q}}) = \frac{1}{2}L_1(\dot{q}_1 + \dot{q}_2)^2 + \frac{1}{2}L_2\dot{q}_2^2 - \frac{q_1^2}{2C_1} - \frac{q_2^2}{2C_2} \tag{4.173}$$

Hence we have:

$$\frac{d}{dt}\nabla L_{\dot{q}}(\mathbf{q}, \dot{\mathbf{q}}) - \nabla L_q(\mathbf{q}, \dot{\mathbf{q}}) = \frac{d}{dt}\begin{bmatrix} \frac{\partial L}{\partial \dot{q}_1} \\ \frac{\partial L}{\partial \dot{q}_2} \end{bmatrix} - \begin{bmatrix} \frac{\partial L}{\partial q_1} \\ \frac{\partial L}{\partial q_2} \end{bmatrix}$$

$$= \frac{d}{dt}\begin{bmatrix} L_1(\dot{q}_1 + \dot{q}_2) \\ L_1(\dot{q}_1 + \dot{q}_2) + L_2\dot{q}_2 \end{bmatrix} - \begin{bmatrix} -\frac{q_1}{C_1} \\ -\frac{q_2}{C_2} \end{bmatrix}$$

$$= \begin{bmatrix} -L_1\dot{i}_1 + \frac{q_1}{C_1} \\ -L_1\dot{i}_1 + L_2\dot{i}_2 + \frac{q_2}{C_2} \end{bmatrix} \tag{4.174}$$

Therefore Lagrange's equations give us:

$$-L_1\dot{i}_1 + \frac{q_1}{C_1} = 0$$

$$-L_1\dot{i}_1 + L_2\dot{i}_2 + \frac{q_2}{C_2} = 0 \tag{4.175}$$

Notice Lagrange's equations give rise to the KVL equations for the circuit.

By applying a (Legendre) transformation Lagrange's Eq. (4.170), we get the **Hamiltonian**. We first define $\boldsymbol{\psi}$ as the conjugate momenta to $\mathbf{q}$:

$$\boldsymbol{\psi} \overset{\triangle}{=} \nabla_{\dot{q}} L(\mathbf{q}, \dot{\mathbf{q}}) \tag{4.176}$$

Notice that $\boldsymbol{\psi}$ has the same number of components as $\mathbf{q}$, i.e., N_C components while there are N_L inductor fluxes: $\boldsymbol{\phi} = \nabla_i \mathscr{E}_L(\mathbf{i})$. In general, $\psi_n \neq \phi_n$. Specifically, using the chain rule:

$$
\begin{aligned}
\boldsymbol{\psi} &= \nabla_{\dot{q}} L(\mathbf{q}, \dot{\mathbf{q}}) \\
&= \nabla_{\dot{q}} \mathscr{E}_L(\mathbf{A}\dot{\mathbf{q}}) \\
&= \nabla_{\dot{q}} (\mathbf{A}\dot{\mathbf{q}}) \nabla_{\mathbf{A}\dot{\mathbf{q}}} \mathscr{E}_L(\mathbf{A}\dot{\mathbf{q}}) \\
&= \mathbf{A}^T \nabla_i \mathscr{E}_L(\mathbf{i}) \qquad\qquad\qquad\qquad\qquad (4.177)
\end{aligned}
$$

Thus: $\boldsymbol{\psi} = \mathbf{A}^T \boldsymbol{\phi}$. The Hamiltonian formalism requires that the **Hamiltonian function** H be expressed in terms of the generalized coordinates $\mathbf{q}$ (capacitor charges) and their conjugate momenta $\boldsymbol{\psi}$:

$$
H(\mathbf{q}, \boldsymbol{\psi}) = \mathscr{E}_L(\boldsymbol{\psi}) + \mathscr{E}_C(\mathbf{q}) \qquad\qquad (4.178)
$$

Hamilton's equations are hence given by:

$$
\dot{\mathbf{q}} = \nabla_{\psi} H(\mathbf{q}, \boldsymbol{\psi}) \qquad\qquad\qquad (4.179)
$$

$$
\dot{\boldsymbol{\psi}} = -\nabla_q H(\mathbf{q}, \boldsymbol{\psi}) \qquad\qquad\qquad (4.180)
$$

Example 4.5.2 Write system equations for the circuit in Fig. 4.47 using the Hamiltonian, for $t \geq 0$. Assume again the inductors have initial current $i_1(0)$, $i_2(0)$ and capacitors have initial charge $q_1(0)$, $q_2(0)$ at $t = 0$.

Solution In Example 4.5.1, we derived the Lagrangian as:

$$
L(\mathbf{q}, \dot{\mathbf{q}}) = \frac{1}{2} L_1(\dot{q}_1 + \dot{q}_2)^2 + \frac{1}{2} L_2 \dot{q}_2^2 - \frac{q_1^2}{2C_1} - \frac{q_2^2}{2C_2} \qquad (4.181)
$$

We can now find the conjugate momenta:

$$
\begin{aligned}
\psi_1 &= \frac{\partial L}{\partial \dot{q}_1} \\
&= L_1(\dot{q}_1 + \dot{q}_2) \\
&= -L_1 i_1 \\
&= -\phi_1 \qquad\qquad\qquad\qquad\qquad\qquad (4.182)
\end{aligned}
$$

(continued)

Example 4.5.2 (continued)

$$\psi_2 = \frac{\partial L}{\partial \dot{q}_2}$$

$$= L_1(\dot{q}_1 + \dot{q}_2) + L_2 \dot{q}_2$$

$$= -L_1 i_1 + L_2 i_2$$

$$= -\phi_1 + \phi_2 \tag{4.183}$$

The total energy is:

$$\mathcal{E}_L + \mathcal{E}_C = \frac{\phi_1^2}{2L_1} + \frac{\phi_2^2}{2L_2} + \frac{q_1^2}{2C_1} + \frac{q_2^2}{2C_2} \tag{4.184}$$

The Hamiltonian is:

$$H(\mathbf{q}, \boldsymbol{\psi}) = \frac{\psi_1^2}{2L_1} + \frac{(\psi_2 - \psi_1)^2}{2L_2} + \frac{q_1^2}{2C_1} + \frac{q_2^2}{2C_2} \tag{4.185}$$

Hamilton's equations give:

$$\dot{q}_1 = \frac{\partial H}{\partial \psi_1}$$

$$= \frac{\psi_1}{L_1} - \frac{\psi_2 - \psi_1}{L_2} \tag{4.186}$$

$$\dot{q}_2 = \frac{\partial H}{\partial \psi_2}$$

$$= \frac{\psi_2 - \psi_1}{L_2} \tag{4.187}$$

$$\dot{\psi}_1 = -\frac{\partial H}{\partial q_1}$$

$$= -\frac{q_1}{C_1} \tag{4.188}$$

$$\dot{\psi}_2 = -\frac{\partial H}{\partial q_2}$$

$$= -\frac{q_2}{C_2} \tag{4.189}$$

It is trivial to verify that Eqs. (4.186) through (4.189) give rise to KCL and KVL.

4.6 Miscellaneous Topics

We would like to wrap up this chapter by discussing three important and fundamental concepts.

4.6.1 Reciprocity

The reciprocity theorem appears in various fields of science and engineering: physics, mechanics, acoustics, electromagnetic waves, and electric circuits [12]. Roughly speaking, it deals with the symmetric role played by the input and output of a physical system. In electric circuits, reciprocity holds for a subclass of linear and nonlinear circuits. In this section, we will only focus on linear time-invariant circuits. Reciprocity with respect to memristors is an active area of research, see [16]. We will simply give three statements of the theorem and illustrate an application of one of the statements with an example [14].

Consider a linear time-invariant network $\mathcal{N}$ which consists of resistors, inductors, mutual inductors, capacitors, and transformers only. $\mathcal{N}$ is in steady-state and not degenerate. Connect four wires to $\mathcal{N}$ obtaining two pairs of terminals 1, 1' and 2, 2'.

Theorem 4.10 (Reciprocity Theorem Statement 1) *Connect a voltage source $e_0(\cdot)$ to terminals 1, 1' and observe the zero-state current response $j_2(\cdot)$ in a short circuit connected to 2, 2' (see Fig. 4.48a). Next, connect the same voltage source $e_0(\cdot)$ to terminals 2, 2' and observe the zero-state current response $\hat{j}_1(\cdot)$ in a short circuit connected to 1, 1' (see Fig. 4.48b). The reciprocity theorem asserts that whatever the topology and element values of $\mathcal{N}$ and whatever the waveform $e_0(\cdot)$, $j_2(t) = \hat{j}_1(t) \ \forall t$.*

In the statement above, we are essentially saying that if the voltage source is interchanged for a zero-impedance ammeter, the reading of the ammeter will not change.

Theorem 4.11 (Reciprocity Theorem Statement 2) *Connect a current source $i_0(\cdot)$ to terminals 1, 1' and observe the zero-state voltage response $v_2(\cdot)$ in an open circuit connected to 2, 2' (see Fig. 4.48c). Next, connect the same current source $i_0(\cdot)$ to terminals 2, 2' and observe the zero-state voltage response $\hat{v}_1(\cdot)$ in an open circuit connected to 1, 1' (see Fig. 4.48d). The reciprocity theorem asserts that whatever the topology and element values of $\mathcal{N}$ and whatever the waveform $i_0(\cdot)$, $v_2(t) = \hat{v}_1(t) \ \forall t$.*

In the statement above, we are observing open circuit voltages.

Theorem 4.12 (Reciprocity Theorem Statement 3) *Connect a current source $i_0(\cdot)$ to terminals 1, 1' and observe the zero-state current response $j_2(\cdot)$ in a short circuit connected to 2, 2' (see Fig. 4.48e). Next, connect a voltage source $e_0(\cdot)$ to*

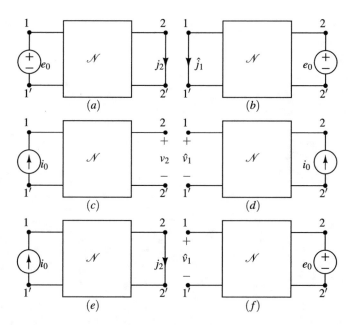

Fig. 4.48 (**a**), (**b**): Reciprocity theorem statement 1, (**c**), (**d**): Reciprocity theorem statement 2, (**e**), (**f**): Reciprocity theorem statement 3

terminals 2, 2' and observe the zero-state voltage response $\hat{v}_1(\cdot)$ in an open circuit connected to 1, 1' (see Fig. 4.48f). The reciprocity theorem asserts that whatever the topology and element values of $\mathcal{N}$, whenever $i_0(t) = e_0(t)$, $\hat{v}_1(t) = j_2(t)$ $\forall t$.

In the statement above, for both measurements, there is an "infinite impedance" connected to 1, 1' and a "zero impedance" connected to 2, 2'. The reader should have noticed that since the reciprocity theorem deals exclusively with the zero-state response (including steady-state response as $t \to \infty$) of a linear time-invariant network, it is convenient to describe it in terms of network functions. We will illustrate the idea in Example 4.6.1 for statement 3 from Theorem 4.12.

Example 4.6.1 Confirm if statement 3 of the reciprocity theorem is true for the circuit shown in Fig. 4.49.

Solution We have defined the ports 1, 1' and 2, 2' as shown in Fig. 4.49b and c, respectively. We are going to find the impulse response and since we do not have any other source, we know at steady state all voltages and currents **must** tend to zero (this will serve as a "sanity check"). By node analysis and

(continued)

Example 4.6.1 (continued)

Laplace transform in Fig. 4.49b we obtain:

$$\begin{bmatrix} 0.2 + s + \frac{1}{s} & -\frac{1}{s} \\ -\frac{1}{s} & 1 + \frac{1}{s} \end{bmatrix} \begin{bmatrix} V_1(s) \\ V_2(s) \end{bmatrix} = \begin{bmatrix} 1 \\ 0 \end{bmatrix} \qquad (4.190)$$

Hence:

$$V_2(s) = \frac{1/s}{(0.2 + s + 1/s)(1 + 1/s) - (1/s)^2} = \frac{1}{s^2 + 1.2s + 1.2} \qquad (4.191)$$

Taking the inverse Laplace transform (using reliable online tables) and noting that $j_2(t) = 1 * v_2(t)$, we obtain:

$$j_2(t) = 1.09e^{-0.6t} \sin(0.916t) \; t \geq 0 \qquad (4.192)$$

For the network in Fig. 4.49c, we will set up circuit equations in terms of $\hat{I}_1(s)$, $\hat{I}_2(s)$ (this is called mesh analysis). The matrix equations are:

$$\begin{bmatrix} 5 + \frac{1}{s} & -\frac{1}{s} \\ -\frac{1}{s} & 1 + s + \frac{1}{s} \end{bmatrix} \begin{bmatrix} \hat{I}_1(s) \\ \hat{I}_2(s) \end{bmatrix} = \begin{bmatrix} 0 \\ 1 \end{bmatrix} \qquad (4.193)$$

Thus:

$$\hat{I}_1(s) = \frac{1/s}{(5 + 1/s)(s + 1 + 1/s) - (1/s)^2} \qquad (4.194)$$

Since $\hat{v}_1(t) = 5\hat{i}_1(t)$, we have:

$$\begin{aligned} \hat{V}_1(s) &= \frac{5}{(5s + 1)(s + 1 + 1/s) - 1/s} \\ &= \frac{5}{5s^2 + 6s + 6} \\ &= \frac{1}{s^2 + 1.2s + 1.2} \end{aligned} \qquad (4.195)$$

Recognizing this function of s to be the transform of $j_2(t)$, we use previous calculations and conclude that:

$$\hat{v}_1(t) = 1.09e^{-0.6t} \sin(0.916t) \; t \geq 0 \qquad (4.196)$$

Thus, the two responses are equal, as required by the reciprocity theorem.

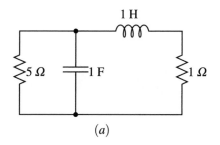

(a)

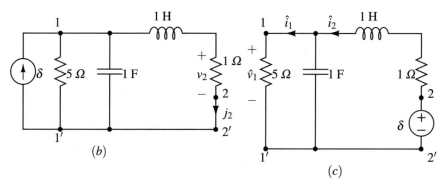

(b)

(c)

Fig. 4.49 Circuit(s) for Example 4.6.1

4.6.2 Synthesis of Higher-Order Circuit Elements

Recall from Chap. 1 that we defined (α, β) circuit elements as a "natural extension" of the four fundamental circuit elements. We have reproduced Fig. 1.40 in Fig. 4.50 for ease of discussion.

In order to give some physical meaning to each higher order element $\mathcal{E}$, it is convenient [5] to examine its **small-signal** behavior about an operating point Q on the associated $v^{(\alpha)} - i^{(\beta)}$ curve. Assuming that $\mathcal{E}$ is characterized by $v^{(\alpha)} = f(i^{(\beta)})$, the small-signal behavior of $\mathcal{E}$ about Q is described by:

$$\delta v^{(\alpha)}(t) = m_Q \delta i^{(\beta)}(t) \tag{4.197}$$

where m_Q denotes the slope $f'(i^{(\beta)})$ at Q. We can define the AC small-signal impedance $Z(j\omega)$ associated with Eq. (4.197) by taking the Laplace transform of Eq. (4.197) and letting $s = j\omega$:

$$\mathcal{L}\{\delta v\} = Z(s)\mathcal{L}\{\delta i\} \tag{4.198}$$

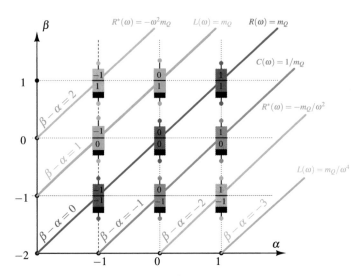

Fig. 4.50 The "periodic table" of all two-terminal (α, β) elements, with a frequency based interpretation

where:

$$Z(j\omega) = (j\omega)^{\beta - \alpha} m_Q \qquad (4.199)$$

Notice we obtained Eq. (4.199) by simply understanding the fact that each derivative constitutes one $j\omega$. We can interpret Eq. (4.199) as the impedance of an associated linearized element $\mathcal{E}_Q$. Since $(\beta - \alpha)$ can be any positive, zero, or negative integer, there are four interesting cases to consider[22]:

- **Case 1:** $\beta - \alpha = \pm 2n$, n = even integer

  In this case, $Z(j\omega) = \omega^{\beta - \alpha} m_Q \overset{\triangle}{=} R(\omega)$ is a **real positive function** and hence $\mathcal{E}_Q$ is **purely resistive**. We can interpret, therefore, $\mathcal{E}_Q$ as a frequency-dependent resistor (red) in Fig. 4.50.

---

[22]The following interpretations are meaningful only for small-signal sinusoidal excitations at a fixed frequency. Such interpretations however often provide valuable information for circuit designers in their analysis of physical nonlinearities. The main point is: depending on the operating point and the operating frequency, the small-signal model of a device may be either resistive, inductive, or capacitive.

- **Case 2:** $\beta - \alpha = \pm 2n$, $n = $ odd integer

  In this case, $Z(j\omega) = -\omega^{\beta-\alpha} m_Q \triangleq R^*(\omega)$ is a **real negative function** and hence $\mathscr{E}_Q$ is **purely resistive**. We can interpret, therefore, $\mathscr{E}_Q$ as a frequency-dependent negative resistor (orange) in Fig. 4.50.
- **Case 3:** $\beta - \alpha = (-1)^n(2n + 1)$, $n = 0, 1, 2, \cdots$

  In this case, $Z(j\omega) = j\omega L(\omega)$ is an imaginary number where

$$L(\omega) \triangleq \begin{cases} \omega^{2n} m_Q, & \text{when } n \text{ even} \\ \omega^{-2(n+1)} m_Q, & \text{when } n \text{ odd} \end{cases} \tag{4.200}$$

and hence $\mathscr{E}_Q$ is **purely inductive**, provided $m_Q > 0$. We can interpret, therefore, $\mathscr{E}_Q$ as a frequency-dependent inductor (BlueGreen) in Fig. 4.50.
- **Case 4:** $\beta - \alpha = (-1)^{n+1}(2n + 1)$, $n = 0, 1, 2, \cdots$

  In this case, $Z(j\omega) = -j\left(\frac{1}{\omega C(\omega)}\right)$ is an imaginary number where

$$C(\omega) \triangleq \begin{cases} \frac{\omega^{2n}}{m_Q}, & \text{when } n \text{ even} \\ \frac{\omega^{-2(n+1)}}{m_Q}, & \text{when } n \text{ odd} \end{cases} \tag{4.201}$$

and hence $\mathscr{E}_Q$ is **purely capacitive**, provided $m_Q > 0$. We can interpret, therefore, $\mathscr{E}_Q$ as a frequency-dependent capacitor (OliveGreen) in Fig. 4.50.

Two applications of the interpretation above: a memristor $\mathscr{M}$ characterized as a $(-1, -1)$ element is classified as a frequency-dependent resistor (red) because the area of the pinched-hysteresis $v - i$ loop is a function of frequency [7]. A second application is in interpreting $(0, -2)$ element as a frequency-dependent negative resistor (orange) or FDNR, see [15].

We will however use the time domain to synthesize the particular higher-order element $(0, -2)$, motivated by the fact that we need $i = \ddot{v}$ for the Duffing oscillator implementation in Sect. 5.5, **not** $i = -\ddot{v}$ as given by an FDNR. Consider the schematic in Fig. 4.51. The concept behind Fig. 4.51 is rooted in

Fig. 4.51 A mutator for synthesizing $(0, -2)$ from a $(0, -1)$ element (capacitor C_2 at port 2)

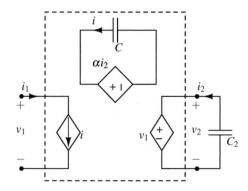

Sect. 2.5.4, specifically Fig. 2.42. Instead of connecting a nonlinear resistor, we have
used a linear capacitor C_2. Based on Sect. 2.5.4, we want the following two-port
relationship:

$$v_1 = v_2$$

$$i_1 = k\frac{d^2}{dt^2}v_2 \tag{4.202}$$

By the VCVS at port 2, we trivially obtain $v_2 = v_1$. From the VCCS across C, we
have:

$$i = -\alpha C\frac{di_2}{dt} \tag{4.203}$$

Dimensionally, $[\alpha] = \Omega$. By the CCCS at port 1, we trivially obtain: $i_1 = i$. Using
the expression for i above and the fact the $v_1 = v_2$, we get the desired relationship
at port 1:

$$i_1 = -\alpha C\frac{di_2}{dt}$$

$$= \alpha C C_2\frac{d^2}{dt^2}v_2$$

$$= \alpha C C_2\frac{d^2}{dt^2}v_1 \tag{4.204}$$

We will synthesize the mutator in Fig. 4.51 by using two opamps and one
CFOA in Sect. 5.5. For the general concept for synthesizing higher-order nonlinear
elements, the interested reader can refer to [10].

4.6.3 Limit Cycles

In Sect. 4.2.1.6, we have already seen how a simple first-order opamp circuit could
burst into a relaxation oscillation. Our analysis of this phenomenon depends on a key
assumption, namely, the jump rule. Our objective in this final section is to justify this
rule.

Every electronic oscillator requires at least two energy-storage elements and
at least one nonlinear element [12]. We will therefore begin with the simplest
nonlinear oscillator circuit, analyze its qualitative behavior, and then examine how

Fig. 4.52 Basic oscillator circuit

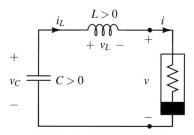

the oscillation waveform varies as we tune a parameter, say the inductance.[23] We will then show that as the inductance decreases, the oscillation changes from a nearly "sinusoidal" waveform into a nearly "discontinuous" waveform. In the limit, when the inductance tends to zero, the waveform becomes discontinuous and we obtain the jump rule. Fig. 4.52 shows the basic structure of an important class of electronic oscillators. Since both the inductor and capacitor are linear and passive (i.e., $L > 0$, $C > 0$), we claim that the resistive one-port $\mathcal{N}_R$ must be active (i.e., the DP characteristic contains at least some points in the second and/or fourth quadrant of the $v - i$ plane) in order for oscillation to be possible.

To see why $\mathcal{N}_R$ must be active, suppose it is strictly passive so that $v(t)i(t) > 0$ for all t; then the energy will continually enter $\mathcal{N}_R$, only to be dissipated in the form of heat.[24] This dissipated energy must of course come from the initial energy stored in the capacitor and inductor. Hence, as $t \to \infty$, the total energy stored in the capacitor and inductor will decrease continuously till it becomes completely dissipated. Since the instantaneous energy stored in the capacitor and inductor is $\mathcal{E}_C(t) = \frac{1}{2}Cv_C^2(t)$, $\mathcal{E}_L(t) = \frac{1}{2}Li_L^2(t)$ (recall Sect. 4.5), it follows that:

$$\text{Total energy} = \frac{1}{2}Cv_C^2(t) + \frac{1}{2}Li_L^2(t) \to 0 \quad \text{as } t \to \infty \tag{4.205}$$

Hence both $v_C(t)$ and $i_L(t)$ must eventually tend to zero and no sustained oscillation is possible.

A typical active resistive one-port has already been described by the three-segment PWL negative resistance characteristic in Fig. 4.26b. In general, any continuous nonmonotonic current-controlled $v - i$ characteristic described by $v = \hat{v}(i)$ satisfying the conditions:

$$\hat{v}(0) = 0$$

$$\hat{v}'(0) < 0$$

---

[23]This change in the steady-state dynamic behavior of a circuit as one (or more) parameters are varied is called a **bifurcation**. The parameter that is being varied is called the **bifurcation parameter**. A detailed study of bifurcations is beyond the scope of this book.

[24]Although the circuit could theoretically oscillate when $\mathcal{N}_R$ is a short circuit (passive but not strictly passive), no oscillation is possible in practice because the connecting wire always has some small but nonzero resistance.

$$\hat{v}(i) \rightarrow \infty \ \text{as } i \rightarrow \infty$$

$$\hat{v}(i) \rightarrow -\infty \ \text{as } i \rightarrow -\infty \qquad (4.206)$$

would cause the circuit in Fig. 4.52 to oscillate. This statement can be proved rigorously, see [12].

Indeed, the conditions in Eq. (4.206) are satisfied by many electronic circuits. For example, the DP characteristic in Fig. 4.54 of the twin-tunnel-diode circuit in Fig. 4.53 clearly satisfies the conditions in Eq. (4.206).

We will now consider the physical mechanisms of oscillation in the simple series $\mathcal{N}_R LC$ circuit from Fig. 4.52. We can write the normal form equations for the circuit by inspection (Fig. 4.54):

$$\dot{v}_C = \frac{-i_L}{C} \ \overset{\triangle}{=} \ f_1(v_C, i_L)$$

$$\dot{i}_L = \frac{v_C - \hat{v}(i_L)}{L} \ \overset{\triangle}{=} \ f_2(v_C, i_L) \qquad (4.207)$$

Fig. 4.53 A negative resistance twin-tunnel-diode circuit

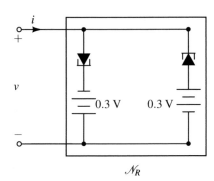

Fig. 4.54 Typical DP characteristic of the circuit in Fig. 4.53

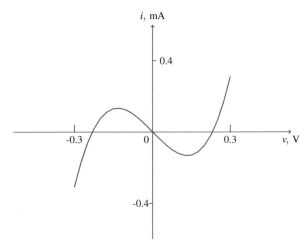

Assuming only that $\hat{v}(i_L)$ satisfies the conditions in Eq. (4.206) it is possible to derive general qualitative behaviors for this circuit. Indeed, equating $f_1(\cdot)$ and $f_2(\cdot)$ to zero in Eq. (4.207), we get the unique equilibrium point located at the origin: $v_{CQ} = 0, i_{CQ} = 0$.

Now, in order to determine if $(0, 0)$ is a stable or unstable equilibrium point, we can perform a small-signal analysis of the circuit about the DC operating point Q (in this case, $(0, 0)$). But, we will now take the opportunity to introduce the concept of the **Jacobian matrix**: if we linearize the RHS of Eq. (4.207) (or any nth-order normal form equations) and ignore the quadratic and other higher order terms, we would get the following result (given for a 2nd-order system such as Eq. (4.207)):

$$\begin{bmatrix} \dot{\bar{x}}_1 \\ \dot{\bar{x}}_2 \end{bmatrix} = \begin{bmatrix} a_{11} & a_{12} \\ a_{21} & a_{22} \end{bmatrix} \begin{bmatrix} \bar{x}_1 \\ \bar{x}_2 \end{bmatrix} \tag{4.208}$$

where $\bar{x}_1 \overset{\triangle}{=} x_1 - x_{1Q}$ and $\bar{x}_2 \overset{\triangle}{=} x_2 - x_{2Q}$ represent the small-signal deviation from the operating point. From Taylor series, we know:

$$\begin{bmatrix} a_{11} & a_{12} \\ a_{21} & a_{22} \end{bmatrix} = \begin{bmatrix} \frac{\partial f_1}{\partial x_1} & \frac{\partial f_1}{\partial x_2} \\ \frac{\partial f_2}{\partial x_1} & \frac{\partial f_2}{\partial x_2} \end{bmatrix}_{\mathbf{x}=\mathbf{x}_Q} \tag{4.209}$$

The matrix on the RHS of Eq. (4.209) is the Jacobian matrix $\mathbf{J}$. For any nth-order system in normal form, we can generalize $\mathbf{J}$ to:

$$\mathbf{J} = \begin{bmatrix} \frac{\partial f_1}{\partial x_1} & \frac{\partial f_1}{\partial x_2} & \cdots & \frac{\partial f_1}{\partial x_n} \\ \frac{\partial f_2}{\partial x_1} & \frac{\partial f_2}{\partial x_2} & \cdots & \frac{\partial f_2}{\partial x_n} \\ \cdot & \cdot & \cdot & \cdot \\ \cdot & \cdot & \cdot & \cdot \\ \cdot & \cdot & \cdot & \cdot \\ \frac{\partial f_n}{\partial x_1} & \frac{\partial f_n}{\partial x_2} & \cdots & \frac{\partial f_n}{\partial x_n} \end{bmatrix}_{\mathbf{x}=\mathbf{x}_Q} \tag{4.210}$$

For the second order Jacobian in Eq. (4.209), we gather from the Hartman-Grobman theorem [12], that the qualitative behavior (stable, unstable) of the associated nonlinear system will be "similar" to the linearized system about an equilibrium point.

For the oscillator described by Eq. (4.207), the Jacobian matrix evaluates to:

$$\mathbf{J} = \begin{bmatrix} 0 & \frac{-1}{C} \\ \frac{1}{L} & \frac{-\hat{v}'(0)}{L} \end{bmatrix} \tag{4.211}$$

We know from our basic calculus courses that the general solution of the linear ODE in Eq. (4.208) is given by:

$$\bar{\mathbf{x}}(t) = (k_1 e^{\lambda_1 t})\boldsymbol{\eta}_1 + (k_2 e^{\lambda_2 t})\boldsymbol{\eta}_2 \tag{4.212}$$

where λ_1, λ_2 are the eigenvalues of $\mathbf{J}$ and η_1, η_2 are the associated eigenvectors. Also from our basic calculus courses, we know that if $\text{Re}(\lambda_1) < 0$, $\text{Re}(\lambda_2) < 0$ the system associated with Eq. (4.208) is stable, etc. Since instead of using the eigenvalues, we can utilize the trace and determinant of $\mathbf{J}$:

$$T = a_{11} + a_{22} = \frac{-\hat{v}'(0)}{L}$$

$$\Delta = a_{11}a_{22} - a_{12}a_{21} = \frac{1}{LC} \tag{4.213}$$

Since $\Delta > 0$ and by the second condition in Eq. (4.206), $T > 0$, we have the following relation for the equilibrium point (origin) of the oscillator to be an **unstable**:

$$\frac{1}{LC} > \frac{1}{4}\left[\frac{-\hat{v}'(0)}{L}\right]^2 \tag{4.214}$$

or equivalently:

$$|\hat{v}'(0)| < 2\sqrt{\frac{L}{C}} \tag{4.215}$$

So all trajectories starting near the origin would diverge from it and head toward infinity. But, just like the relaxation oscillator we studied earlier, $\mathcal{N}_R$ is eventually passive (i.e. the $v - i$ characteristic must lie in the first and third quadrants beyond a certain finite distance from the origin). Thus, in view of conditions 3 and 4 in Eq. (4.206), $\mathcal{N}_R$ will start absorbing energy from the external world—the capacitor and inductor in this case.

Consequently, the energy initially supplied by the "active" $\mathcal{N}_R$ (when the (v_C, i_L) is near the "unstable" origin) to propel the trajectory toward infinity eventually fizzles as $\mathcal{N}_R$ becomes passive and begins to absorb energy instead. Therefore, the initial outward motion of the trajectory will be damped out by losses due to power dissipated inside $\mathcal{N}_R$ when the trajectory is sufficiently far out. Soon, the trajectory must "grind to a halt" and start "falling" back toward the origin.

The above scenario is depicted in Fig. 4.55, where we have included a cubic $v(i) = \frac{i^3}{3} - i$. The parametric plot of $(i(t), v(t))$ is called a **phase portrait**, so named because the $x - y$ plane is historically called the phase plane.

Observe that since the circuit has only one equilibrium state, and since it is unstable, there is no point where any trajectory could come to rest. Therefore all trajectories must continue to move at all times. Since they cannot stray too far

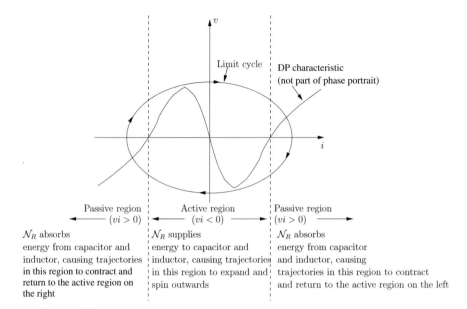

Passive region
$(vi > 0)$
$\mathcal{N}_R$ absorbs energy from capacitor and inductor, causing trajectories in this region to contract and return to the active region on the right

Active region
$(vi < 0)$
$\mathcal{N}_R$ supplies energy to capacitor and inductor, causing trajectories in this region to expand and spin outwards

Passive region
$(vi > 0)$
$\mathcal{N}_R$ absorbs energy from capacitor and inductor, causing trajectories in this region to contract and return to the active region on the left

Fig. 4.55 Physical mechanism for oscillation

beyond the active region and since no trajectory of any autonomous state equation can intersect itself,[25] except at equilibrium points, each trajectory must eventually tend toward some limiting orbit,[26] henceforth called a **limit cycle**. Note that a limit cycle is a periodic trajectory that is unique to a nonlinear system. By definition, linear oscillations are **not** limit cycles, because linear oscillations are a continuum of orbits. A limit cycle Γ must contain no other closed trajectories in a small band around Γ.

Specifically, let us now discuss the phase portrait of the typical Van der Pol oscillator, which helps us derive the generic jump rule. Suppose we choose $\hat{v}(i) = \frac{i^3}{3} - i$. Then Eq. (4.207) reads:

$$\dot{v}_C = \frac{-i_L}{C}$$

$$\dot{i}_L = \frac{v_C - \left(\frac{1}{3}i_L^3 - i_L\right)}{L} \tag{4.216}$$

---

[25]If a trajectory were to intersect itself at $(\hat{x}_1, \hat{x}_2)$, then its slope $\frac{dx_2}{dx_1}$ would have two different values at $(\hat{x}_1, \hat{x}_2)$. This is impossible since our system of equations is deterministic, not stochastic.

[26]Our reasoning does not prove that all trajectories must tend towards a unique limit cycle, although this is actually the case for the particular $v - i$ characteristic. The particular question of the number of limit cycles for a second order autonomous ODE is unsolved and is famously referred to as "Hilbert's sixteenth problem."

Fig. 4.56 Simulated
(ProcessBlue) and physical
(Red) limit cycles from an
implementation (to be
discussed in Sect. 5.1) of the
Van der Pol oscillator

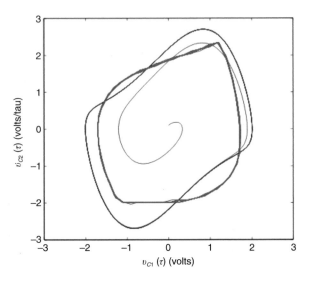

For fixed values of L and C, we could use a computer to generate the phase portrait
of Eq. (4.216). One such phase portrait is shown in Fig. 4.56 But how does the phase
portrait change (or bifurcate) as we vary parameters L and C? In more complicated
state equations, this question can only be answered in general by a brute-force
computer simulation method. But, we can often reduce the number of parameters
without loss of generality by writing the equations in terms of **dimensionless
variables**. For the Van der Pol oscillator, let us introduce the following "scaled"
time variables:

$$\tau \triangleq \frac{1}{\sqrt{LC}}t \tag{4.217}$$

Note that since $\sqrt{LC}$ has the dimensions of time, τ is dimensionless and will
henceforth be called "dimensionless time." Note that this τ is unrelated to the time
constant that we had defined earlier.

Observe that:

$$\dot{v}_C = \frac{dv_C}{d\tau}\frac{d\tau}{dt} = \frac{1}{\sqrt{LC}}\frac{dv_C}{d\tau}$$

$$\dot{i}_L = \frac{di_L}{d\tau}\frac{d\tau}{dt} = \frac{1}{\sqrt{LC}}\frac{di_L}{d\tau} \tag{4.218}$$

Substituting Eq. (4.216) into Eq. (4.218), we obtain the following equivalent state
equation in terms of dimensionless time variable τ:

$$\frac{dv_C}{d\tau} = -\frac{1}{\epsilon}i_L$$

$$\frac{di_L}{d\tau} = \epsilon \left[v_C - \left(\frac{1}{3} i_L^3 - i_L \right) \right] \tag{4.219}$$

where

$$\epsilon \overset{\Delta}{=} \sqrt{\frac{C}{L}} \tag{4.220}$$

Observe that Eq. (4.219) now contains only one parameter, ϵ, as defined by Eq. (4.220). In fact, Fig. 4.56 uses the dimensionless time form of the Van der Pol equation. The dimensionless form not only reduces the number of parameters, but also has the added advantage for computer simulation of scaling. In the case of time for instance, by going from say μs to s, we can use a more realistic time step for the numerical algorithm to avoid convergence issues. We will further explore dimensionless normal form in Chap. 5.

Suppose $\epsilon \to \infty$, Eq. (4.220) implies $L \to 0$. But, from the physical Van der Pol Eq. (4.216), we see that $L \to 0$ implies $\frac{di_L}{dt} \to \infty$. In other words we will have a **vertical jump** in the $v - i$ plane, assuming i is the vertical axis, just as we discussed in Sect. 4.2.1.6.

Let us now consider the general Eq. (4.207) of a series oscillator:

$$\overset{\bullet}{v}_C = -\frac{i_L}{C} \tag{4.221}$$

$$\overset{\bullet}{i}_L = \frac{1}{L}[v_C - \hat{v}(i_L)] \tag{4.222}$$

The function $\hat{v}(\cdot)$ representing the nonlinear resistor characteristic can be quite arbitrary except that it satisfies the conditions in Eq. (4.206). This class, as discussed earlier, includes the negative resistance opamp relaxation oscillator.

Dividing Eq. (4.222) by Eq. (4.221), we obtain the slope:

$$m(P) \overset{\Delta}{=} \frac{di_L}{dv_C} = -\frac{C}{L} \left[\frac{v_C - \hat{v}(i_L)}{i_L} \right] \tag{4.223}$$

of the tangent vector at any point $P \overset{\Delta}{=} (v_C, i_L)$ on a trajectory in the $v_C - i_L$ plane. Thus, we have:

1. As $L \to 0$ in Eq. (4.223), the limiting slope $|m(P)| \to \infty$, as long as $v(C) \neq \hat{v}(i_L)$. Thus all trajectories, **except on the DP characteristic**, will tend to vertical line segments as $L \to 0$. In particular, at the impasse points, we will have a jump discontinuity.

2. Note that from Eq. (4.222), $\overset{\bullet}{i}_L > 0$, if $v_C > \hat{v}(i_L)$ and vice versa. In other words, this gives us the condition for the dynamic route derived in Sect. 4.2.1.6.

3. To complete our analysis of the jump phenomenon, we must estimate the amount of time it takes a trajectory line segment to go from one branch of the DP plot

to another. This is easily found from the velocity along the vertical direction as specified by Eq. (4.222), namely, $\lim_{L \to 0} \left| \dfrac{di_L}{dt} \right| \to \infty$, provided again we are not on the DP characteristic. In particular, the trajectory through each impasse point must execute a vertical instantaneous jump as $L \to 0$.

We have thus formally justified the introduction of the jump rule in Sect. 4.2.1.6.

4.7 Conclusion

As a concluding note to this chapter, let us recall the overall idea behind this chapter was to analyze dynamic nonlinear networks. We essentially had three approaches: time domain, frequency domain, and energy. We restricted our discussion of frequency domain techniques to linear time-invariant circuits but learned about the powerful concepts of phasors and Laplace transforms. The mindful reader would have noticed that many of the ideas involved studying an associated linear system about a particular operating point. Although much insight can be gained for first and second order systems via the linearization technique, third and higher order systems exhibit extremely complicated nonperiodic phenomena, generally known as chaos. Thus, chaos is a phenomena that cannot be fully studied by linearization and is hence a property unique to nonlinear circuits. Therefore, Chap. 5 appropriately concludes the book by incorporating a plethora of ideas encountered throughout the book.

Because of the large body of material in this chapter, we have summarized concepts below, instead of specific formulae:

1. The order of complexity of a dynamic network is the minimum number of initial conditions that must be specified in terms of circuit variables, in order to determine the full behavior of the network.
2. When possible, the dynamic nonlinear network equations should be expressed in normal form.
3. Dual circuits help us reduce the enormous solution space of dynamic nonlinear networks.
4. We learned the following from time domain analysis of nth-order nonlinear networks:

   a. Current through a linear inductor and voltage across a linear capacitor cannot change instantaneously across discontinuities.
   b. Discontinuities in other circuit variables in the network occur because of the constraint in (a) above.
   c. Circuits exhibiting impasse points indicate that we need to augment the circuit model, most likely with parasitics.
   d. MNA, Tableau and Small Signal analysis can be easily extended to include dynamic networks.

5. An alternative to time domain analysis is frequency domain analysis. The advantage of this approach when applied to linear time-invariant circuits is that time domain differential equations are mapped to algebraic equations involving complex variables in the frequency domain. The main ideas discussed were:

   a. We use complex numbers to define a phasor, in order to obtain the steady-state response when the network is excited by a sinusoid of a particular frequency ω.

   b. Differential equations in the time domain can be converted to algebraic equations in the phasor domain, and hence techniques covered in Chap. 3 such as nodal analysis, tableau analysis, superposition, and Thévenin-Norton theorems are applicable to circuits in the phasor domain.

   c. For general excitation, we use the Laplace transform.

   d. To calculate the time response, we need to use partial fraction expansion and then use a table of inverse Laplace transforms.

   e. Laplace transforms can be used to find both the transient and steady-state responses.

6. For memristor networks:

   a. We discussed the Flux-Charge Analysis Method (FCAM). The advantage of this method is a reduction in the number of ODEs for the associated memristive network.

   b. Memristors display a distinct pinched-hysteresis $v - i$ characteristic under sinusoidal excitation.

   c. Due to physical parasitics, a memristor's $v - i$ characteristic may become unpinched at the origin.

   d. We described small-signal AC characteristics of memristive devices.

7. A third approach to studying (dynamic nonlinear) networks is energy. We discussed formulation of system equations from both the Lagrangian and Hamiltonian. The main ideas discussed are:

   a. Inductors store the mechanical equivalent of "kinetic energy" via the current flowing through them (or the flux-linkage across them). Capacitors store the mechanical equivalent of "potential energy" via the voltage across them (or the charge stored in them).

   b. Lagrangian formalism is in terms of the difference between kinetic and potential energies. Hamiltonian formalism is in terms of the sum of kinetic and potential energies.

8. Reciprocity helps us understand the symmetric role played by the input and output of a physical system.

9. Higher-order circuit elements in general can be synthesized using higher order mutators. We showed how to synthesize a particular type of higher order mutator for $i = \ddot{v}$.

10. Limit cycles are an exclusive steady-state behavior of nonlinear oscillators, that usually arise due to unstable equilibrium points.

Lab 4: Relaxation Oscillator (Transient Simulation) and High-Pass filter (AC Simulation)

Just like lab 3, we encourage the reader to perform simulation in QUCS first, so they can verify their answers to the appropriate problems via simulation.

Objective To understand time domain (transient) simulation and frequency response (AC simulation)in QUCS

Theory There are two steps to this lab: in the first step, you construct a relaxation oscillator. In the second step, you go through the QUCS online workbook to simulate a high-pass filter. For the relaxation oscillator, you will be performing a **transient analysis** or a time domain simulation. Please do not confuse transient analysis, as defined by circuit simulators, with the concept of transient response discussed in the text!

   To perform sinusoidal steady-state analysis, the terminology used by circuit simulators is **AC simulation**. We will use a simple RC circuit to illustrate the idea of **filtering** signals. A discussion of filtering is beyond the scope of this book, but the reader is encouraged to go through the appropriate material in an excellent reference such as [12]. Moreover, as the reader simulates the high-pass filter, they are encouraged to modify the circuit to understand its functionality.

Lab Exercise

1. For this step, construct the circuit shown in Fig. 4.57.
2. Once you enter the appropriate parameters, simulating the circuit should result in Fig. 4.58. Compare your result with the discussion of relaxation oscillators in this chapter (see also Exercise 4.7).
3. For this step, simulate the circuit under "AC simulation - A simple RC highpass" in the QUCS online workbook. Make sure you understand the results. If necessary, construct the circuit physically.

Exercises

4.1 Consider the memristor circuits in Figs. 4.5 and 4.6 from Example 4.1.3. What is the order of complexity for the two circuits if the memristive devices are replaced with ideal memristors?

4.2 For the circuit shown in Fig. 4.59, calculate $v_0(t)$ for $t \geq 0$, given $i_L(0) = 2$ A.

4.3 Consider the circuit shown in Fig. 4.60a where the inductor is nonlinear and is given by the $i - \phi$ characteristic shown.

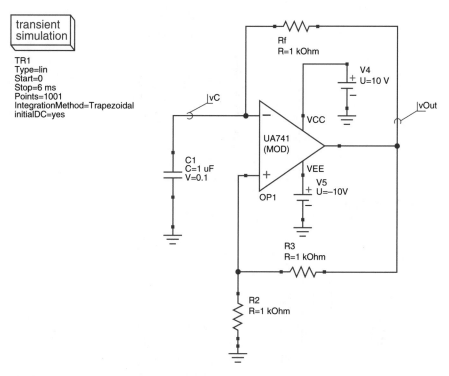

Fig. 4.57 An opamp based relaxation oscillator

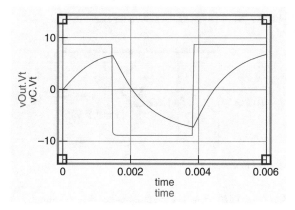

Fig. 4.58 Steady-state $v_C(t)$ and $v_{\text{out}}(t)$ for the circuit in Fig. 4.57

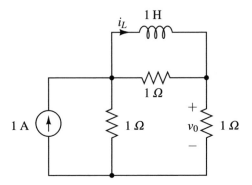

Fig. 4.59 Circuit for Exercise 4.2

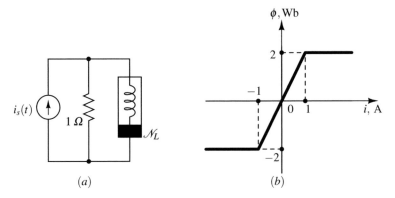

Fig. 4.60 (**a**) Circuit and (**b**) nonlinear characteristic for Exercise 4.3

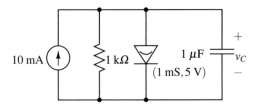

Fig. 4.61 Circuit for Exercise 4.4

1. Let $i_s(t) = 3u(t)$ and $i(0^-) = -1$ A. Determine the current $i(t)$ for $t \geq 0$.
2. What is the amount of energy delivered to the inductor for $t \geq 0$?

4.4 For the circuit shown in Fig. 4.61, calculate and sketch $v_C(t)$ for $t > 0$. Assume $v_C(0) = 0$ V.

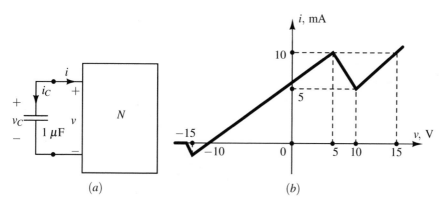

Fig. 4.62 (a) Circuit and (b) DP characteristic for Exercise 4.5

4.5 Consider the circuit shown in Fig. 4.62a, where N is described by the $v - i$ characteristic shown in Fig. 4.62b.

1. Indicate the dynamic route. Label all equilibrium points and state whether they are stable or unstable.
2. Suppose $v_C(0) = 15\,\text{V}$. Find and sketch $v_C(t)$ and $i_C(t)$ for $t \geq 0$. Indicate all pertinent information on the sketches.

4.6 Consider the circuit shown in Fig. 4.63a, where N is described by the $v - i$ characteristic shown in Fig. 4.63b.

1. Indicate the dynamic route. Label all equilibrium points and state whether they are stable or unstable.
2. Suppose $i_L(0) = 20\,\text{mA}$. Find and sketch $v(t)$ and $i(t)$ for $t \geq 0$. Indicate all pertinent information on the sketches.

4.7 Determine closed form expressions and sketch $v_C(t)$ and $v_o(t)$ waveforms for the relaxation oscillator in Fig. 4.26a.

4.8 Write the modified node equations for the circuit shown in Fig. 4.64.

4.9 The roots of a general cubic equation in X may be viewed (in the $X - Y$ plane) as the intersections of the X-axis with the graph of a cubic of the form:

$$Y = X^3 + AX^2 + BX + C \qquad (4.224)$$

1. Show that the point of inflection of the graph occurs at $X = -\frac{A}{3}$.
2. Deduce (algebraically and geometrically) that the substitution $X = \left(x - \frac{A}{3}\right)$ will reduce the above equation to the form $Y = x^3 + bx + c$.

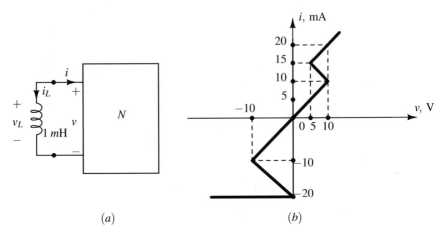

Fig. 4.63 (a) Circuit and (b) DP characteristic for Exercise 4.6

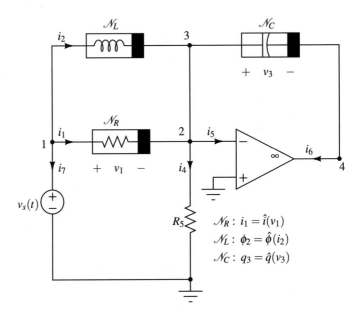

Fig. 4.64 Circuit for Exercise 4.8

4.10 Reconsider the cubic: $x^3 = 3px + 2q$. To derive the general formula for the cubic:

1. Make the inspired substitution $x = s + t$ and deduce that x solves the cubic if $st = p$, $s^3 + t^3 = 2q$.
2. Eliminate t between the two equations above, thereby obtaining a quadratic in s^3.

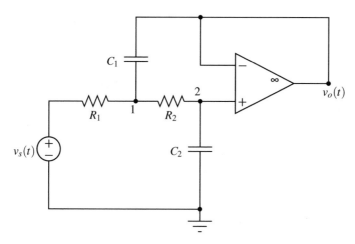

Fig. 4.65 Circuit for Exercise 4.12

3. Solve the quadratic to obtain two possible values of s^3. By symmetry, what are the possible values of t^3?
4. Given that we know $s^3 + t^3 = 2q$, deduce the formula for x in Eq. (4.91).

4.11 Algebraically (**and/or** geometrically) prove the following:

1. $|z| = \sqrt{x^2 + y^2}$
2. $z\bar{z} = |z|^2$
3. $\frac{1}{x+jy} = \frac{x}{x^2+y^2} - j\frac{y}{x^2+y^2}$
4. $(1 + j)^4 = -4$
5. $(1 + j)^{13} = -2^6(1 + j)$
6. $(1 + j\sqrt{3})^6 = 2^6$

4.12 Write nodal equations in the phasor domain for the circuit shown in Fig. 4.65. Use the nodal equations to find the ratio $V_o(j\omega)/V_s$.

4.13 Reconsider the system S from Exercise 1.9. If the input to S is a sinusoidal signal of frequency ω, is the frequency of the output signal still ω?

4.14 Prove the differentiation property of Laplace transforms for nth-order derivatives:

$$\mathcal{L}\{\frac{d^n}{dt^n} f(t)\} = s^n F(s) - s^{n-1} f(0^-) - s^{n-2} f'(0^-) \cdots - f^{n-1}(0^-) \qquad (4.225)$$

4.15 Show that if the initial conditions were not zero in Example 4.3.6, then we would have obtained:

$$I_L(s) = \frac{\omega_0^2}{s^2 + 2\alpha s + \omega_0} I_s(s) + \frac{(s + 2\alpha)i_L(0^-) + \dot{i}_L(0^-)}{s^2 + 2\alpha s + \omega_0^2} \qquad (4.226)$$

Fig. 4.66 Circuit for
Exercise 4.16

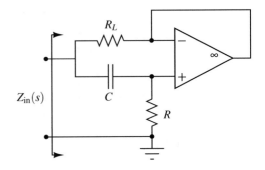

$Z_{in}(s)$

Fig. 4.67 Circuit for
Exercise 4.18

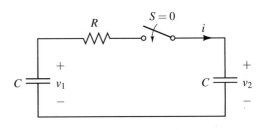

4.16 Determine $Z_{in}(s)$ for the circuit in Fig. 4.66. Show that the circuit functions as a physical implementation of a gyrator.

4.17 Derive the small-signal model for the thermistor from Sect. 4.4.2.

4.18 The circuit shown in Fig. 4.67 is made of linear time-invariant elements. Prior to time 0, the left capacitor is charged to V_0 volts, and the right capacitor is uncharged. The switch is closed at time 0. Calculate the following:

1. The current i for $t \geq 0$.
2. The energy dissipated in the interval $(0, T)$.
3. The limiting values for $t \rightarrow \infty$ of (a) the capacitor voltages v_1 and v_2, (b) the current i and (c) the energy stored in the capacitor and the energy dissipated in the resistor.
4. Is there any relation between the energies? If so, state what it is.
5. What happens when $R \rightarrow 0$?

4.19 We have encountered many resistive circuits having multiple equilibrium points. For example, the tunnel-diode circuit in Fig. 3.4a from Chap. 3 has three operating points. This answer seems to contradict the fact that a single laboratory measurement on the corresponding physical circuit can only give one operating point.

We are now in a position to resolve the so-called **operating point paradox**. Basically, the tunnel-diode circuit in Fig. 3.4a is **not** a realistic model of the **physical** circuit. In any physical circuit, as we have discussed numerous times, there always exist parasitic effects. In circuits having unique solution, these effects can often

Fig. 4.68 Realistic model for a biased tunnel-diode circuit

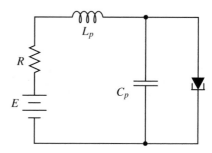

be neglected without discernible errors. In circuits exhibiting multiple solutions, however, some of these parasitic elements cannot be neglected.

Consider the realistic tunnel-diode circuit shown in Fig. 4.68. The three operating points in the **resistive** circuit can now be interpreted as **equilibrium** points in the remodeled **dynamic** circuit. Show that in Fig. 3.4b:

1. Q_2 is an **unstable** equilibrium point.
2. Use numerical simulation and phase portraits to show different initial conditions will give rise to either Q_1 or Q_3 as the operating point.

4.20 NOTE: This is an open-ended problem

Going through this chapter, the reader should have realized that there are three main approaches to studying circuits: time domain, frequency domain, and the energy approach. With respect to circuit simulation programs such as QUCS, they readily implement the time domain and frequency domain approaches.

So, a natural question is: what about energy based approaches? That is, can we supplement QUCS to compute Lagrangian and Hamiltonian for a specified circuit? And, how would one go about interpreting the results?

We would recommend investigating the questions above as a capstone project.

References

1. Ambelang, S., Muthuswamy, B.: From Van der Pol to Chua: an introduction to nonlinear dynamics and chaos for second year undergraduates. In: Proceedings of the 2012 IEEE International Symposium on Circuits and Systems, pp. 2937–2940 (2012)
2. Calaprice, A. (ed.): The Ultimate Quotable Einstein. Princeton University Press/The Hebrew University of Jerusalem, Princeton/Jerusalem (2011)
3. Chua, L.O.: Introduction to Nonlinear Network Theory. McGraw-Hill, New York (1969)
4. Chua, L.O.: Memristor - the missing circuit element. IEEE Trans. Circuit Theory **18**(5), 507–519 (1971)
5. Chua, L.O.: Device modeling via basic nonlinear circuit elements. IEEE Trans. Circuits Syst. **27**(11), 1014–1044 (1980)
6. Chua, L.O.: Dynamic nonlinear networks: state-of-the-art. IEEE Trans. Circuits Syst. **27**(11), 1059–1087 (1980)
7. Chua, L.O.: Nonlinear circuit foundations for nanodevices, part I: the four-element torus (invited paper). Proc. IEEE **91**(11), 1830–1859 (2003)

8. Chua, L.O.: Supplementary Lecture Notes on First-Order Circuits. University of California, Berkeley (Fall 2008), pp. EE100. Available, online: http://inst.eecs.berkeley.edu/~ee100/fa08/lectures/EE100supplementary_notes_13.pdf. Accessed 29 Dec 2017

9. Chua, L.O., Kang, S.M.: Memristive devices and systems. Proc. IEEE **64**(2), 209–223 (1976)

10. Chua, L.O., Szeto, E.W.: Synthesis of higher order nonlinear circuit elements. IEEE Trans. Circuits Syst. **31**(2), 231–235 (1984)

11. Chua, L.O., Tseng, C.: A Memristive circuit model for pn junction diodes. Int. J. Circuit Theory Appl. **4**(2), 367–389 (1976)

12. Chua, L.O., Desoer, C.A., Kuh, E.S.: Linear and Nonlinear Circuits. McGraw-Hill, New York (1987)

13. Corinto, F., Forti, M.: Memristor circuits: flux-charge analysis method. IEEE Trans. Circuits Syst. Regul. Pap. **63**(11), 1997–2009 (2016)

14. Desoer, C.A., Kuh, E.S.: Basic Circuit Theory. Tata McGraw Hill, New York (1969)

15. Elwakil, A.S., Kennedy, M.P.: Chaotic oscillator configuration using a frequency dependent negative resistor. J. Circuits Syst. Comput. **9**(3,4), 229–242 (1999)

16. Georgiou, P.S., et al.: On Memristor ideality and reciprocity. Microelectron. J. **45**(11), 1363–1371 (2014)

17. Hamill, P.: A Student's Guide to Lagrangians and Hamiltonians. Cambridge University Press, Cambridge (2013)

18. Havil, J.: Gamma : Exploring Euler's Constant. Princeton University Press, Princeton (2003)

19. Jeltsema, D., Scherpen, J.M.A.: Multidomain modeling of nonlinear networks and systems. IEEE Control. Syst. **29**(4), 28–59 (2009)

20. Kennedy, M.P., Chua, L.O.: Hysteresis in electronic circuits: a circuit theorist's perspective. Int. J. Circuit Theory Appl. **19**, 471–515 (1991)

21. Lin, D., Hui, S.Y.R., Chua, L.O.: Gas discharge lamps are volatile memristors. IEEE Trans. Circuits Syst. Regul. Pap. **61**(7), 2066–2073 (2014)

22. Marszalek, W.: On the action parameter and one-period loops of oscillatory memristive circuits. Nonlinear Dyn. **82**(1–2), 619–628 (2015)

23. Muthuswamy, B., Chua, L.O.: Simplest chaotic circuit. Int. J. Bifurcation Chaos **20**(5), 1567–1680 (2010)

24. Muthuswamy, B., et al.: Memristor modelling. In: Proceedings of the 2014 IEEE ISCAS, pp. 490–493 (2014)

25. Needham, T.: Visual Complex Analysis. Oxford University Press, Oxford (2000)

26. Penfield, P.P.: Varactor Applications, p. 513. M.I.T. Press, Cambridge (1962). Available, online: https://catalog.hathitrust.org/Record/001619307

27. Riaza, R., Tischendorf, C.: Semistate models of electrical circuits including memristors. Int. J. Circuit Theory Appl. **39**(6), 607–627 (2011)

28. Sah, M.P., et al.: A generic model of memristors with parasitic components. IEEE Trans. Circuits Syst. Regul. Pap. **62**(3), 891–898 (2015)

29. Weisstein, E.W.: Second Fundamental Theorem of Calculus (2018). Available, online: http://mathworld.wolfram.com/SecondFundamentalTheoremofCalculus.html. Accessed 1 Jan 2018

Chapter 5
Chaos

Simulated chaotic attractor from the Muthuswamy-Chua system [19, 22]

Abstract So far we have studied the fundamentals of (nonlinear) circuit theory. We have encountered a variety of multi-terminal elements and circuit analysis techniques. In this final chapter, we will discuss the fascinating mathematical concept of chaos. Notice we use the word mathematical: chaos has been largely studied by mathematicians and scientists. Yet we conclude this book on circuit theory with an advanced mathematical topic because chaos will prove invaluable in **integrating** a majority of the concepts discussed in this book. Chaos is also fundamentally restricted to nonlinear circuits, linear networks do not exhibit chaos. So it is appropriate to conclude this book with a chapter on an exclusive property of nonlinear circuits. For those mathematically familiar with chaos, this chapter takes an "experimentalist" approach when discussing a variety of chaotic circuits.

5.1 An Introduction to Chaos

Chaotic circuits provide excellent examples for utilizing nonlinear elements in **topologically simple circuits**,[1] to study an interesting phenomenon.

Since a thorough treatment of chaos requires a book on its own, in this chapter we will instead mainly focus on a fundamental idea discussed throughout the book:

---

[1] That is, most of the normal form equations can be derived by inspection.

© Springer International Publishing AG, part of Springer Nature 2019
B. Muthuswamy, S. Banerjee, *Introduction to Nonlinear Circuits and Networks*,
https://doi.org/10.1007/978-3-319-67325-7_5

Fig. 5.1 Circuit model of the
Van der Pol oscillator

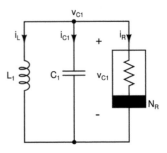

the concept of device modeling. Hence a recurring theme throughout this chapter
would be to first discuss a circuit that has been **systematically designed to exhibit
chaotic behavior**, and then discuss a circuit that **exhibits chaos because of physical
nonlinearities**. We will also focus on the PWL approximation technique, because
as we learned throughout the book, we can synthesize any PWL characteristic using
opamps, etc. We will design chaotic circuits using nonlinear resistors, capacitors,
and inductors. We will also discuss memristor and transistor based chaotic circuits.

We will be performing simulations of the (chaotic) circuits in QUCS and
normal form equations in SageMath. We will show implementation results only
for the Muthuswamy-Chua chaotic circuit,[2] in order to encourage the reader to
investigate the physical implementation of other chaotic circuits, and hence "learn
by experimenting." Also, many end-of-chapter exercises are essentially capstone
design problems. In other words the ideas discussed in this chapter should lead to
interesting research problems for the motivated reader, perhaps even resulting in
"good" publications.

We will start by revisiting the Van der Pol oscillator from Sect. 4.6.3. Reconsider
the schematic of the Van der Pol oscillator in Fig. 5.1. Note that we are using
the dual [3] of the series $LC\mathcal{N}_R$ circuit from Sect. 4.6.3. We now have a voltage-
controlled $\mathcal{N}_R$:

$$i_R(v_{C1}) = a v_{C1} + b v_{C1}^3 \tag{5.1}$$

One could implement the cubic nonlinearity in Eq. (5.1) using the twin-tunnel-diode
circuit in Fig. 4.53 or by using analog multipliers as shown in Fig. 5.2 From Fig. 5.2
(see Exercise 5.1), we get:

$$i_R(v_{C1}) = \left[-v_{C1} + \left(1 + \frac{R_5}{R_4} \right) \frac{(v_{C1})^3}{100} \right] \frac{1}{R_3} \tag{5.2}$$

However, as stated earlier, we will use a PWL approximation for the cubic
nonlinearity. Consider the QUCS schematic in Fig. 5.3. U1 and U2 are QUCS

---

[2]We picked this circuit to implement because it requires a memristor emulator that has the most
number of components of all the chaotic circuits discussed in this chapter.

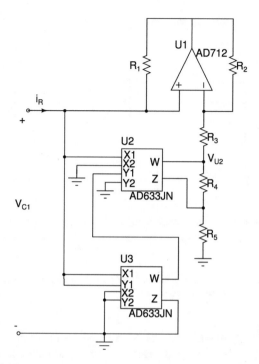

Fig. 5.2 A circuit implementation of $\mathcal{N}_R$ from Fig. 5.1. All power supplies are ± 15 V. Opamp U1 acts as a current inverter when $R_1 = R_2$, U2 and U3 are analog multipliers

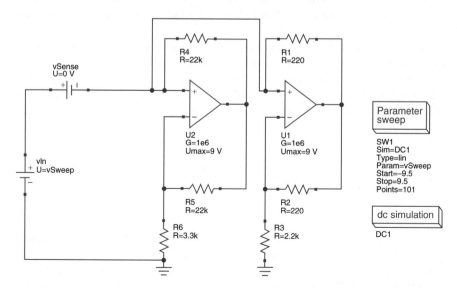

Fig. 5.3 PWL approximation of the cubic nonlinearity. This circuit is called the "Chua diode" [6]

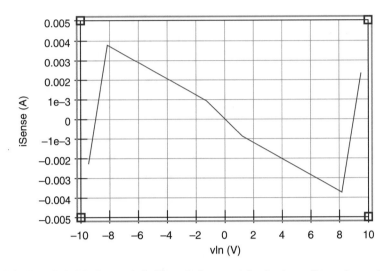

Fig. 5.4 Chua diode DP characteristic. iSense is the current flowing (according to the passive sign convention) through the vSense voltage source in Fig. 5.3

nonlinear model[3] of opamps. We have specified an open-loop gain of 1e6 and $E = \pm 9$ V. We will perform a parameter sweep simulation and plot the DP characteristic. The result is shown in Fig. 5.4. Notice that opamp circuit is a parallel combination of two nonlinear resistors. Each opamp is the voltage-controlled dual of the opamp negative impedance converter from Fig. 2.41. Because of the parallel combination of the resistors, we have two additional breakpoints[4] in the negative resistance region.

Next, we will simply add a resistor and capacitor (to make the system three dimensional,[5] in other words, the order of complexity is now three) as shown in Fig. 5.5, to obtain **Chua's circuit** [6]. The purpose of adding just two components will be clear from simulating Chua's circuit. The complete QUCS schematic for simulation is shown in Fig. 5.6. Simulating the circuit in Fig. 5.6, we get the results in Fig. 5.7.

---

[3]These can be found under **Nonlinear Components** in the components tab of QUCS. Note that simulating chaotic circuits with a physical opamp (like μA741 models) from the Opamp QUCS Library may not cause the simulation to converge for some circuits.

[4]The reason for breakpoints is to obtain chaos in a three-dimensional extension of the Van der Pol oscillator. The justification is beyond the scope of this book, for details refer to [5].

[5]The minimal dimension of a continuous time chaotic system is usually said to be three because of the Poincaré-Bendixson theorem. But there are unusual systems of lower order that violate this theorem and exhibit chaos, see [27]. Moreover, even one-dimensional discrete-time systems (maps) can exhibit chaotic behavior. Such maps will not be discussed in this book, although they appear in very simple electric circuits, see [26].

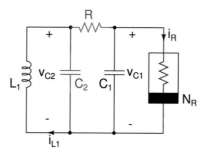

Fig. 5.5 Circuit model for Chua's circuit. The only elements added to the Van der Pol oscillator in Fig. 5.1 are highlighted in red

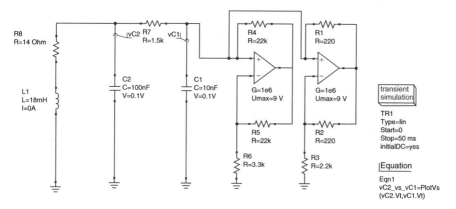

Fig. 5.6 Chua's circuit in QUCS setup for transient analysis

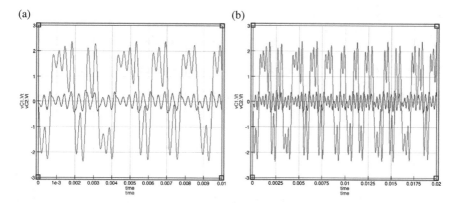

Fig. 5.7 Time domain plots of the voltages across the two capacitors. (**a**) $t = 0$ to $t = 10$ ms. (**b**) $t = 0$ to $t = 20$ ms

An important point is the issue of convergence in circuit simulation of nonlinear (chaotic) circuits. Although this topic is beyond the scope of this text (see [23]), one needs to pay attention to the log messages from the simulator to investigate the source of the issue, and if the convergence error can be safely ignored. If the simulation does not converge at all or seems to converge to incorrect results, the warnings should be closely studied.

In this case, the log messages show:

Listing 5.1 QUCS log from simulating Chua's circuit

```
1   Output:
2   -------
3
4   Starting new simulation on Thu 11. Jan 2018 at 11:33:19:169
5
6   creating netlist... done.
7   Starting /usr/local/bin/qucsator
8
9   project location:
10  modules to load: 0
11  factorycreate.size() is 0
12  factorycreate has registered:
13  parsing netlist...
14  checking netlist...
15  checker notice, variable `vC1.Vt' in equation `vC2_vs_vC1'
        not yet defined
16  checker notice, variable `vC2.Vt' in equation `vC2_vs_vC1'
        not yet defined
17  netlist content
18       2 C instances
19       1 L instances
20       8 R instances
21       2 OpAmp instances
22       1 TR instances
23  creating netlist...
24  checker notice, variable `vC1.Vt' in equation `vC2_vs_vC1'
        not yet defined
25  checker notice, variable `vC2.Vt' in equation `vC2_vs_vC1'
        not yet defined
26  NOTIFY: TR1: average time-step 2.16797e-06, 4326 rejections
27  NOTIFY: TR1: average NR-iterations 3.10133, 947 non-
        convergences
28
29  Simulation ended on Thu 11. Jan 2018 at 11:33:20:959
30  Ready.
```

We can see that there are non-convergence warnings, in this case, they can be safely ignored.

Comparing Fig. 5.7a, b, we may be tempted to conclude the transient response is periodic. But closely examining the two time domain waveforms, we can see that

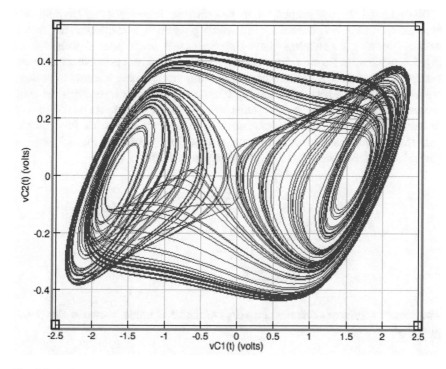

Fig. 5.8 A phase portrait (plot) of $(v_{C1}(t), v_{C2}(t))$ from Fig. 5.7, plotted from $t = 0$ to $t = 50$ ms

may not be true. For instance, they clearly show the number of "oscillating maxima and minima" in each ms interval is different.

A more insightful picture is the phase portrait (plot) discussed in Sect. 4.6.3. The $v_{C1} - v_{C2}$ phase plot is shown Fig. 5.8. We now see that there is a "structure," called a **chaotic attractor**, in phase space. It turns out the chaotic attractor **is the steady-state** response for this circuit.

The structure is so named because it tends to "attract" points in a "basin of attraction" into the attractor (see Exercise 5.2). The term "chaotic" or "chaos" was coined by James Yorke and T.Y. Li [18]. Nevertheless, there is no agreed upon definition of chaos, although researchers generally concur that a chaotic system should satisfy the following properties:

1. Boundedness[6]
2. Aperiodicity
3. Sensitive dependence on initial conditions

---

[6]We add this property because without it, linear one-dimensional unstable systems could be considered "chaotic" since they satisfy properties 2 and 3.

We will not discuss property 1 in this chapter. Property 2 we have already encountered in Fig. 5.7. Before we investigate property 3, a mindful reader should have noticed that we seemingly have trajectories that are "crossing" each other in Fig. 5.8, whereas in Sect. 4.6.3 we commented that this is not possible because our system is deterministic. The reason why the trajectories seem to cross is because we are looking at a projection on the 2D plane! To further investigate the third seminal property of chaos and the actual structure of the chaotic attractor, it would be helpful to invoke the idea of dimensionless scaling from Sect. 4.6.3. First, we can easily write down the normal form equations for Chua's circuit in Fig. 5.5 (assume parasitic series resistance R_8 of L is 0, see Exercise 5.3):

$$\frac{dv_{C1}}{dt} = \frac{1}{C_1}\left[\frac{v_{C2} - v_{C1}}{R} - g(v_{C1})\right]$$

$$\frac{dv_{C2}}{dt} = \frac{1}{C_2}\left[\frac{v_{C1} - v_{C2}}{R} + i_L\right]$$

$$\frac{di_L}{dt} = \frac{-v_{C2}}{L} \tag{5.3}$$

The general nonlinear characteristic $g(v_R)$ of the Chua diode, shown in Fig. 5.9, is given by:

$$g(v_R) = G_b v_R + \frac{1}{2}(G_a - G_b)(|v_R + B_p| - |v_R - B_p|) \tag{5.4}$$

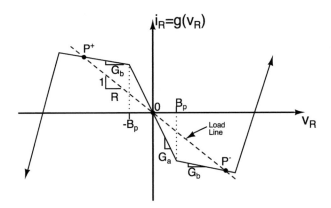

Fig. 5.9 The five segment PWL characteristic from Fig. 5.4, with $v_{C1} = v_R$. Note that $\mathcal{N}_R$ is not passive since we have segments in the second and fourth quadrant, which correspond to the active regions of $\mathcal{N}_R$. The linear resistor R from Fig. 5.5 has been plotted as a load line at steady state. For particular values of R, notice we get three equilibrium points, $P^+, 0, P^-$. This justifies our choice of connecting two voltage-controlled nonlinear resistors in parallel in Fig. 5.3. Note that $\mathcal{N}_R$ eventually becomes strictly passive. Those segments are due to opamp saturation and should not be included when deriving Eq. (5.4)

We will now scale the equations from Chua's circuit into dimensionless, as shown in Example 5.1.1.

Example 5.1.1 Derive the dimensionless form for Eq. (5.3).

Solution Following the procedure in Sect. 4.6.3, let $\tau \triangleq \frac{t}{RC_2}$. Replacing t in Eq. (5.3), we get:

$$\frac{dv_{C1}}{d\tau} = \frac{C_2}{C_1}[(v_{C2} - v_{C1}) - Rg(v_{C1})]$$

$$\frac{dv_{C2}}{d\tau} = [(v_{C1} - v_{C2}) + Ri_L]$$

$$\frac{di_L}{d\tau} = RC_2\left(\frac{-v_{C2}}{i_L}\right) \qquad (5.5)$$

We will take the dimensionless time form in Eq. (5.5) and make the state variables dimensionless as well. Hence we will finally get a dimensionless state equation. To do this, consider the first equation, replacing $g(v_{C1})$ from Eq. (5.4):

$$\frac{dv_{C1}}{d\tau} = \frac{C_2}{C_1}\left[(v_{C2} - v_{C1}) - \{RG_bv_{C1} + \frac{R}{2}(G_a - G_b)(|v_{C1} + B_p| - |v_{C1} - B_p|)\}\right]$$

$$(5.6)$$

Factoring out B_p ($B_p > 0$), we get:

$$\frac{dv_{C1}/B_p}{d\tau} = \frac{C_2}{C_1}\left[\left(\frac{v_{C2}}{B_p} - \frac{v_{C1}}{B_p}\right) - Rg\left(\frac{v_{C1}}{B_p}\right)\right] \qquad (5.7)$$

Let: $x \triangleq \frac{v_{C1}}{B_p}, y \triangleq \frac{v_{C2}}{B_p}, a \triangleq G_aR, b \triangleq G_bR, \alpha \triangleq \frac{C_2}{C_1}$. Equation (5.7) simplifies to:

$$\frac{dx}{d\tau} = \alpha(y - x - f(x)) \qquad (5.8)$$

where $f(x) = bx + \frac{1}{2}(a - b)(|x + 1| - |x - 1|)$.

(continued)

Example 5.1.1 (continued)

Notice all the parameters defined above are dimensionless. Multiplying and dividing $\frac{dv_{C2}}{d\tau}$ in Eq. (5.5) by B_p:

$$\frac{dv_{C2}/B_p}{d\tau} = \left[\left(\frac{v_{C1}}{B_p} - \frac{v_{C2}}{B_p}\right) + \frac{Ri_L}{B_p}\right] \tag{5.9}$$

Let: $z \triangleq \frac{i_L R}{B_p}$. We thus have:

$$\frac{dy}{d\tau} = x - y + z \tag{5.10}$$

Finally, if we define $\beta \triangleq \frac{R^2 C_2}{L}$, $\frac{di_L}{d\tau}$ in Eq. (5.5) becomes:

$$\frac{dz}{d\tau} = -\beta y \tag{5.11}$$

Hence, the dimensionless form of Chua's circuit equations are:

$$\frac{dx}{d\tau} = \alpha(y - x - f(x))$$
$$\frac{dy}{d\tau} = x - y + z$$
$$\frac{dz}{d\tau} = -\beta y \tag{5.12}$$

Some observations from the dimensionless form:

1. Example 5.1.1 illustrates a semi-systematic procedure for obtaining dimensionless normal form: we always start by scaling the time variable. The justification is that there are three choices for scaling to dimensionless time: $\tau = \frac{t}{R_n C_n}$, $\tau = \frac{t}{L_n/R_n}$ or $\tau = \frac{t}{\sqrt{L_n C_n}}$. After scaling time, the actual scaling of the state variables, parameters, and nonlinearities depend on the particular system. Hence, the procedure is "semi-systematic."
2. Equation (5.12) has four parameters that can be tuned: α, β, a, b. In contrast, Eq. (5.3) has seven parameters: $R, C_1, C_2, L, G_a, G_b, B_p$. Hence it is always convenient to scale circuit equations to dimensionless form.

SageMath simulation results for Eq. (5.12) with two different initial conditions ([0.1, 0, 0.1], [0.1, 0, 0.01]) are shown in Fig. 5.10. SageMath code is shown in Listing 5.2.

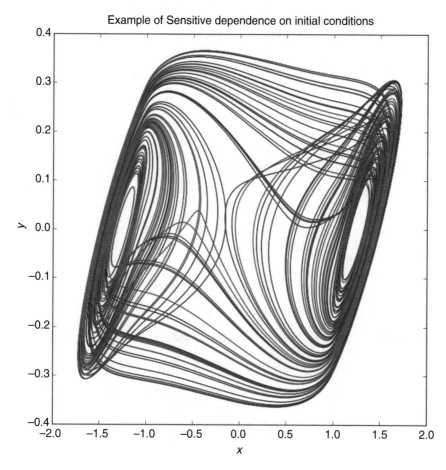

Fig. 5.10 The result of simulating with two different initial conditions are shown in different colors. Notice that the trajectories do not superimpose

Listing 5.2 SageMath code

```
1  # Simulate the autonomous Chua oscillator (dimensionless form
       )
2  from scipy.integrate import odeint
3  from matplotlib import pyplot as plt
4  plt.rcParams['figure.figsize'] = (8.0,8.0)
5  plt.rc('text', usetex=True)
6  plt.rc('font', family='serif')
7  from mpl_toolkits.mplot3d import Axes3D
8  # Circuit parameters
9  C1=10e-9
10 C2=100e-9
11 L=18e-3
12 R=1514
```

```
13   Bp=1
14   Ga=(0.864e-3+0.864e-3)/(-1.14-1.14)
15   Gb=(2.94e-3-0.864e-3)/(-8.36+1.14)
16   #dimensionless parameters
17   a=Ga*R
18   b=Gb*R
19   alpha=C2/C1
20   beta=((R**2)*C2)/L
21   # system
22   def f(x,a,b):
23       return b*x+0.5*(a-b)*(abs(x+1)-abs(x-1))
24   def chuaDimensionless(previousState,t):
25       # Let previousState = [x(t-dt),y(t-dt),z(t-dt)].
26       # Hence, we are going to return the normal form equations
                to be integrated
27       # by odeint:
28       # x(t) = f_1(x(t-dt),y(t-dt),z(t-dt))
29       # y(t) = f_2(x(t-dt),y(t-dt),z(t-dt))
30       # z(t) = f_3(x(t-dt),y(t-dt),z(t-dt))
31       x,y,z=previousState
32       return (alpha*(y-x-f(x,a,b)),x-y+z,-beta*y)
33   # setup and run simulation
34   times=srange(0,500,0.01)
35   ics=[0.1,0,0.1]
36   chuaDimensionlessSolIC1=odeint(chuaDimensionless,ics,times,
         rtol=1e-14,atol=1e-13)
37   chuaDimensionlessSolIC2=odeint(chuaDimensionless,[0.1,0,0.01
         ],times,rtol=1e-14,atol=1e-13)
38   # make sure we obtain STEADY STATE values of (x,y,z)
39   x1=chuaDimensionlessSolIC1[30000:45000,0]
40   y1=chuaDimensionlessSolIC1[30000:45000,1]
41   z1=chuaDimensionlessSolIC1[30000:45000,2]
42   x2=chuaDimensionlessSolIC2[30000:45000,0]
43   y2=chuaDimensionlessSolIC2[30000:45000,1]
44   z2=chuaDimensionlessSolIC2[30000:45000,2]
45   # 2D plot
46   plt.plot(x1,y1,'b',x2,y2,'r')
47   plt.xlabel('$x$',fontsize=16)
48   plt.ylabel('$y$',fontsize=16)
49   plt.title('Example of sensitive dependence on initial
         conditions')
50   plt.show()
51   # 3D plot
52   fig = plt.figure()
53   ax = fig.add_subplot(111, projection='3d')
54   ax.view_init(20,70)
55   plt.xlabel('$x$',fontsize=16)
56   plt.ylabel('$y$',fontsize=16)
57   plt.ylabel('$z$',fontsize=16)
58   plt.plot(x,y,z)
59   plt.show()
```

Figure 5.11 shows a 3D plot. Notice how the chaotic attractor trajectories clearly will not self-intersect in three dimensions, consistent with our explanation in

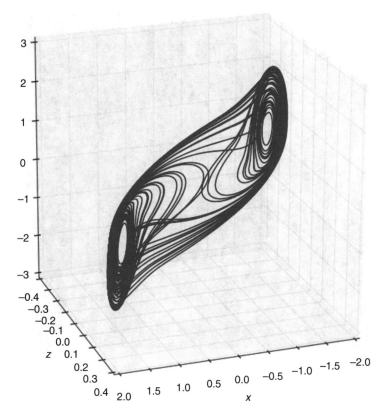

Fig. 5.11 Chaotic attractor in 3D

Sect. 4.6.3. In fact, the chaotic attractor has a **fractional Kaplan-Yorke dimension** between 2 and 3. For details on computing measures for chaotic systems such as the (Kaplan-Yorke) dimension, Lyapunov exponents, etc., see [27]. Browsing through the simulation code shows a very important point: **we have to be careful before declaring a system to be chaotic from simulation results alone**. It could be the steady state solution is (very) long-term periodic, or we may be looking at a transient response. Although a variety of mathematical techniques exist that can be used to rigorously prove chaos, they are beyond the scope of this book. For a very good overview of the different techniques available, with a circuit theoretic emphasis, see related chapters in [1]. However, we can easily avoid the trap of **misidentifying a transient response as the steady state solution by simply plotting the phase portrait with different time ranges**. In this case, we chose to plot the range [30,000 : 45,000].

Another point to note are the parameters we chose. Since the parasitic series resistance of L was chosen to be zero, we added 14 Ω to R because at **DC, we get a load line using $R = 1514 \, \Omega$, consistent with the circuit in Fig. 5.6.**

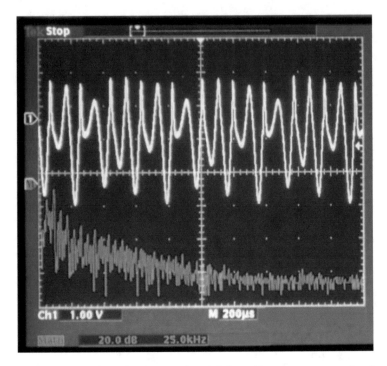

Fig. 5.12 Plot (in red) of the Fourier transform of the voltage across a capacitor from the Muthuswamy-Chua circuit (to be discussed in Sect. 5.4.1). Notice how the signal has content across a wide range of frequencies

A very important property of chaos that is beyond the scope of this book is the frequency content of chaotic signals. It turns out that chaotic signals possess a **wideband** frequency content. An example is shown in Fig. 5.12. As a result chaos can be easily confused with noise. Hence, before the advent of computer simulations, chaos was observed but not identified in a variety of circuits and systems. We will now look at a very brief history of chaos and see circuits where chaos was observed but not identified.

5.2 A History of Chaos in Circuit Theory

In 1922, Armstrong invented the (super)regenerative circuit as a detector with high sensitivity and selectivity as compared to other types of receivers [16]. In the early days of radio engineering, this type of detection was frequently used. Nowadays, regenerative devices are still used as predetection systems when very high frequencies (e.g., microwave communication) are involved. The regenerative detector

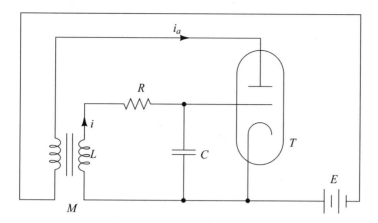

Fig. 5.13 The grid oscillator, T is a three-terminal nonlinear vacuum tube. An incoming signal is modeled as a sinusoidal forcing of the form $A\cos(\omega t)$. For circuit equations and detailed analysis, see [16]

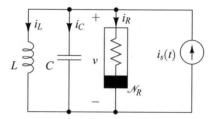

Fig. 5.14 Sinusoidally forced Van der Pol oscillator that displays chaotic behavior, $i_s(t) = A\cos(\omega t)$, see [27]

is favorably used in applications where simplicity and compactness outweigh the need for low noise reception. These circuits use a three terminal vacuum tube, as a receiver as well as in the transmitter, with inductive coupling. Figure 5.13 shows the grid oscillator, a good model for studying chaos in Armstrong's circuit. It turns out that in a simplified model of the circuit above, one can show that the current i behaves chaotically during a small period in time after which the circuit becomes an oscillator. Armstrong was not aware of the circuits' chaotic behavior, but reported "strange irregular startups of the oscillator." It also turns out that during the period in which irregularities appear, the amplification of the circuit is maximal. Hence Armstrong's circuit is an example application of chaos to signal amplification.

Van der Pol in fact also observed similar phenomena, i.e. "irregular noise," when he forced his oscillator with a sinusoidal signal (Fig. 5.14). Unfortunately, he also dismissed chaos as "noise" and did not study the phenomenon further. Note that the nonautonomous Van der Pol equations for the circuit in Fig. 5.14:

$$\frac{di_L}{dt} = \frac{v}{L}$$
$$\frac{dv}{dt} = \frac{-i_L - i_R + A\cos(\omega t)}{C} \tag{5.13}$$

can be put it into autonomous normal form by a simple change of variables. If $z \overset{\triangle}{=} \omega t$, we get:

$$\frac{i_L}{dt} = \frac{v}{L}$$
$$\frac{dv}{dt} = \frac{-i_L - i_R + A\cos(z)}{C}$$
$$\frac{dz}{dt} = \omega \tag{5.14}$$

A variety of "near-misses" also occurred with respect to chaos when investigating nonlinear circuits. Ueda studied combinations such as the Duffing[7]-Van der Pol oscillator by means of analog and digital computers as early as 1961 while he was a graduate student but did not publish results [27]. Many investigations into chaotic circuits in the 1970s were focused on the nonautonomous type, where an external (usually sinusoidal) forcing function was used.

In 1983, while on a visit to Dr. Matsumoto's lab in Waseda University, Dr. Leon O. Chua witnessed his colleague unable to reproduce chaos in a physical circuit implementation of the Lorenz chaotic system. Chua realized that the issue at hand was the use of analog multipliers, which were not reliable in the early 1980s [5]. As a result of the failure, Dr. Chua **systematically designed** an **autonomous** circuit that could potentially reproduce chaotic behavior physically. The core concept was to make use of the $\mathcal{N}_R$ shown in Fig. 5.9 such that at equilibrium, the circuit possessed three **unstable** equilibrium points, as Example 5.2.1 shows.

Example 5.2.1 Using the dimensionless formulation of Chua's circuit from Eq. (5.12) in Example 5.1.1, determine the equilibrium points for the parameter values from Fig. 5.6 and classify them as unstable or stable.

Solution The equilibrium points are simply found by setting the derivatives equal to zero and solving the resulting system of nonlinear equations. Thus, if

(continued)

---

[7]We will study the Duffing oscillator in Sect. 5.5.

Example 5.2.1 (continued)
the equilibrium points are (x^*, y^*, z^*), we have:

$$\alpha(y^* - x^* - f(x^*)) = 0$$

$$x^* - y^* + z^* = 0$$

$$\beta y^* = 0 \qquad (5.15)$$

Simplifying:

$$x^* = -f(x^*)$$

$$x^* = -z^*$$

$$y^* = 0 \qquad (5.16)$$

Solving the above equations, we get the equilibrium points $(0, 0, 0)$, $(+1.261, 0, -1.261)$, $(-1.261, 0, 1.261)$. When $|x(t)| < 1$, the nonlinear function is: $f(x) = ax$. Thus the Jacobian matrix J_0 is:

$$\mathbf{J}_0 = \begin{bmatrix} -\alpha - \alpha \cdot a & \alpha & 0 \\ 1 & -1 & 1 \\ 0 & -\beta & 0 \end{bmatrix} \qquad (5.17)$$

When $|x(t)| > 1$, the nonlinear function is: $f(x) = bx \pm (a - b)$. the Jacobian matrices $J_{\pm 1}$ are both:

$$\mathbf{J}_{\pm 1} = \begin{bmatrix} -\alpha - \alpha \cdot b & \alpha & 0 \\ 1 & -1 & 1 \\ 0 & -\beta & 0 \end{bmatrix} \qquad (5.18)$$

Notice $\mathbf{J}_0$ and $\mathbf{J}_{\pm 1}$ do not depend on the values of the equilibrium points, but only the parameters. For $\mathbf{J}_0$, the eigenvalues are $\lambda_1 \approx 2.659$, $\lambda_{2,3} \approx -1.092 \pm 2.423j$. For $\mathbf{J}_{\pm 1}$, the eigenvalues are $\lambda_1 \approx -6.937$, $\lambda_{2,3} \approx 0.143 \pm 3.217j$.

Notice how the Jacobian shows all three equilibrium points are unstable. But, since $\mathcal{N}_R$ in Fig. 5.9 is eventually passive, the circuit variables cannot arbitrarily increase. Hence, the circuit eventually settles into a strange attractor.[8]

---

[8]There is another very important criterion for a chaotic attractor in Chua's circuit—the homoclinic orbit. For details, see [1].

Although Chua's circuit has more components than required for a chaotic electronic circuit, its significance is the fact that since it was systematically designed, a rigorous proof of chaos was quickly possible within only 2 years after its invention [1]. Moreover, systematically understanding how chaos is produced in Chua's circuit will help in approaching the study of chaos in other "simpler" electronic circuits. This is because of the fact that Chua used PWL analysis in designing the Chua diode. So, when possible, we encourage the reader to use a PWL approximation of a nonlinear function in order to not only implement the function physically, but also to aid in the mathematical analysis of the underlying differential equations.

For instance, in the next section, we will list two circuits which have very simple topologies, but they produce "rich" chaotic behavior. Also, the underlying nonlinear device model that gives rise to chaotic behavior is still a subject of active research.

5.3   Chaos from Physical Nonlinearities: *pn*-Junctions and PWL Inductors

5.3.1   RLD Chaotic Circuit

Consider the QUCS schematic of the RLD circuit in Fig. 5.15. A chaotic time-domain waveform is shown in Fig. 5.16. Investigations into the physical source of chaos in the diode for the circuit from Fig. 5.15 focus on the nonlinear junction capacitance. In fact, [20] has an elegant PWL model for the nonlinear junction capacitance. Hence a circuit that uses a nonlinear capacitor to produce chaotic

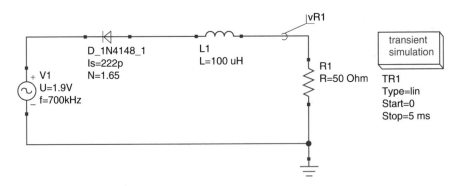

Fig. 5.15 Chaotic circuit from a forced diode resonator

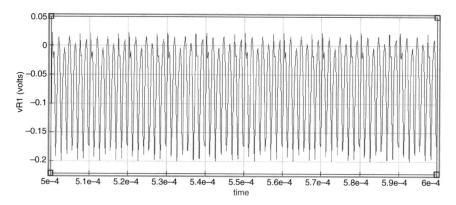

Fig. 5.16 The voltage across the 50 Ω resistor from Fig. 5.16 as a function of time

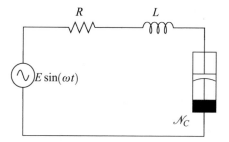

Fig. 5.17 PWL capacitor used to model the diode in the RLD chaotic circuit

behavior is shown in Fig. 5.17. The circuit equations for the PWL model are:

$$\frac{dq}{dt} = i$$

$$\frac{di}{dt} = -\frac{1}{L}\left(iR + f(q) - E\sin(\omega t)\right) \qquad (5.19)$$

where:

$$f(q) = a|q| + bq + E_0 \qquad (5.20)$$

$a = \frac{C_2 - C_1}{2C_1 C_2}, b = \frac{C_1 + C_2}{2C_1 C_2}$. Chaos has been observed by fixing $R = 60\,\Omega, L = 100\,\mu\mathrm{H}, C_1 = 0.1\,\mu\mathrm{F}, C_2 = 400\,\mathrm{pF}, \omega/2\pi = 700\,\mathrm{kHz}, E_0 = 0.1\,\mathrm{V}$ and varying E from 0 to 2.0 V. We leave the exploration to the reader.

But, there are a variety of nonlinearities present in the junction diode, besides capacitance. For example, the conductivity modulation effect present in diodes is due to memristance [7]. Hence an interesting avenue for further research would be to examine if the memristor's nonlinearity plays any role in chaos in the RLD circuit (under appropriate range of parameters).

5.3.2 PWL Inductor Circuit

If we make only the inductor nonlinear in an RLC circuit by introducing hysteresis (example: iron core inductors), chaos can occur [8]. The PWL schematic for the circuit is shown in Fig. 5.18. $\mathcal{N}_L$ is defined by the following PWL function:

$$i(\phi) = \begin{cases} \frac{\phi - \phi_1}{L_1} & \text{for } \phi > \phi_0 \\ \frac{\phi}{L_0} & \text{for } |\phi| < \phi_0 \\ \frac{\phi + \phi_1}{L_1} & \text{for } \phi < -\phi_0 \end{cases} \tag{5.21}$$

where

$$\phi_1 = \phi_0 \left(1 - \frac{L_1}{L_0} \right) \tag{5.22}$$

Practically, $0 < L_1 < L_0$. The circuit equations are:

$$\frac{d\phi}{dt} = \frac{R_1 R_2}{R_1 + R_2} i(\phi) - \frac{R_2}{R_1 + R_2} (v - E\cos(\omega t))$$
$$\frac{dv}{dt} = \frac{1}{C} \left[\frac{R_2}{R_1 + R_2} i(\phi) - \frac{1}{R_1 + R_2} (v - E\cos(\omega t)) \right] \tag{5.23}$$

Circuit parameters used for simulation are: $\omega/2\pi = 50\,\text{Hz}$, $R_1 = 50\,\Omega$, $R_2 = 10\,k\Omega$, $C = 1.69\,\mu\text{F}$, $L_0 = 33.33\,\text{H}$, $L_1 = 1.28\,\text{H}$, $\phi_0 = 0.92\,\text{Vs}$. Varying E should produce chaotic behavior, we also leave this exploration to the reader. What is interesting however is the physical mechanism of chaos in this circuit is still unknown.

Fig. 5.18 A nonlinear resonant circuit

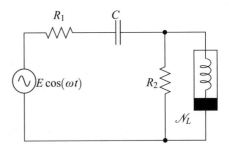

Fig. 5.19 Canonical Chua's oscillator with a flux-controlled memristor

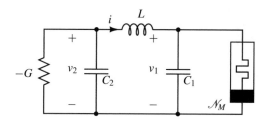

5.4 Memristor Based Chaotic Circuits

We will now discuss chaotic circuits using the fourth fundamental circuit element, the memristor. Consider the canonical Chua's circuit[9] in Fig. 5.19, with $\mathcal{N}_R$ replaced by $\mathcal{N}_M$ [13]. However, one has to use caution when deriving the circuit equations. This is because simply writing the equations in terms of current and voltage would give:

$$\frac{di}{dt} = \frac{1}{L}(v_2 - v_1)$$

$$\frac{dv_2}{dt} = \frac{1}{C_2}(Gv_2 - i)$$

$$\frac{dv_1}{dt} = \frac{1}{C_1}(i - W(\phi_1)v_1)$$

$$\frac{d\phi_1}{dt} \overset{\triangle}{=} v_1 \tag{5.24}$$

Exercise 5.5 asks you to rewrite the system equations in terms of charge and flux, so the number of ODEs is reduced by one.

A variety of chaotic attractors have been derived for the circuit in Fig. 5.19. What is interesting however is that we **only need one** capacitor, **one** inductor, and **one** memristor to obtain a chaotic circuit, as the next section illustrates.

5.4.1 Muthuswamy-Chua Circuit

Let us play the same "trick" that we used in Chua's circuit, of replacing $\mathcal{N}_R$ with $\mathcal{N}_M$, but in the topologically simpler (series) Van der Pol oscillator. We consider the series implementation in this section because Muthuswamy and Chua **systematically** obtained chaos [22] in the circuit shown in Fig. 5.20, called the Muthuswamy-Chua circuit [19]. Assuming a current-controlled $\mathcal{N}_M$ with only one

---

[9]This circuit is slightly different from the circuit we discussed in Sect. 5.1.

Fig. 5.20 The
Muthuswamy-Chua circuit
[22]

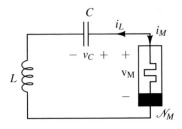

internal state z,[10] the system equations for the circuit can be trivially derived:

$$\frac{dv_C}{dt} = \frac{i_L}{C}$$

$$\frac{di_L}{dt} = -\frac{1}{L}(v_C + R(z, i_L)i_L)$$

$$\frac{dz}{dt} \triangleq f(z, i) \tag{5.25}$$

The significance of the Muthuswamy-Chua circuit is that it is the **simplest known chaotic circuit** that uses only the **fundamental circuit elements**. Also, only the memristor is nonlinear.[11] Consider the following specific system equations derived from Eq. (5.25). We have assumed $x = v_C$, $y = i_L$.

$$\frac{dx}{dt} = \frac{y}{C}$$

$$\frac{dy}{dt} = -\frac{1}{L}\left(x + \beta(z^2 - 1)y\right)$$

$$\frac{dz}{dt} = -y - \alpha z + yz \tag{5.26}$$

Inspired by Rössler's intuitive arguments in deriving his namesake chaotic equation, the memristance and state functions in Eq. (5.26) were systematically derived by Dr. Muthuswamy for producing chaotic behavior. Assuming $\beta > 0$, $R(z) = \beta(z^2 - 1)$ is negative for $|z| < 1$. Hence when we "power on" the circuit in Fig. 5.20, since initial memristor state variable will naturally be assumed to be close to zero, we have a

---

[10]The system of equations is autonomous and the minimum number of state variables to obtain chaos in a continuous time autonomous system is three. Hence we need only one internal state for the memristor to have a three-dimensional autonomous ODE model of the Muthuswamy-Chua circuit.

[11]Of course, we know by definition a linear memristor is simply a resistor.

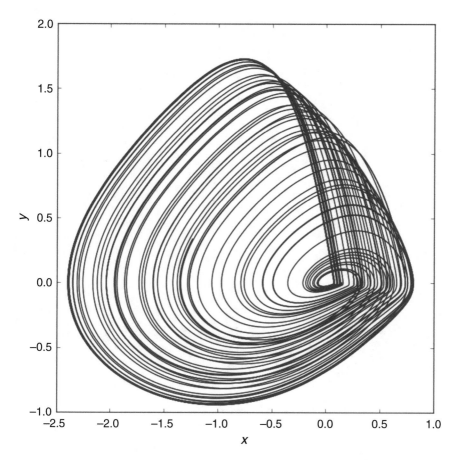

Fig. 5.21 $x - y$ phase plot resulting from simulating Eq. (5.26) with parameters $C = 1, L = 3.3, \beta = 1.7, \alpha = 0.2$. We chose an appropriate interval to avoid transient and plot the "steady-state" chaotic attractor

negative memristance. Hence the circuit is **unstable** and the voltage, current values start increasing. In the RHS of the $\dot{z}$ Eq. (5.26), we can see the product yz also starts increasing. But, if $|z| > 1$, the memristance is positive and hence the circuit becomes **stable**. Furthermore, the $-y - \alpha z$ will also eventually cause trajectories in z to head back to the origin, until the circuit becomes unstable again. This alternating unstable and stable behavior leads to limit cycles (similar to our discussion in Sect. 4.6.3) for some parameter values, and chaos for other (systematically chosen) parameter values. In fact, it has been rigorously proved [12] that the chaotic attractor shown in Fig. 5.21 is topologically the same as the Rössler attractor.

We will now discuss the implementation of the Muthuswamy-Chua circuit in detail since we have to emulate the memristor. Consider the schematic Fig. 5.22.

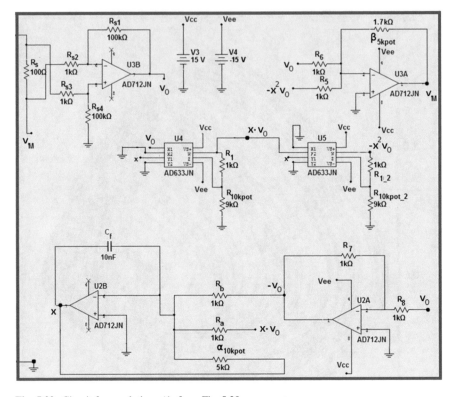

Fig. 5.22 Circuit for emulating $\mathcal{N}_M$ from Fig. 5.20

The first step is to sense the current using a "sense" resistor R_s. This resistor must be "small enough" so it does not affect the dynamics of the circuit. In our case we have $R_s = 100\,\Omega$ connected to the difference amplifier U3B. Hence the output of U3B is:

$$v_0 = \frac{R_{s1}}{R_{s2}} R_s i_M = -R_{\text{scale}} i_L \qquad (5.27)$$

Thus we have scaled and mapped the current into a voltage v_0, so that we can easily use components such as analog multipliers, which are voltage based.

The next step is to realize the memristor function $R(x) = \beta(x^2 - 1),$[12] using opamp U3A, multipliers U4, U5. Using the datasheets of the multipliers and the

---

[12]We will use x instead of z for the memristor state in the implementation discussion, to be consistent with the original publication [22].

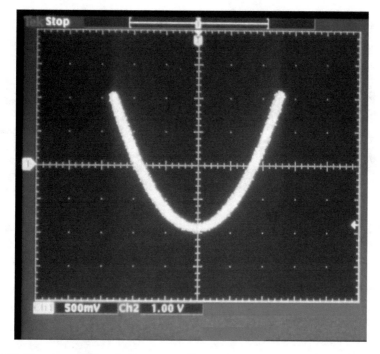

Fig. 5.23 Experimental plot of $R(x) = 1.5(x^2 - 1)$. Horizontal axis scale is 0.5 V/division; vertical axis scale is 1.00 V/division. The experimental curve crosses the horizontal axis at -1 V and approximately 0.9 V

connections shown in the schematic, we can infer that:

$$v_M = -\frac{\beta_{5kpot}}{R_6} v_0 - \frac{\beta_{5kpot}}{R_5}(-x^2 v_0) \tag{5.28}$$

We will choose $R_5 = R_6 = R = 1k$ and $\beta \triangleq \frac{\beta_{5kpot}}{R}$. Replacing v_0 in Eq. (5.28) from Eq. (5.27) and simplifying, we get:

$$v_M = \beta R_{scale}(x^2 - 1)i_L \tag{5.29}$$

A plot of $R(x)$ obtained from the circuit is shown in Fig. 5.23. The reader should have noticed that R is not PWL. A good avenue for further research would be to consider PWL versions of R.

The final step is to realize the memristor internal state equation. This is done by means of opamps U2B, U2A. The output x of opamp U2B is given by:

$$\frac{dx}{dt} = \frac{1}{C_f}\left[\frac{v_0}{R_b} - \frac{x}{\alpha_{10kpot}} - \frac{x v_0}{R_a}\right] \tag{5.30}$$

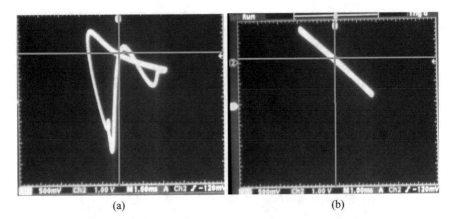

Fig. 5.24 Memristor pinched hysteresis loop (Lissajous figures). Axes scales are 0.5 V/division for horizontal axis (i_M), 1.00 V/division for vertical axis (v_M). (**a**) 3 kHz. (**b**) 35 kHz

Substituting for v_0 from Eq. (5.27), we get:

$$\frac{dx}{dt} = \frac{1}{C_f}\left[-\frac{R_{\text{scale}}i_L}{R_b} - \frac{x}{\alpha_{10\text{kpot}}} + \frac{R_{\text{scale}}i_L x}{R_a}\right] \tag{5.31}$$

Let us check memristor pinched-hysteresis $v - i$ characteristics, based on our discussions in Sect. 4.4.2. Results are shown in Fig. 5.24a, b. The DC characteristic has already been verified in Fig. 4.41. Notice that as $\omega \to \infty$, the hysteresis loop degenerates to that of a linear resistor, as required. Let us now consider the memristor emulator connected to a physical C_n and L_n. Circuit equations are:

$$\frac{dv_C}{dt} = \frac{i_L}{C_n}$$

$$\frac{di_L}{dt} = -\frac{1}{L_n}\left[v_C + \beta R_{\text{scale}}(x^2 - 1)i_L + R_s i_L\right]$$

$$\frac{dx}{dt} = \frac{1}{C_f}\left[-\frac{R_{\text{scale}}i_L}{R_b} - \frac{x}{\alpha_{10\text{kpot}}} + \frac{R_{\text{scale}}i_L x}{R_a}\right] \tag{5.32}$$

Notice that we have included the effect of the sense resistor R_s in the $\dot{i}_L$ equation above.

We finally need to convert the circuit equations into the system Eq. (5.26). To do this, we will first perform the time scaling as $\tau \triangleq T_s t = 10^5 t$. In this case, we **do not perform a dimensionless scaling** using $\tau = \frac{t}{\sqrt{LC}}$. The reason is that choosing this time scale would eventually make $L = C$ in Eq. (5.26). We will also scale $y(\tau)$ to hundreds of microamps, for physical implementation. Thus, we have:

$$x(\tau) \overset{\triangle}{=} v_C(t)$$

$$y(\tau) \overset{\triangle}{=} R_{scale} i_L(t)$$

$$z(\tau) \overset{\triangle}{=} x(t) \tag{5.33}$$

To be clear, the **dimension of** x, y, z **are all volts**. Substituting the definitions above into Eq. (5.32) and simplifying using the component values from the emulator we get:

$$\frac{dx}{d\tau} = \frac{y}{C}$$

$$\frac{dy}{d\tau} = -\frac{1}{L}\left(x + \beta(z^2 - 1)y + 0.01y\right)$$

$$\frac{dz}{d\tau} = -y - \alpha z + yz \tag{5.34}$$

where:

$$C = R_{scale} C_n T_s$$

$$L = \frac{L_n T_s}{R_{scale}}$$

$$\beta = \frac{\beta_{5kpot}}{R}$$

$$\alpha = \frac{1}{T_s C_f \alpha_{10kpot}} \tag{5.35}$$

Choosing $C_n = 1\,\text{nF}$, $L_n = 330\,\text{mH}$ and $\beta_{5kpot} = 1.7k$, $\alpha_{10kpot} = 5k$, we get: $C = 1$, $L = 3.3$, $\beta = 1.7$, $\alpha = 0.2$. The attractor obtained from the physical circuit is shown in Fig. 5.25. Based on our experience with the memristor emulator, a natural follow-up question is: are there chaotic circuits where physical memristor nonlinearities cause chaos? As of the writing of this book (January 2018), no one has **explicitly** found such a circuit. But, there are a variety of candidates. One promising candidate is Theodorchik's oscillator [2] shown in Fig. 5.26. Anischenko et. al. [2] do not explicitly state the $R(T)$ modeled by a thermistor, is a memristor. They rather assume the memristance to be a linear function of temperature. They modify the

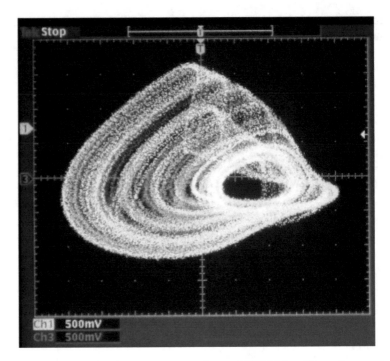

Fig. 5.25 Experimental chaotic attractor, axes scales are 0.5 V/division. We used a current probe to measure the current i_L through the inductor. In the experimental plot, $(0, 0)$ has been shifted to the right for clarity on the oscilloscope. Compare to Fig. 5.21

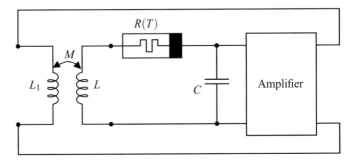

Fig. 5.26 Theodorchik's oscillator with "inertial nonlinearity"

oscillator in Fig. 5.26 by adding more amplifier stages, assume the amplifier to be nonlinear and obtain chaos. It would be interesting for the reader to investigate this problem further (see Exercise 5.7).

5.5 Implementing the Duffing Oscillator Using a Higher-Order Element

Having obtained chaotic circuits by using all four fundamental circuit elements, we will next propose a chaotic circuit using a higher-order circuit element. We alluded to this in earlier chapters, recall the Duffing oscillator equation from Chap. 1:

$$\ddot{v} + c\dot{v} + v(b + a \cdot v^2) = i(t) \tag{5.36}$$

We discussed that we will use a $(0, -2)$ element to implement $\ddot{v}$ and proposed a schematic in Sect. 4.6.2, reproduced in Fig. 5.27. Consider now the circuit in Fig. 5.28. From the U2 voltage follower, we get: $v_1 = v_2$.

The voltage drop across R is $i_2 R$, since the current into the noninverting input of U3 is zero. Also, since the noninverting input of U3 is at virtual ground, we have the **noninverting input voltage of** U1 to be equal to $i_2 R$. Hence:

$$i_3 = -C_3 \frac{di_2 R}{dt}$$

$$= RC_2C_3 \frac{d^2 v_2}{dt^2}$$

$$= RC_2C_3 \frac{d^2 v_1}{dt^2} \tag{5.37}$$

Since the current into the noninverting input of U2 is also zero, the CFOA ensures $i_1 = i_3$. So we finally have:

$$i_1 = RC_2C_3 \frac{d^2 v_1}{dt^2} \tag{5.38}$$

Fig. 5.27 A mutator for synthesizing $(0, -2)$ from a $(0, -1)$ element (capacitor C_2 at port 2)

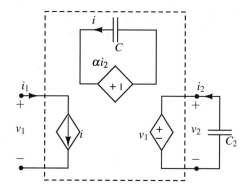

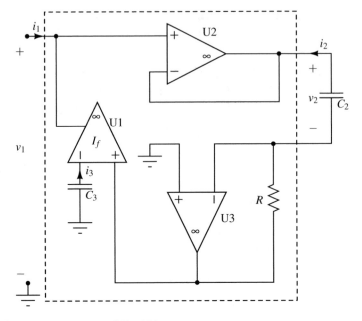

Fig. 5.28 One implementation of Fig. 4.51

Notice the equation above is dimensionally consistent. Exercise 5.4 asks you to complete the implementation of the Duffing oscillator using the higher-order element we implemented above.

5.6   Transistor Based Chaotic Circuits

Consider[13] the Colpitts Chaotic Oscillator[14] QUCS schematic shown in Fig. 5.29. Simulation results are shown in Fig. 5.30. The PWL equivalent circuit is shown in Fig. 5.31. The circuit equations based on the PWL model are:

$$\frac{dV_{CE}}{dt} = \frac{1}{C_1}(I_L - I_C)$$

$$\frac{dV_{BE}}{dt} = -\frac{1}{C_2}\left(\frac{V_{EE} + V_{BE}}{R_{EE}} + I_L + I_B\right)$$

$$\frac{dI_L}{dt} = \frac{1}{L}(V_{CC} - V_{CE} + V_{BE} - I_L R_L) \qquad (5.39)$$

---

[13]This is another example of a circuit where a physical nonlinearity is the cause of chaos.

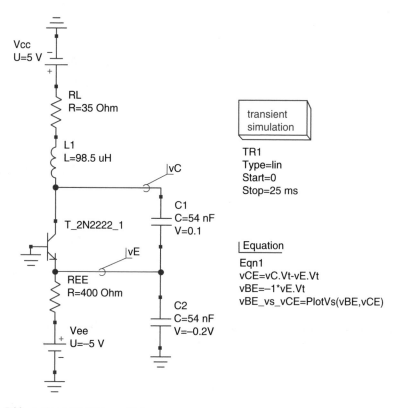

Fig. 5.29 A chaotic Colpitts oscillator

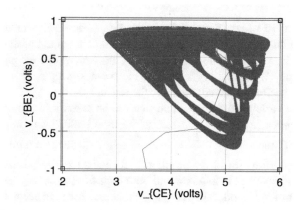

Fig. 5.30 QUCS simulated chaotic attractor. y-axis is v_{BE}, x-axis is v_{CE}. Notice the initial transient settling into the chaotic attractor

Fig. 5.31 PWL model of the
Colpitts chaotic oscillator

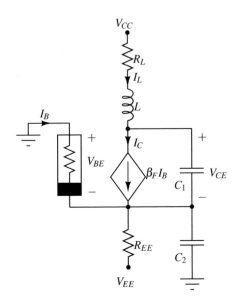

Experimentally, it has been observed that the transistor is operating in either forward
active or cutoff. Consequently, the following PWL linear function for I_B is used
(compare to Fig. 2.18):

$$I_B = \begin{cases} 0 & \text{if } V_{BE} \leq V_{TH} \\ \frac{V_{BE}-V_{TH}}{R_{ON}} & \text{if } V_{BE} > V_{TH} \end{cases} \tag{5.40}$$

$$I_C = \beta_F I_B \tag{5.41}$$

where $V_{TH} \approx 0.75$ V is the threshold voltage, R_{ON} is the small-signal on-resistance
of the base-emitter junction, and β_F is the forward current gain of the device.

We would like to conclude this chapter by illustrating the elegance of dimensionless scaling. We will only highlight the main concepts, leaving the actual
dimensionless scaling of the equations to the reader.

There are a variety of concepts that need to be taken into account for dimensionless scaling of Eq. (5.39). The first is the time scale. There is every possible
combination of time scaling: $\tau \stackrel{\Delta}{=} \frac{t}{R_L C1}$, $\tau \stackrel{\Delta}{=} \frac{t}{L/R_L}$, etc. in the circuit. So let us take
a step back and **understand** the problem.[14] It seems like the primary mechanism
of chaos should involve the inductor and **both** capacitor(s), as they are the dynamic
elements in our PWL model. Hence a logical choice for time scale should involve
some combination of L, C_1, C_2. Again from experience, the reader will realize by

---

[14]The reader has hopefully been applying the steps to problem solving elucidated in Example 1.10.2 throughout the book.

looking at the form of the RHS in Eq. (5.39) that:

$$\tau \triangleq \sqrt{\frac{LC_1C_2}{C_1 + C_2}} \tag{5.42}$$

is a "good" choice. Next, let us examine the I_B nonlinearity. Notice it can be rewritten as:

$$I_B = \begin{cases} 0 & \text{if } V_{BE}/V_{TH} \leq 1 \\ \frac{V_{TH}(V_{BE}/V_{TH}-1)}{R_{ON}} & \text{if } V_{BE}/V_{TH} > 1 \end{cases} \tag{5.43}$$

The justification for doing so is to define (for the second state equation in Eq. (5.39)) $y \triangleq \frac{V_{BE}}{V_{TH}}$. But notice we can carry the simplification above for the PWL one step further:

$$\frac{I_B R_{ON}}{V_{TH}} = \begin{cases} 0 & \text{if } V_{BE}/V_{TH} \leq 1 \\ V_{BE}/V_{TH} - 1 & \text{if } V_{BE}/V_{TH} > 1 \end{cases} \tag{5.44}$$

The LHS of the equation above gives us a hint that:

$$z \triangleq \frac{I_L R_{ON}}{V_{TH}} \tag{5.45}$$

Moreover, the nonlinear function simply becomes:

$$f(y) = \begin{cases} 0 & \text{if } y \leq 1 \\ y - 1 & \text{if } y > 1 \end{cases} \tag{5.46}$$

We have now the dimensionless definitions for time, all state variables **and** the nonlinearity!

5.7 Conclusion to This Book

In this book we have covered lumped circuit theory. A reader who is probably familiar with classic linear circuit theory should hopefully now appreciate the advantage of following a "top-down" general approach to circuit theory: it enables them to properly analyze a very broad class of circuits. For example, consider our discussion of on opamps in Sect. 2.5. The reason we were able to properly analyze negative **and** positive feedback circuits is because we clearly (and correctly) separate static behavior from dynamic circuit behavior. Thus, the concept of instability in positive feedback circuits (opamp or otherwise) **requires** us to introduce (parasitic) dynamic $v - i$ components such as capacitors and inductors (perhaps even the memristor for a hitherto undiscovered circuit). In this chapter, we saw how all the concepts integrated together in the form of chaotic circuits using fundamental circuit elements, opamps and transistors.

So, having been armed with the proper approach to circuit theory, where does a reader go from here? An answer to this question is for the reader to follow up on particular concepts of interest. Some (by no means, exhaustive) examples:

1. Recall the Simultaneity Postulate from Sect. 1.3. This postulate dictates when the techniques in this book are valid. Hence, a natural follow-up for the interested reader would be on distributed circuits,[15] where the simultaneity postulate is inapplicable.
2. Another approach would be to pick up books that exhaustively cover the ideas introduced here. For example, the graph theoretic approach to circuits is extensively covered in [4]. Another excellent very recent volume on the topic is [24].
3. Chaotic systems are the subject of many excellent books. An excellent starting point is [27]. Sprott has a chapter devoted exclusively to chaotic electrical circuits.

Exercises

5.1 Show that Fig. 5.2 (assuming $R_1 = R_2$) gives Eq. (5.2).

5.2 NOTE: This is an open-ended problem
Systematically change the values of the initial conditions; inductor, capacitor(s) values and the DP characteristic of $\mathcal{N}_R$, to obtain different behaviors in Chua's circuit. What happens in Fig. 5.6 if the parasitic series resistance for the inductor is removed? Does the simulation converge?

The reader should notice that it is much easier to simulate the dimensionless form in a mathematical package such as SageMath, rather than obtaining results from the circuit simulator. Why do you think this is the case? **Think** about **all** reasons. Also think about advantages of simulating the circuit equations.

5.3 Derive the circuit Eq. (5.3) for Fig. 5.5. Use Sect. 1.9.1.2 to derive the $\mathcal{N}_R$ characteristics in Eq. (5.4) from Fig. 5.9.

---

[15] We (Dr. Muthuswamy and Dr. Banerjee) are planning to write such a follow-up volume to this book, tentatively titled: "Advanced Nonlinear Circuits and Networks." In the follow-up book, we plan to first discuss rigorously how circuit theory is an approximation of electromagnetic field theory. This would then set us up nicely to discuss distributed circuits.

5.4 NOTE: This is an open-ended problem
Synthesize the Duffing oscillator from Sect. 5.5. We recommend approximating the cubic using a simple PWL nonlinearity (realized using one opamp).

5.5 Write system equations for the Chua oscillator from Fig. 5.19 in terms of charge and flux, using the ideas from Sect. 4.4.1.

5.6 NOTE: This is an open-ended problem
Design and implement a memristor simulation library for QUCS.

5.7 NOTE: This is an open-ended problem
Investigate chaotic circuit implementations where the source of chaos is a **physical (not emulated)** memristor's nonlinearity. As a starting point, we know of three devices that can be modeled by memristors: *pn*-junction diodes, thermistors, and discharge tubes. Hence a good approach would be to investigate existing chaotic circuits based on these devices and check if the underlying memristor nonlinearity is the cause of chaos.

Lab 5: Capstone Chaos Project(s)

In this final "lab," we will give some further suggestions for capstone projects. Note again that most of the exercises above are capstone projects.

1. At the turn of the twenty-first century, an active area of research in chaotic circuits (systems) is the notion of classifying chaotic attractors into "self-excited" and "hidden." We have discussed "self-excited" chaotic attractors: those that arise due to unstable equilibrium points. Kuznetsov et. al. coined the notion of "hidden" attractors, so named because they are present in a neighborhood of stable equilibrium points. An excellent starting point is the survey paper by Leonov and Kuznetsov [17]. Physically implementing chaotic circuits that exhibit hidden attractors are tricky because they exist close to stable equilibrium points, for a good example, see [28].
2. Mathematically investigating chaotic circuits is difficult because one has to be well-versed in the theory of dynamical systems. But, excellent works abound online. A good tractable starting point for the curious undergraduate would be the papers on **interval arithmetic** by Galias [11].
3. There has been no experimental confirmation of chaos from a physical (Josephson junction, *pn*-junctions, thermistor, discharge tube) memristor.
4. An energy approach to the study of chaotic systems.
5. Chaotic circuits with time delay, see [27], although a realistic circuit representation of chaotic time delay systems should probably use distributed components such as waveguides.
6. We also encourage the reader to look through some of the references in this chapter for exciting ideas related to chaotic circuits.
7. Further references for projects are [9, 10, 15, 21, 25].

References

1. Adamatzky, A., Chen, G.: Chaos, CNN, Memristors and Beyond: A Festschrift for Leon Chua. World Scientific, Singapore (2011)
2. Anischenko, V.S., Vadivasova, T.E., Strelkova, G. I.: Deterministic Nonlinear Systems: A Short Course. Springer, Berlin (2014)
3. Ambelang, S., Muthuswamy, B.: From Van der Pol to Chua: An Introduction to Nonlinear Dynamics and Chaos for Second Year Undergraduates. In: 2012 IEEE ISCAS. https://doi.org/10.1109/ISCAS.2012.6271932
4. Chua, L.O.: Introduction to Nonlinear Network Theory. McGraw-Hill, New York (1969) (out of print)
5. Chua, L.O.: The genesis of Chua's circuit. ArchivfürElektronik und Ubertragung-stechnik **46**, 250–257 (1992)
6. Chua, L.O.: Chua's circuit (2007). Available, online: http://www.scholarpedia.org/article/Chua_circuit#The_Chua_Diode_is_Locally_Active Last accessed January 10th 2018
7. Chua, L.O., Tseng, C.: A memristive circuit Model for pn junction diodes. Int. J. Circuit Theory Appl. **4**(2), 367–389 (1976)
8. Chua, L.O., et al.: Dynamics of a piecewise-linear resonant circuit. IEEE Trans. Circuits Syst. **CAS-29**(8), 535–547 (1982)
9. Chua, L.O., Wah, C.W., Huang, A., Zhong, G.: A universal circuit for studying and generating chaos - part I : routes to chaos. IEEE Trans. Circuits Syst. **40**(10), 732–744 (1993)
10. Galias, Z.: Numerical study of multiple attractors in the parallel inductor-capacitor-memristor circuit. Int. J. Bifurcation Chaos **27**, 1730036–1730051 (2017)
11. Galias, Z.: List of publications. Available, online: http://www.zet.agh.edu.pl/~galias/publ.html Last accessed January 11th 2018
12. Ginoux, J.M., Letellier, C., Chua, L.O.: Topological analysis of chaotic solution of a three-element memristive circuit. Int. J. Bifurcation Chaos **20**(11), 3819–3827 (2010)
13. Itoh, M., Chua, L.O.: Memristor oscillators. Int. J. Bifurcation Chaos **18**(11), 3183–3206 (2008)
14. Kennedy, M.P.: Chaos in the colpitts oscillator. IEEE Trans. Circuits Syst. I, Fundam. Theory Appl. **41**(11), 771–774 (1994)
15. Kevorkian, P.: Snapshots of dynamical evolution of attractors from Chua's oscillator. IEEE Trans. Circuits Syst. I, Fundam.Theory Appl. **40**(10), 762–780 (1993)
16. Leenaerts, D.M.W.: Chaotic behavior in super regenerative detectors. IEEE Trans. Circuits Syst. I, Fundam. Theory Appl. **43**(3), 169–176 (1996)
17. Leonov, G.A., Kuznetsov, N.V.: Hidden attractors in dynamical systems. From hidden oscillations in Hilbert-Kolmogorov, Aizerman, and Kalman problems to hidden chaotic attractor in Chua circuits. Int. J. Bifurcation Chaos **23**(1), 1330002-1–1330002-69 (2013)
18. Li, T.Y., Yorke, J.A.: Period three implies chaos. Am. Math. Mon. **82**, 985 (1975)
19. Llibre, J., Valls C.: On the integrability of a Muthuswamy-Chua system. J. Nonlinear Math. Phys. **19**(4), 477–488 (2012)
20. Matsumoto, T., Chua, L.O., Tanak, S.: Simplest chaotic nonautonomous circuit. Phys. Rev. A **30**(2), 1155–1158 (1984)
21. Matsumoto, T., Chua, L.O., Tokunaga, R.: Chaos via torus breakdown. IEEE Trans. Circuits Syst. **CAS-34**(3), 240–253 (1987)
22. Muthuswamy, B., Chua, L.O.: Simplest chaotic circuit. Int. J. Bifurcation Chaos **20**(5), 1567–1680 (2010)
23. Parker, T.S., Chua, L.O.: Practical Numerical Algorithms for Chaotic Systems. Springer, Berlin (1989)
24. Parodi, M., Storace, M.: Linear and Nonlinear Circuits: Basic and Advanced Concepts, vol. 1. Springer, Berlin (2018)
25. Pivka, L., Wu, C.W., Huang, A.: Chua's Oscillator : a compendium of chaotic phenomena. J. Frankl. Inst. **331**(6), 705–741 (1994)

26. Sharkovsky, A.N., Chua, L.O.: Chaos in some 1-D discontinuous maps that appear in the analysis of electrical circuits. IEEE Trans. Circuits Syst. I, Fundam. Theory Appl. **40**(10), 722–731 (1993)

27. Sprott, J.C.: Elegant Chaos : Algebraically Simple Chaotic Flows. World Scientific, New York (2010)

28. Wei, Z., et al.: Hidden hyperchaos and electronic circuit application in a 5D self-exciting homopolar disc dynamo. Chaos Interdisciplinary J. Nonlinear Sci. **27**, 033101 (2017). https://doi.org/10.1063/1.4977417

Appendix A
Installing QUCS

In this appendix, we will discuss how to install QUCS [1]. Note that QUCS has a lot of components, many of which we will not use. Nevertheless, we will install all components for completeness.

A.1 Windows

Please install the official Windows QUCS package from the download section in the QUCS homepage [1].

A.2 OS X

Please install the official OS X QUCS package from the download section in the QUCS homepage [1].

A.3 Linux

If you are using Linux, make sure you have a reliable internet connection, as we will be installing from source. If you are on a Linux platform, we will assume that you are comfortable with basic command line tools such as `tar`, `apt-get`, etc. and have `sudo` access.

The instructions below are specifically for Ubuntu 14.04 distribution, but they should be applicable to any of the popular Linux distributions.

© Springer International Publishing AG, part of Springer Nature 2019   353
B. Muthuswamy, S. Banerjee, *Introduction to Nonlinear Circuits and Networks*,
https://doi.org/10.1007/978-3-319-67325-7

1. The first step is to download the latest QUCS tarball (v0.0.19 as of this writing) from [1] in your home folder.
2. Extract and unzip the tarball:

```
1   $ tar xvzf qucs-0.0.19.tar.gz
```

3. Change into the QUCS directory and go through README.md.
4. You may need to install missing dependencies via the Debian package manager. In our case, we had to install the following packages:

```
1   $ sudo apt-get install gperf libxml-libxml-perl libxml2
        libxml2-dev libgd-perl octave texlive-math-extra texlive-
        science build-essential libqt4-dev libqt4-qt3support qt4-
        dev-tools libqt4-opengl-dev octave-epstk automake libtool
        gperf flex bison git cmake
```

5. Next, we need to install ADMS. To do, clone the repository from github into your root folder, configure and install:

```
1   $ git clone https://github.com/Qucs/ADMS.git
2   $ export LD_LIBRARY_PATH=/usr/local/lib
3   $ cd ADMS
4   $ sh bootstrap.sh
5   $ ./configure --enable-maintainer-mode
6   $ make
7   $ sudo make install
8   $ sudo ldconfig
```

6. Configure, make and install QUCS:

```
1   $ cd ~/qucs-0.0.19/
2   $ ./configure
3   $ make
4   $ sudo make install
```

Reference

1. QUCS Project: Quite Universal Circuit Simulator. Available online. http://qucs.sourceforge.net Cited 24 May 2017

Solutions

For step-by-step solutions to all problems, please visit online material at: http://www.youtube.com/user/bharathberkeley/IntroToNonlinearCircuitsAndNetworks.

© Springer International Publishing AG, part of Springer Nature 2019

B. Muthuswamy, S. Banerjee, *Introduction to Nonlinear Circuits and Networks*,
https://doi.org/10.1007/978-3-319-67325-7

Index

© Springer International Publishing AG, part of Springer Nature 2019
B. Muthuswamy, S. Banerjee, *Introduction to Nonlinear Circuits and Networks*,
https://doi.org/10.1007/978-3-319-67325-7

Printed in the United States
By Bookmasters